Advances in Technology-Rich Science Education

Volume 3

This book series provides a comprehensive and authoritative overview of current research, trends, and innovations that leverage educational technology to improve science teaching and learning. Contributions from a global community of scholars, educators, policymakers, and leaders address formal, non-formal, and informal science education contexts.

It provides a platform for sharing research, practice, development, and policy, highlighting emerging innovations, and examining how technology influences engagement, learning, and workforce readiness. Additionally, it also addresses key challenges and opportunities related to access, equity, inclusion, and digital literacies while fostering critical perspectives and discussion about technology's evolving role in science education.

Focused on the enterprise of science education, the series explores the design and use of technology-enhanced learning environments, applications of artificial intelligence, simulation, and computation, and innovations in online and blended learning, including MOOCs and virtual laboratories. It also considers augmented reality, authentic digital experiences, strategies for supporting diversity and inclusion, research methodologies, workforce development, and the creation and critique of educational policy and leadership.Book proposals for this series may be submitted to the Publishing Editor: Claudia Acuna E-mail: Claudia.Acuna@ Springer.com

Xiaoming Zhai • Kent J. Crippen
Editors

Advancing AI in Science Education

Envisioning Responsible and Ethical Practice

Springer

Editors
Xiaoming Zhai
University of Georgia
Athens, GA, USA

Kent J. Crippen
University of Florida
Gainesville, FL, USA

ISSN 3091-339X ISSN 3091-3403 (electronic)
Advances in Technology-Rich Science Education
ISBN 978-3-032-16870-2 ISBN 978-3-032-16871-9 (eBook)
https://doi.org/10.1007/978-3-032-16871-9

This Springer imprint is published by the registered company Springer Nature Switzerland AG
The registered company address is: Gewerbestrasse 11, 6330 Cham, Switzerland

Foreword

Science education is entering a new era—an era shaped by artificial intelligence (AI). Across every branch of science, AI is changing how scientists make discoveries, interpret evidence, and even think about asking questions. And it's not just transforming scientific research—it's transforming how we learn. For those of us who have spent our careers helping students make sense of the natural world, this is an extraordinary moment. Technology and pedagogy are coming together in ways that make us rethink what it really means to teach and learn science in the twenty-first century.

That's why I find *Advancing AI in Science Education: Envisioning Responsible and Ethical Practice* so timely and valuable. Edited by Xiaoming Zhai and Kent Crippen, and written by a diverse group of international scholars, the book goes far beyond showing how teachers can use AI in classrooms. It presents a picture of how researchers and teachers can responsibly and ethically weave AI into science teaching and learning. It reminds us that innovation should never come at the expense of fairness, curiosity, creativity, agency, or human connection. Those must remain at the heart of what we do.

What I love about this book is its balance. The authors embrace the promise of AI—its potential to personalize learning, expand access, and make science more authentic and inquiry-driven—while also acknowledging its risks. They talk openly about bias, inequity, and the danger of over-automation. They remind us that technology alone has never improved education. What matters is *how* we make use of it and *why* we use it.

A significant contribution of this book is the Responsible and Ethical Principles (REP) framework, developed by Zhai, Crippen, and Lee. REP serves as a moral compass and a practical guide for using AI. It distinguishes between responsible principles—such as fairness, transparency, privacy, and human oversight — and ethical principles —such as beneficial use, scientific integrity, and respect for human values. To me, this framework is a game-changer. It asks us to look beyond what AI *can* do and instead focus on what it *should* do. Every new technology in education should be judged not just by its efficiency, but by whether it builds trust, justice, and dignity, and I would add, enhances our capabilities.

The book's structure mirrors its depth. Each section—on goals, procedures, materials, assessment, and outcomes—offers a fresh lens on how AI is reshaping science teaching and learning. Taken together, the various chapters show that AI is not just changing the tools we use to teach; it's changing the very purposes of science education.

One chapter introduces Science Literacy Plus, an expanded vision of scientific literacy that weaves AI literacy into the fabric of science learning. I found this idea powerful. Students today need to do more than understand scientific ideas—they also need to understand how AI systems shape the creation and sharing of knowledge. They must learn to use AI to explore data and build models while also questioning its outputs and considering their ethical and social implications. That's how science education becomes preparation for living—and leading—in an AI-infused world.

The book also reimagines teaching. AI can help with routine work, analyze student thinking, and generate real-time feedback. But rather than reducing the teacher's role, it makes teachers even more essential. Teachers become interpreters and ethical stewards, and can focus their time where it is most needed. Teachers turn data into dialogue, determine strategies to use, and provide opportunities for student growth. AI should *amplify* what teachers do best—fostering curiosity, empathy, and the joy of doing science—not replace it.

I found the discussion about how AI can make learning materials more dynamic and inclusive valuable to changing how teachers teach and students learn. Virtual labs, intelligent simulations, and adaptive tools provide new ways for students to explore complex ideas. Generative AI can even help design lessons that reflect cultural and linguistic diversity, offering new entry points for learners who have historically not taken part in science. But, as the book's authors remind us in several chapters, equity doesn't just happen because technology is available. Responsible design requires intentional human oversight and cultural awareness. In this vision, AI becomes a partner in broadening participation in science—not through automation, but through thoughtful, human-guided innovation.

Assessment is another area where AI's impact is profound. The authors show how AI can make assessment more meaningful by focusing on how students think, argue, and collaborate, rather than just what they can recall. AI can analyze students' written responses, drawn models, and even their group discussions to provide teachers with timely insights. But they also warn against letting AI take over judgment. When teachers or students rely too heavily on machine scoring or feedback, we lose human involvement, reflection, and dialogue that make assessment such a powerful part of learning.

As someone who's spent much of my career developing curriculum materials and assessments, I found this part especially meaningful. Assessment isn't just about measuring learning—it's about supporting it. Well-designed assessments help students make their thinking visible and help teachers see what students understand and where they're struggling. The ideas in this book connect deeply with that vision. AI, used responsibly, can give educators richer, more immediate insights into student reasoning and can help us design curriculum materials that truly foster sensemaking and inquiry. I find that exciting because it reinforces what I have been striving for: curriculum materials and assessments that are coherent, centered on students'backgrounds and ideas, and enhance wonderment.

As the book draws to a close, the authors return to the heart of the matter: the human side of teaching and learning. Even in AI-rich classrooms, what matters most are the things AI can't do—care, imagine, or wonder. AI can generate text, but it can't feel joy when a student's eyes light up with understanding. It can simulate inquiry, but it can't build community. Those human qualities—curiosity, compassion, agency, and creativity—will always be the essence of outstanding science teaching and the values we hope to instill in learners.

That's what makes *Advancing AI in Science Education* such a significant contribution. It doesn't treat AI as a disruptor or a magic that will solve all the challenges we face in science education. Instead, it treats AI as a cognitive tool we can use to change how teaching and learning occur. Will we design AI systems that chase automation and efficiency, or ones that nurture curiosity, creativity, and usable knowledge? Will we let technology drive education, or will we use it to strengthen our shared vision for meaningful science learning so that learners can make sense of phenomena and solve challenging problems they face in their lives?

For me, this book reaffirms something I've always believed: science education is about helping people make sense of the world, developing usable knowledge to solve problems, and empowering individuals to shape it for the better. If used wisely, AI can help us do that. It can support teachers, engage students, and bring new kinds of learning experiences to life.

Advancing AI in Science Education shows us that future. It challenges us to innovate boldly, but with integrity. It reminds us that technology should serve people, not the other way around. And it invites us—all of us—to help shape a future where science education continues to inspire curiosity, critical thinking, and care for the world we share.

Lappan-Phillips Professor of Science Education
University Distinguished Professor
Michigan State University,
East Lansing, MI, USA

Joe Krajcik

Preface

This book was catalyzed by the *Research in Artificial Intelligence-Involved Science Education (RAISE)* Research Interest Group that we co-founded within the NARST organization. The RAISE community has created a space that facilitates numerous ongoing dialogues and collaborations about using artificial intelligence (AI) to enhance the implementation of the *Framework for K-12 Science Education* and the *Next Generation Science Standards*. Although substantial progress has been made, educators continue to seek ways to realize the potential of AI and understand how to make three-dimensional science learning accessible and meaningful in diverse classrooms. During the course of writing this book, new questions have arisen about how students, teachers, and policymakers engage with AI in science learning.

Supported by the U.S. National Science Foundation as a conference grant from the DRK-12 Program under Grant No. 2332964, we formed a committee, *Advancing Artificial Intelligence for Science Education* (AASE), bringing together a group of 14 experts internationally recognized in the field with the charge to examine the *changes* that AI brings to science education and *how* to meet the changes responsibly and ethically. To meet the goal, AASE committee members met numerous times online and conducted an in-person meeting in Athens, Georgia, in February 2025. They collaboratively and systematically examined how AI impacts five aspects of science education: educational goals, procedures, learning materials, assessment, and outcomes. Based on the outcomes of that process, they use the chapters of this book to recommend frameworks and directions for research, practice, and policy.

This book does not simply document the uses of AI in science education, but instead provides a framework for how these technologies can be adopted responsibly, ethically, and with human values at the forefront. This book serves as both a reflection of the field's current moment and an invitation to shape what comes next.

The book reflects the committee's collective charge and vision. Although chapters were authored individually, we made a deliberate decision for the volume to speak with one voice, that of the committee. To achieve this, the committee worked toward a common structure and shared commitments across chapters. Each chapter is designed to be both accessible and innovative, striking a balance between conceptual clarity and forward-looking insights. We sought readability without sacrificing rigor, and inclusivity without diluting disciplinary depth.

The contributors to this volume, the members of the committee, bring a wide range of expertise, spanning teacher education, learning sciences, classroom practice, educational technology, AI, and policy, and each is recognized as a leader in their domain of science education. This diversity is reflected throughout the book, ensuring a breadth of perspectives while also highlighting the depth of reputation and scholarship each author contributes.

As co-chairs of the AASE committee, we are deeply grateful to the authors for their care and expertise, to colleagues and students whose insights sharpened our thinking, and to the communities that inspire our commitment to ethical and responsible practice. We also gratefully acknowledge the National Science Foundation for its support of this work, and especially the collegiality and guidance of Program Officer Asli Sezen-Barrie. Any opinions, findings, and conclusions or recommendations expressed in this material are those of the author(s) and do not necessarily reflect the views of the National Science Foundation. We also thank Jennifer Kleiman, graduate student, who serves as manager for the AASE conference, and Christina Chhin, who proofread all the chapters.

Xiaoming Zhai
Kent Crippen

Contents

Part I Introduction

1 **The Landscape of AI in Science Education: What Is Changing and How to Respond** . . . 3
Xiaoming Zhai and Kent Crippen

2 **AI in Contemporary Scientific Practice: Implications for AI-Integrated Science Education** . . . 27
Ross H. Nehm and Marcus Kubsch

3 **Responsible and Ethical Principles for the Practice of AI-Supported Science Education** . . . 49
Kent J. Crippen, Xiaoming Zhai, and Okhee Lee

Part II Educational Goals

4 **AI-Enhanced Science Literacy: Science Literacy Plus** . . . 83
Xiufeng Liu and Jennifer N. Tripp

5 **AI Transformations of Science Education Goals to Prepare Students as Lifelong Investigators** . . . 107
Marcia C. Linn, Libby Gerard, and Allison Bradford

Part III Educational Procedures

6 **Science Teaching and Learning with Artificial Intelligence** . . . 133
Knut Neumann, Xiaoming Zhai, and Marcus Kubsch

7 **Capitalizing on Artificial Intelligence to Create Critical Entry Points That Support Science Teacher Learning and the Work of Science Teaching** . . . 163
Jamie N. Mikeska, Natalie S. King, and Yizhu Gao

Part IV Learning Materials

8 Transforming Science Learning Materials in the Era of Artificial Intelligence 193
Xiaoming Zhai and Kent Crippen

Part V Assessment

9 Addressing Challenges of Science Assessment with Artificial Intelligence 219
Lei Liu, Kevin Haudek, Xiaoming Zhai, Ross H. Nehm, and Knut Neumann

10 AI-Based Science Assessment: Potential Pitfalls and Responsible and Ethical Practices 247
Gyeonggeon Lee, Xiaoming Zhai, and Okhee Lee

Part VI Educational Outcomes

11 Preparing Students for Living in an AI Infused World: Revisiting the Outcomes of STEM Education 273
Marcus Kubsch, Paul Hur, Knut Neumann, and Juan-Carlos Aguilar

Part VII Conclusion

12 Charting the Future of AI-Supported Science Education: A Human-Centered Vision 295
Xiaoming Zhai and Kent Crippen

About the Contributors

AASE Committee Co-Chairs

Xiaoming Zhai Associate Professor in Science Education and AI, (courtesy) Associate Professor in Computer Science and in Statistics, serves as Director of the AI4STEM Education Center and the National GENIUS Center at University of Georgia. His research pioneers the use of AI to transform STEM education. He has published widely, serves as guest editor of various journals, chaired the 2022 and 2025 AI in STEM conferences, and co-leads NARST's RAISE group. He received the AERA TACTL Early Career Award, Humboldt Fellowship, and NAEd/Spencer Award. Dr. Zhai is a leading voice in shaping the future of AI for STEM education.

Kent J. Crippen Professor of STEM Education at University of Florida. His research advances a resilient STEM workforce by designing environments that immerse students in authentic science and engineering practices. The Crippen Research Group investigates learning and participation across interdisciplinary contexts, including extended reality, AI and data science, engineering design, and molecular modeling. With strong funding and national honors, including as AAAS Fellow, Crippen's scholarship emphasizes innovation, integrity, impact, and teamwork. He serves as Editor-in-Chief of the *Journal of Science Education and Technology*, founding editor of the *Advances in Technology-Rich Science Education* book series, and co-leads NARST's RAISE RIG.

Committee Members

Natalie S. King PhD, is an Associate Professor of Science Education at Georgia State University. Her research advances Black girls in STEM, community-based programming, and equity-centered curriculum. She received the 2023 NSF Alan T. Waterman Award, the nation's highest honor for early-career scientists and engineers, and an NSF CAREER Award. Her work challenges traditional narratives about girls' participation in STEM while centering identity, brilliance, and belonging. Dr. King served as the PI for an NSF Noyce project diversifying the STEM teaching workforce. She is Founder and Executive Director of I AM STEM Foundation, expanding culturally responsive STEM opportunities for Black and Brown youth.

Marcus Kubsch Assistant Professor of Physics Education Research at Freie Universität Berlin, is the co-founder and chair of the AI in Science Education Research working group in the German physics, chemistry, and biology education research professional organizations. His research uses a diverse range of AI methods and techniques to better understand and support teaching and learning in physics. Further, his work strives to develop the methodological foundations for effective AI use in science education research. His honors include numerous awards, including best paper awards from the NARST RAISE group and the NARST Early Career Research Award.

Knut Neumann Full Professor, serves as the Director of the Department of Physics Education at the IPN–Leibniz Institute for Science and Mathematics Education. His research interests are centered around the question of how students can be best supported in developing competence in science. He has contributed to the development of learning progressions for several core science concepts. More recently, he has focused on the potential of AI to analyze individual students' learning trajectories and to individually support students for an optimized progression toward competence in science. His work has been published in numerous high-quality, peer-reviewed journals.

Lei Liu is Research Director at ETS Research Institute and Adjunct Professor at the University of Pennsylvania. Her research advances technology-enhanced assessment and learning innovation, with a recent focus on integrating artificial intelligence into educational measurement to capture and foster students' reasoning, argumentation, collaboration, and other durable skills. She leads multiple federally funded and philanthropic projects and contributes to large-scale operational programs such as NAEP and state assessment systems. Widely published in leading journals and edited volumes, her work bridges learning sciences, science education, educational technology, and assessment to advance evidence-centered and insight-driven innovations in learning and measurement.

Xiufeng Liu Chair Professor of STEM Education in the Faculty of Education, University of Macau, and SUNY Distinguished Professor Emeritus of the University at Buffalo, State University of New York. His research interests include measurement and evaluation in STEM education, student conceptual progression of cross-cutting concepts (i.e., matter and energy), and student and teacher STEM identity measurement and development. He led multiple NSF and NIH funded projects to support teachers and students to learn science and to measure their learning. Currently, he is developing and validating instruments and AI-enhanced tools to measure the processes and outcomes of STEM literacy comprising creative interdisciplinary problem-solving, STEM identity, and nature of STEM.

Gyeonggeon Lee is Assistant Professor at Natural Sciences and Science Education Department, National Institute of Education, Nanyang Technological University. His research focuses on integrating artificial intelligence (AI) into science education, encompassing automatic assessment, human–AI collaboration in learning, and multimodality. Building on the interdisciplinarity enabled by AI, he aims to fulfill whole-person education. His honors include two doctoral dissertation awards from KERA and KCS, early-career research award and paper award from KASE, and Jhumki Basu Fellowship from NARST. He serves on editorial boards of JRST and JOST.

Marcia C. Linn Evelyn Lois Corey Professor of Instructional Science, School of Education, University of California, Berkeley. Member: National Academy of Education, International Academy of Education. Fellow: American Association for the Advancement of Science (AAAS), American Psychological Association (APA), Association for Psychological Science (APS), American Educational Research Association (AERA), and the International Society of the Learning Sciences (ISLS). President: International Society of the Learning Sciences (ISLS), Chai: AAAS Education Section. Board member: AAAS, Educational Testing Service Graduate Record Examination, McDonnell Foundation Cognitive Studies in Education Practice, and National Science Foundation Education and Human Resources Directorate.

Juan-Carlos Aguilar is Director of the Innovative Programs and Research Office at the Georgia Department of Education (GaDOE). He manages and coordinates research-based programs supporting effective state education initiatives aligned with the agency's strategic goals. He also serves as a liaison between GaDOE, statewide organizations, and the University System of Georgia on science education matters. Previously, he served as Science Program Manager, leading efforts to include selected CTAE courses for the fourth science credit, coordinating K-12 science curriculum development, co-directing Georgia's K-12 STEM initiative, and overseeing alignment of state assessments with the Georgia Performance Standards for Science.

Jamie Mikeska is Managing Principal Research Scientist in ETS's Research Institute and Co-director of the Understanding Teaching Pathways research program. Her research focuses on studying the use of measures designed to assess STEM teachers' content knowledge for teaching and instructional quality as summative and formative assessment tools within teacher education and examining how artificial intelligence can be used to support STEM teacher learning of key teaching competencies. She is nationally recognized for research using digital teaching simulations to support teachers' learning and has been principal investigator on several National Science Foundation research grants (#1445641, #1621344, #1813254, #1813476, #2037983, #2032179).

Okhee Lee is a professor and Peter L. Agnew Professor of Education in the Steinhardt School of Culture, Education, and Human Development at New York University. She is committed to advancing research, policy, and practice that promote STEM and language learning for all students. Lee was a member of the Next Generation Science Standards writing team and served as leader for the Diversity and Equity Team. Her work involves integrating science, language, and computational thinking with a focus on multilingual learners. Her latest work addresses justice-centered STEM education with multilingual learners by integrating multiple STEM subjects to address societal challenges.

Kevin Haudek is Associate Professor in the Department of Biochemistry and Molecular Biology and the CREATE for STEM Institute at Michigan State University. His work focuses on discipline-based education research, with an emphasis on enhancing the use of formative assessments in STEM education to support student learning. He employs AI-driven methods to identify emergent ideas in student writing and to develop models that automatically assess student responses, aiming to support students in solving complex scientific problems. He is also interested in how AI tools can be applied to diverse responses while tailoring individualized support for students' developmental learning.

Yizhu Gao is Assistant Professor in AI and Science Education at the University of Georgia. Her work focuses on developing technology to enhance STEM learning and assessment. Using AI-based and big data techniques (data mining, learning analytics, computational psychometrics), she specializes in capturing metrics that show how students learn and what they know by analyzing multimodal data. These metrics drive her to develop explainable digital agents that personalize feedback for students and provide teachers with real-time, actionable alerts. Her work has been published in numerous high-quality, peer-reviewed journals in science education and educational technology.

Ross H. Nehm is Professor in the Department of Ecology and Evolution and a member of the graduate program in Applied Mathematics and Statistics at Stony Brook University, State University of New York. His lab uses AI and other tools to investigate and assess biology concept learning.

Part I
Introduction

Chapter 1
The Landscape of AI in Science Education: What Is Changing and How to Respond

Xiaoming Zhai and Kent Crippen

1.1 Science Education Is at a Crossroads

Science education currently stands at a pivotal crossroads, increasingly shaped by the rapid convergence of longstanding pedagogical traditions alongside the rise of a disruptive innovation—artificial intelligence (AI). AI is no longer confined to isolated industries; its transformative impact now permeates virtually every sector of society such as healthcare, finance, transportation, manufacturing, communication, and including science education (Guo et al., 2024b; Lee et al., 2025). These widespread changes are altering the nature of scientists' work yet are also redefining how scientific knowledge is generated, spread, and applied (Herdliska & Zhai, 2024) (see Chap. 2). Scientists are leveraging AI tools to expedite complex research processes such as data mining, pattern recognition, and simulation in fields like genomics, neuroscience, and environmental science (Eraslan et al., 2019; Gil et al., 2014). This technological evolution is having a cascading effect on the workforce, necessitating new competencies that extend far beyond traditional disciplinary knowledge (Shi, 2025). Employers increasingly value critical thinking, data literacy, adaptability, and interdisciplinary collaboration—skills that are often underdeveloped in conventional educational frameworks (Cheng et al., 2025; OECD, 2021). As the societal fabric becomes more intertwined with AI capabilities, science education must evolve in parallel to prepare learners not only to participate in this new landscape but also to lead within it.

X. Zhai (✉)
AI4STEM Education Center, University of Georgia, Athens, GA, USA
e-mail: Xiaoming.zhai@uga.edu

K. Crippen
School of Teaching and Learning, University of Florida, Gainesville, FL, USA

X. Zhai, K. J. Crippen (eds.), *Advancing AI in Science Education*, Advances in Technology-Rich Science Education 3,
https://doi.org/10.1007/978-3-032-16871-9_1

These far-reaching societal and occupational shifts demand a fundamental rethinking of science education to ensure that students emerge as informed, capable contributors in both civic and professional realms. AI's integration into daily life intensifies this need, exposing the inadequacy of educational systems still rooted in standardized science curricula and outdated assessment practices. For over a decade, researchers have drawn attention to persistent limitations in traditional science education, including declining student engagement, limited differentiation, and a troubling disconnect between classroom instruction and real-world application (OECD, 2017; Zhai & Pellegrino, 2023). Moreover, structural inequities continue to manifest through significant achievement gaps along socioeconomic lines, as many under-resourced schools lack access to current materials, qualified educators, and interactive learning environments (Hu & Morgan, 2024). When viewed against the backdrop of an AI-driven society, these deficits signal an urgent need for systemic reform—one that prioritizes equity, responsiveness, and the cultivation of future-ready skills through science education (Varsik & Vosberg, 2024). Building on this need for transformation, AI is increasingly recognized as a driving force capable of bridging the gap between outdated science education models and contemporary societal demands.

Positioned as a catalyst for innovation, AI introduces a suite of tools and frameworks that directly confront long-standing challenges in science education. These technologies include AI-supported scoring of open-ended written explanations and arguments or drawn models that delivers rubric-aligned scores and formative feedback at scale (Zhai, 2024), allowing students to practice modeling, explaining, and arguing from evidence while teachers track growth with reliability approaching human ratings. Adaptive learning platforms that tailor instruction to individual students' cognitive and emotional profiles, promote differentiated learning experiences that foster deeper understanding (Gao & Zhang, 2025). Intelligent Tutoring Systems (ITS), such as AutoTutor and EER-Tutor, replicate aspects of one-on-one instruction by offering real-time, customized feedback based on student inputs, thereby enhancing engagement and learning efficiency (Latif et al., 2025). Generative AI further enables automatic creation of instructional materials, such as lesson plans and assessments (Lee & Zhai, 2024), reducing teachers' routine workload and freeing them to concentrate on higher-order instructional tasks that more directly support science learning. In addition, AI-supported learning analytics can provide educators with fine-grained, real-time insights into student behavior, comprehension levels, and progression, enabling data-informed instructional decisions for science learning (Gašević et al., 2022; Sajja et al., 2025). These analytics help identify learning bottlenecks, predict at-risk students, and support timely interventions. Empirical studies show that AI-supported platforms like the Web-based Inquiry Science Environment (WISE) (https://wise.berkeley.edu/) significantly improve academic performance, especially among students who benefit from personalized pacing and targeted scaffolding (Li et al., 2024). Through such capabilities, AI is not merely supporting existing science pedagogical strategies but fundamentally re-engineering the teaching and learning process to align with the demands of a future-oriented, AI-driven society. Science education is at a crossroads due to the significant ongoing changes brough by AI, and understanding these changes better equips us to respond thoughtfully and productively.

1.2 An Educational Transformation Model in the Era of AI

Extending the impact of AI technologies beyond tools and platforms, these innovations are fundamentally altering the aims and structure of science education. Educational goals, the broad, guiding purposes that articulate what education seeks to achieve, are shifting from rote memorization toward fostering essential twenty--first-century scientific competencies that ensure graduates can succeed in AI environments (Zhai, 2023). With the integration of AI-supported learning environments to facilitate instructional activities, students can feasibly engage in science and engineering practices that cultivate disciplinary core ideas and crosscutting concepts. AI can enhance the development of critical thinking and inquiry-based reasoning by enabling adaptive simulations, intelligent tutoring, and real-time feedback that personalize exploration (Mafarja et al., 2025). Students build AI literacy by learning to navigate, interpret, and critically assess AI-supported tools, algorithms, and data systems (Cheng et al., 2025). These changes align with broader societal needs for a scientifically literate workforce capable of tackling interdisciplinary challenges in biotechnology, climate science, energy systems, and data-intensive research (National Research Council, 2012). By aligning goals with the realities of an AI-driven society, science education is moving toward preparing students not only to understand scientific content but also to apply scientific knowledge to solve pressing global problems.

The transformation of *educational goals* in science education reflects a paradigm shift in what it means to be scientifically literate. Scientists are increasingly relying on AI for significant discoveries (Bianchini et al., 2022; Kitano, 2016), and these innovations need to be introduced to students if we expect them to succeed in contemporary science and achieve scientific literacy in the era of AI (Vartiainen et al., 2021). Instead of emphasizing factual recall and algorithmic problem-solving, the focus has shifted toward cultivating higher-order thinking skills, the ability to interpret and analyze scientific data, and the capacity to engage critically with emerging technologies such as AI, gene editing, and the ever expanding applications of computational modeling (Sinha et al., 2023). Students are also being prepared to grapple with the ethical and societal implications of scientific innovation, ensuring that they can act as responsible citizens and professionals in a rapidly changing world (Akgün & Krajcik, 2024). This transformational goal situates science education at the core of equipping learners with the intellectual agility required for lifelong learning and cross-disciplinary collaboration.

Equally important is the transformation of *educational procedures* in science classrooms, which are being redesigned to promote inquiry, experimentation, and collaborative problem-solving. Educational procedures refer to the structured methods, routines, and pedagogical practices educators use to organize and deliver instruction. With the integration of AI technologies, pedagogical models such as flipped classrooms, problem-based learning, and inquiry-driven laboratory work are continuing to gain in relevance, and prominence. AI-supported platforms provide real-time feedback, simulate experimental conditions, and personalize learning sequences, enabling students to design and test hypotheses, analyze experimental data, and refine their understanding of scientific concepts (Krenn et al., 2022; Lee et al., 2024). More importantly, the instructional and learning process can be more efficient with the support of AI. For example, intelligent tutoring systems can guide students through complex topics such as molecular dynamics or climate modeling, offering targeted support where misconceptions arise and scaffolding deeper conceptual growth (Gobert et al., 2024). This transformation inevitably requires teachers to develop new knowledge and skills to accommodate the changes (Shi & Choi, 2024), thus improving the overall learning experience for students.

Learning materials in science education, the resources designed or adapted to support the learning process, are also undergoing substantial transformation, increasingly evolving into interactive, adaptive, and multimodal forms. Traditional static textbooks are being replaced by virtual laboratories, AI-supported simulations, and augmented reality environments that allow students to engage directly with scientific phenomena (Lee et al., 2024). For instance, platforms such as Google Teachable Machine provide opportunities for learners to manipulate variables and train algorithms for scientific inquiry (Herdliska & Zhai, 2024), while generative AI such as GPT offers opportunities to explain graphical representations for real-world phenomena (Bewersdorff et al., 2025). In addition, to mimic scientists' authentic research, these AI-supported materials promote inclusivity and accessibility by adjusting to individual learning preferences, linguistic backgrounds, and cognitive needs, making advanced scientific exploration available to diverse populations of learners worldwide. Moreover, AI is playing an increased role in assisting educators and researchers in creating up-to-date learning materials (Su & Zhong, 2022), freeing curriculum designers from heavy time-consuming work so that they can focus on more intellectual design and development.

Assessment practices in science education, both formative and summative, are similarly being reshaped by AI, moving away from standardized, summative testing toward formative, performance-based, and authentic evaluation (Linn & Gerard, 2024). AI has enabled automatic scoring of students' written explanations and arguments and drawn models, significantly advancing automaticity, and extending the complexity of constructs that assessment can target (Nehm, 2024; Zhai et al., 2020). AI-supported learning analytics can track how students design experiments, interpret data, and communicate scientific arguments, thereby assessing both process and product in scientific inquiry (Pellegrino, 2024). Real-time dashboards for teachers and learners capture patterns of misunderstanding in topics like genetics or thermodynamics, allowing for immediate intervention. In addition, longitudinal assessment models supported by AI generate comprehensive learner profiles, making it possible to trace the development of scientific reasoning skills across years of study (Liu et al., 2024), thereby embedding assessment as a continuous component of the learning cycle rather than as a final endpoint.

Ultimately, these transformations in goals, procedures, materials, and assessment culminate in a redefinition of *educational outcomes* in science education, the knowledge, skills, attitudes, and capacities that learners develop. Outcomes are no longer confined to the mastery of scientific facts or completion of laboratory exercises; they now emphasize creativity in experimental design, collaboration in interdisciplinary contexts, resilience in problem-solving, and ethical reflection on the applications of science and technology, including AI. Students who have mastered AI-supported science education are envisioned as adaptive, lifelong learners who are capable of addressing complex scientific and societal challenges with the responsible and ethical use of AI. This transformation ensures that science education not only conveys disciplinary knowledge but also equips learners with the competencies required to lead innovation in an AI-driven future.

1.3 Obstacles Along the Journey

Yet, despite AI's transformative promise, its integration into science education brings a complex and expansive array of ethical, social, and pedagogical concerns that demand not only closer examination but ongoing critical reflection. At the forefront of these concerns is the issue of fairness (Xing & Li, 2024). While AI holds the potential to enhance learning opportunities equitably, its current deployment may reflect and reinforce systemic inequities already embedded in educational systems. For example, algorithmic bias, stemming from historically skewed, incomplete, or non-representative datasets, can entrench preexisting disparities rather than address them. Educational AI tools, especially those trained predominantly on data from affluent or high-performing schools, may unintentionally marginalize students from under-resourced environments. This marginalization can manifest in various ways—from unfair or inaccurate feedback to faulty predictions that influence academic placement or opportunities, leading to a form of digital gatekeeping that limits student potential and exacerbates existing achievement gaps (Kubsch et al., 2022). In addition, over-cautiousness and pseudo AI bias have emerged as significant factors that prevent the acceptance of AI, leading to expanded disparities among users in different groups (Zhai & Krajcik, 2024). These concerns become especially acute in diverse classrooms where educational equity is already a persistent challenge.

Transparency and explainability present equally significant issues. A troubling number of AI systems function as opaque boxes, generating decisions and recommendations that even developers may struggle to fully interpret (Wu et al., 2024). In educational contexts where AI influences assessment, learning pathways, or instructional resources, students and educators must be able to discern the reasoning behind the system's outputs. When transparency is lacking, the result is a troubling reliance on machine-generated judgments that go unscrutinized, undermining the educational process and placing undue trust in unaccountable algorithms. Furthermore, this opacity leads directly to the issue of accountability (Akgün & Krajcik, 2024). When AI systems make mistakes—whether through flawed programming, biased data, or contextual misinterpretation—who is held responsible? Unfortunately, educational settings often lack clear protocols for assigning accountability, leaving students and teachers without recourse when AI tools cause harm or misjudgement.

Privacy and data security add another dimension to the ethical landscape. AI applications in education rely heavily on gathering vast amounts of data, including not only academic performance but also behavioral patterns, engagement levels, and sometimes even biometric or location-based information (Latif & Zhai, 2025). This extensive data collection raises urgent questions about consent, control, and usage. For example, who owns this data? How long is it stored? Who has access, and under what circumstances? Increasing concerns have emerged about the exploitation of student data for commercial purposes or surveillance, undermining institutional trust and student autonomy. More than a technical challenge, data privacy reflects a broader institutional value system: whether education remains a protected space for intellectual growth or becomes a site of commodified surveillance.

Concerns about human oversight and autonomy further complicate AI's role in the classroom. While AI can offer valuable support by automating routine tasks or generating real-time analytics, it cannot and should not replace the human dimensions of teaching (Shi & Choi, 2024). There is a growing risk that over-reliance on AI will diminish the educator's role, reducing teachers to monitors of algorithm-driven instruction and students to passive recipients of machine-curated content. This shift could erode opportunities for meaningful dialogue, mentorship, and moral development—hallmarks of quality education. Students must be empowered to question AI recommendations and remain active participants in their learning journey, rather than being conditioned to accept technological authority as absolute. The erosion of human judgment and agency in classrooms poses a serious threat to the core educational mission of fostering independent, critical thinkers.

Another pressing concern involves the intended purposes behind AI use in educational environments. While AI is often introduced under the guise of innovation, many implementations serve efficiency-driven objectives, such as standardizing content delivery or streamlining testing procedures. Such applications, without careful attention and thoughtful human-involvement, risk reducing science education to mechanized instruction, stripped of the exploratory, creative, and inquiry-based elements that make scientific learning meaningful. This could be particularly detrimental to historically minoritized populations in STEM fields, for whom sustained engagement and personal relevance are essential for inclusion and long-term success (Nyaaba et al., 2024). The tension between genuine educational enrichment and superficial AI integration calls for deep reflection on whose needs are being prioritized and what pedagogical values are being upheld.

Further complicating the integration of AI into science education are concerns surrounding robustness and scientific integrity. Despite bold claims of precision and adaptability, AI systems remain susceptible to technical limitations. Inaccurate models, poorly curated datasets, or oversimplified algorithms can distort complex scientific concepts or propagate misinformation or feedback (e.g., over inferences or over praise, see Guo et al. (2024a)). The iterative and often ambiguous nature of scientific inquiry—which depends on uncertainty, debate, and evolving evidence—can be flattened by AI systems that privilege deterministic outputs. At worst, students may receive misleading or overly confident explanations that contradict best practices in scientific reasoning. Such distortions undermine science education's foundational commitment to evidence-based thinking and intellectual rigor.

Lastly, there is an overarching concern that AI could erode essential human values in education. As systems increasingly quantify and rank learners through data analytics, there is a risk of reducing students and teachers to mere data points. This process may inadvertently marginalize diverse cultural perspectives and suppress ethical discourse in favor of standardized metrics. Moreover, embedded biases in AI systems may reflect dominant ideologies, leading to exclusions that are subtle yet powerful in shaping who feels seen, valued, and heard within educational spaces. These dynamics raise fundamental questions about moral agency, ethical responsibility, and the kind of society education aims to cultivate.

Without sustained and critical attention to these multifaceted concerns—including fairness, transparency, accountability, privacy, human oversight, engagement, robustness, and respect for human values—AI in science education risks becoming a force of simplification and exclusion rather than empowerment and inclusion. Instead of enhancing learning, it may deepen existing divides, reduce science to algorithmic exercises, and erode the human relationships and reflective capacities that lie at the heart of meaningful education.

1.4 Our Approach: Responsible and Ethical Principles for the Practice of AI-Supported Science Education

The concerns raised in the previous section—*fairness, transparency, accountability, privacy, human oversight, and the preservation of human valu*es—make clear that the integration of AI into science education must be guided by more than technical feasibility or efficiency. Left unchecked, AI could deepen inequities, obscure accountability, or diminish the agency of teachers and students. At the same time, its potential for supporting inquiry, enabling personalization, and widening access to scientific practices is undeniable. Navigating these tensions requires an intentional framework that balances innovation with integrity. It is not enough to ask what AI *can* do in classrooms; we must also ask *how* it should be used and *why* it should be used at all.

To meet this challenge, we turn to the Responsible and Ethical Principles (REP) Framework, which is developed more fully later in the book (See Chap. 3, Crippen et al., 2025). The REP framework makes an important distinction between *responsible principles*—those that focus on the safeguards, practices, and methods for implementing AI (i.e., how)—and *ethical principles*—those that clarify the purposes, values, and broader human aims that justify its use (i.e, why). By holding these two dimensions together, the framework provides a balanced approach: it insists that AI systems be transparent, fair, and accountable, while also reminding us that their ultimate purpose is to promote meaningful learning, uphold scientific integrity, and respect the dignity and diversity of all learners.

In this book, the REP framework is used not as a compliance checklist but as a lens for educational transformation. Applied across the five core areas identified in our model—goals, procedures, materials, assessment, and outcomes—REP helps us see how AI can reshape science education in ways that align practice with values. For example, it frames educational goals around cultivating inquiry and critical reasoning rather than passive knowledge acquisition; it encourages classroom procedures that use AI to amplify rather than replace teacher and student agency; it calls for learning materials that are not only adaptive but also inclusive and culturally relevant; it informs assessment practices that prioritize authenticity and ownership of learning; and it guides the envisioning of outcomes that prepare learners not only to use AI tools but also to critique and shape them as informed citizens and professionals. In this way, REP links ethical commitments with practical action across the entire educational system. Our approach, then, positions AI as a partner in science education rather than a replacement for human judgment or relational teaching. By embedding REP into the design and use of AI, we ensure that innovation does not erode the values that make science education meaningful: equity, inquiry, creativity, integrity, and democratic participation. This orientation invites educators, students, policymakers, and designers to see AI not as a neutral tool but as a technology whose influence must be continually interpreted, critiqued, and guided by shared human purposes.

Below, we weave the eight principles of the REP into a synthesized call for action on how they may unfold in science education (for details see Chap. 3, Crippen, Zhai, & Lee, 2025):

Fairness in Opportunity and Design Science education must ensure that AI expands, rather than restricts, participation. Fairness means more than avoiding overt bias; it requires proactive inclusion. Imagine an AI simulation platform for physics that disproportionately recommends advanced challenges to students whose prior performance signals mastery, but neglects learners still developing foundational skills. Such practices could amplify inequities. To counteract this, teachers and developers must design adaptive systems that scaffold struggling learners without narrowing their opportunities. Inclusive design can be seen in AI-supported lab simulations that provide multilingual scaffolds or accessibility features for students with disabilities, thereby broadening participation in authentic scientific inquiry.

Protecting Privacy and Data Security As AI systems rely heavily on learner data, protecting privacy is both a technical and ethical obligation. Data should not only comply with Family Educational Rights and Privacy Act (FERPA) and the General Data Protection Regulation (GDPR). But also align with contextual sensitivities in the classroom. For example, an AI dashboard tracking student progress should present insights without exposing sensitive personal details to peers. Teachers can model ethical reasoning by engaging students in discussions about consent and data use, making privacy a topic of scientific and civic learning rather than a hidden technical concern. Robust data protections reassure learners that participation in science education does not come at the cost of their dignity or autonomy.

Affirming Human Oversight and Autonomy AI should never displace the expertise of teachers or the agency of learners. For instance, when AI generates a step-by-step solution to a chemistry problem, the teacher's role is to challenge students to analyze, critique, or extend that solution rather than accept it passively. By situating AI as a partner for exploration, learners develop critical stances toward evidence and reasoning. Similarly, teachers should retain decision-making authority in adapting AI-supported recommendations about instruction, ensuring that the unique social and relational dimensions of classrooms are preserved.

Fostering Beneficial and Meaningful Use AI must be deployed with a focus on enriching authentic learning rather than streamlining efficiency alone. In practice, this may mean leveraging AI to support inquiry-based projects where students use AI-supported visualizations of ecological data to form hypotheses and test them through field investigations. Such applications turn AI into a catalyst for curiosity and deeper reasoning, enabling learners to engage more directly in the practice of science. By contrast, systems that automate grading without meaningful feedback risk diminishing the learning process and reducing opportunities for reflection.

Upholding Robustness and Scientific Integrity Scientific accuracy and reliability are non-negotiable in AI-supported learning. Consider a scenario in which an AI chatbot explains photosynthesis but provides oversimplified or inaccurate causal mechanisms. Left unchecked, such outputs undermine scientific literacy. Developers must rigorously test systems for alignment with disciplinary knowledge, and teachers must cultivate classroom practices where AI outputs are treated as hypotheses to be tested, not facts to be memorized. In doing so, learners develop both epistemic humility and scientific reasoning skills.

Respecting Human Values and Cultural Identities Science education must foreground the dignity and cultural identities of learners. AI tools can be designed to incorporate culturally responsive contexts—for example, presenting climate change data using local environmental issues that resonate with students' lived experiences. Teachers can also use AI-supported prompts to facilitate classroom debates about the ethical implications of emerging technologies, thereby strengthening both civic reasoning and cultural sensitivity. In this way, AI enhances, rather than erodes, the human dimensions of science education.

Promoting Transparency and Explainability Complex AI models must be rendered interpretable for teachers and learners. A biology class might use an AI tool that predicts species survival under different climate conditions. Rather than treating predictions as opaque, students should be given insights into the data inputs, assumptions, and limitations behind the model. Teachers can frame these explanations as opportunities for critical engagement with uncertainty and scientific modeling, thereby strengthening data literacy. Transparency allows learners to not only consume AI outputs but also interrogate how those outputs are produced.

Embedding Accountability Across Stakeholders Accountability ensures that responsibility rests with the humans and institutions that design and implement AI, not with learners or the technology itself. For example, if an AI-supported assessment system consistently underrates the performance of students from certain linguistic backgrounds, schools must have clear mechanisms for appeal, review, and correction. Developers, too, must commit to monitoring long-term impacts and communicating failures openly. This shared accountability reinforces trust and ensures that AI use remains aligned with educational goals.

1.5 Overview of the Book: What AI Is Changing and How Should We Respond

In response to these complex dynamics, in this book, the authors undertake a comprehensive analysis of how AI is reshaping the landscape of science education. We critically examine how AI is catalyzing shifts in educational goals toward scientific literacy, such as critical thinking, ethical reasoning, and interdisciplinary problem-solving. We explore the transformation of teaching practices through intelligent systems that support differentiated instruction and real-time scaffolding, and the redesign of curriculum to include AI literacy, data science, and inquiry-driven exploration. We also analyze changes in assessment modalities, highlighting the move from static, summative evaluations to dynamic, formative assessments informed by real-time analytics and predictive feedback. Finally, the authors investigate how these changes collectively redefine learning outcomes, emphasizing the development of agile, collaborative, and reflective learners equipped to thrive in an AI-driven future. Drawing on empirical evidence and case studies, we propose ethically grounded, practically viable strategies for integrating AI in ways that amplify educational equity, foster scientific literacy, and prepare learners for the evolving demands of the twenty-first century. In the subsequent section, we outline the key ideas for each chapter of sections II to VI of the book according to the components of educational transformation: educational goals, educational procedures, learning materials, assessment, and educational outcomes.

1.5.1 Educational Goals

To address the profound changes to science education goals brought by AI, this book includes two dedicated chapters on the topic. Chapter 4 highlights how AI is reshaping science literacy by expanding it across three visions. In *Vision I*, AI transforms core scientific practices such as hypothesis generation, modeling, data analysis, and solution design, requiring students to develop competencies in AI-supported scientific inquiry rather than focusing solely on traditional content mastery. In *Vision II*, AI operates as a collaborative partner in applying science to solve authentic, real-world problems, urging learners to engage with AI tools critically and creatively. *Vision III* extends science education goals further by empowering students as agents of change, where responsible and ethical engagement with AI becomes crucial for addressing socio-political, cultural, and environmental challenges. Collectively, these changes call for an integrative model of *Science Literacy Plus*, where AI literacy is interwoven with science literacy to prepare learners for active participation in an AI-supported world.

To meet these evolving goals responsibly and ethically, Chap. 4 stresses the need to embed ethical frameworks and critical reflection into AI-supported science education. Responsible practices include ensuring equitable access to AI tools, mitigating algorithmic bias, and fostering transparency, accountability, and human oversight. Educators are encouraged to guide students in recognizing AI's limitations, questioning whose perspectives are represented, and reflecting on the broader social and environmental impacts of technological applications. Ethical engagement also involves protecting privacy, respecting Indigenous knowledge, and promoting sustainability in the development and use of AI systems. Ultimately, we argue that science education should cultivate not only technical proficiency but also values of empathy, justice, and collective responsibility, enabling learners to leverage AI for societal good while avoiding harm.

Chapter 5 builds on these insights by emphasizing that AI is fundamentally transforming science education goals from memorization and routine experimentation to sustained inquiry and lifelong investigation. By combining the advances brought by AI with Knowledge Integration (KI) pedagogy, classrooms can become design laboratories where students engage in authentic investigations, interpret complex phenomena, and devise solutions to pressing societal and environmental challenges. In this case, AI scaffolds the inquiry cycle by supporting students as they brainstorm researchable questions, visualize and test models, analyze data, and refine explanations. This shift reorients educational goals toward agency, creativity, and problem-solving, ensuring learners are prepared not only to understand science but also to apply it in addressing real-world dilemmas alongside AI tools.

To ensure these transformations occur responsibly, Chap. 5 underscores the importance of aligning AI-supported inquiry with principles of fairness, transparency, accountability, and human oversight. Ethical use of AI in science education requires that guidance systems honor the diversity of students' ideas without reinforcing bias, while safeguarding privacy and ensuring data security. Partnerships among teachers, students, and designers are essential to critically examine AI outputs, refine instructional uses, and preserve student agency in investigations. Professional development and collaborative curriculum design are also needed so that educators can effectively mediate AI integration while upholding inclusivity, human dignity, and democratic participation. By embedding responsible practices into AI-supported inquiry, science education can leverage AI's opportunities while mitigating risks, preparing learners as autonomous investigators and ethical participants in an AI-driven society.

In summary, Chaps. 4 and 5 together illustrate a comprehensive transformation of science education goals driven by AI. While Chap. 4 reframes science literacy through the lens of AI across three visions, Chap. 5 emphasizes the inquiry-focused practices that AI can scaffold in classrooms. Both chapters highlight that realizing AI's potential in education requires a strong ethical foundation—one that balances technical advancement with values of fairness, inclusivity, sustainability, and democratic participation. Together, these chapters chart a path toward cultivating learners who are not only scientifically and technologically literate but also grounded in the ethical and socially responsible use of AI.

1.5.2 Educational Procedures

Chapters 6 and 7 collectively illuminate the transformative role of AI in reshaping science education by redefining both teacher practice and student learning. These chapters emphasize that the integration of AI is not a peripheral enhancement but a structural shift toward more data-informed, student-centered, and co-orchestrated approaches to teaching and learning.

Chapter 6 introduces three significant changes to the procedures of science education—enhancing the personalization, authenticity, and efficiency of learning processes. Through predictive and generative technologies, AI systems support teachers in tailoring instruction to diverse learners, provide real-time feedback, and create adaptive learning environments such as intelligent tutoring systems and exploratory learning platforms. These tools enable students to engage more actively and autonomously in inquiry-based learning, while also offering teachers synthesized insights about student performance and needs. In addition, AI-supported platforms can create multiple versions of assignments, provide visualizations, and generate scaffolded supports that address students' varied levels of prior knowledge. This flexibility allows educators to more effectively accommodate the wide range of interests, competencies, and cultural backgrounds present in modern classrooms. Moreover, AI reshapes classroom dynamics by functioning not only as a tool but also as an epistemic partner, thereby redistributing roles of authority and responsibility among teachers, students, and technological agents. As a result, science education is evolving into a more individualized, data-informed, and co-orchestrated learning ecosystem that can better address the complexity of student diversity, the challenges of authentic scientific inquiry, and the broader demands of modern scientific literacy.

Addressing these changes responsibly and ethically requires a multi-faceted approach that emphasizes equity, transparency, and human-centered practice. The chapter highlights that while AI tools can amplify instructional precision, they also carry risks related to bias, privacy, and inequitable access, particularly in under-resourced educational contexts. Ethical use demands that teachers and school systems critically interrogate how algorithms are trained, what assumptions underlie their design, and how their outputs are interpreted in relation to student learning. Professional development is thus central, not only for building technical fluency but also for fostering educators' ability to question, contextualize, and adapt AI recommendations within culturally responsive and justice-oriented pedagogies. Equitable implementation also requires systemic commitments—such as providing infrastructure, sustained support, and opportunities for collaboration across institutions—to ensure that AI enhances rather than exacerbates existing disparities. Ultimately, the responsible integration of AI hinges on positioning it as a complement to, rather than a substitute for, teachers' expertise, creativity, and ethical judgment. By doing so, schools can harness AI's efficiency and analytical power to expand inquiry, foster deeper engagement, and cultivate classrooms where technological innovation aligns with core values of inclusivity, fairness, and respect for the human dimensions of learning.

Chapter 7 underscores the transformative impact of AI within science teacher education, where these technologies are reshaping teacher practice by streamlining traditionally time-intensive tasks such as lesson planning, formative assessment design, and classroom orchestration. Through natural language processing and machine learning applications, teachers now have access to automated scoring systems, intelligent dashboards, and personalized feedback mechanisms that highlight student misconceptions in real time. This allows educators to adapt instruction more responsively, aligning it more closely with the multidimensional goals of frameworks like the Next Generation Science Standards (NGSS; NGSS Lead States, 2013). Moreover, the use of AI can support teacher professional growth through simulations and practice-based environments where teachers can rehearse core instructional competencies, such as eliciting student thinking or facilitating scientific argumentation, in reduced-complexity contexts. These developments not only save time but also extend opportunities for teachers to refine their craft with ongoing, targeted, and data-informed insights. In sum, AI integration signifies a shift from labor-intensive processes to more efficient, evidence-based, and student-centered instructional decision-making, positioning teachers to more effectively meet the needs of diverse learners while continuing to advance their professional development.

The transformative impact of AI in science education procedures, however, underscores the necessity of responsible and ethical integration. To safeguard meaningful learning, educators must design instructional strategies that integrate AI with curricula rather than rely on AI as a stand-alone solution. Policies and frameworks are needed to define clear roles for teachers, students, and AI, ensuring that technology complements rather than replaces human judgment. Ethical challenges such as bias in training data, lack of transparency in algorithmic processes, and issues of inclusivity demand systematic attention, alongside efforts to foster AI literacy among both teachers and students. Professional development programs are essential to equip educators with the competencies required for critical evaluation, effective orchestration, and the nurturing of students' epistemic autonomy. Moreover, institutional support is required to ensure that AI adoption does not widen educational inequities but instead contributes to inclusive and equitable opportunities for learning. Teachers and students must learn to critically interrogate AI outputs, engage in reflective practices, and understand both the potentials and the limitations of algorithmic systems. Ultimately, responsibility rests with human actors to ensure that AI tools amplify equity, fairness, transparency, and deep engagement in science learning rather than reduce education to superficial or automated processes.

1.5.3 Learning Materials

Chapter 8 addresses the changes to science learning materials by highlighting AI's profound role in transforming the design, function, and accessibility of the resources used to support learning. AI expands the scope of learning materials beyond static, text-based representations to dynamic, adaptive, and multimodal environments. Unlike conventional textbooks that present fixed facts and limited interactivity, AI-supported platforms integrate authentic scientific practices such as hypothesis testing, large-scale data analysis, and predictive modeling, mirroring how science is conducted in contemporary fields like genomics and climate research. Adaptive systems powered by AI adjust the pace, modality, and complexity of instruction in real time, providing personalized learning trajectories that respond to students' prior knowledge, interests, and cultural backgrounds. Interactive AI simulations, for instance, allow learners to design and manipulate experiments, analyze outcomes, and iteratively refine their models, fostering inquiry-based and evidence-driven learning. At the same time, generative AI facilitates co-creation by enabling teachers to rapidly develop culturally responsive and contextually relevant materials—lesson plans, formative assessments, or analogies to abstract concepts—that evolve alongside both curricular standards and scientific discoveries. AI also plays a critical role in advancing accessibility by providing translation, text simplification, and multimodal adaptations for neurodiverse and multilingual learners, thereby supporting more equitable participation in science classrooms. Taken together, these changes reimagine science education as more participatory, responsive, and reflective of the computational, data-intensive nature of modern science.

Nevertheless, the transformative potential of AI in science learning materials is inseparable from pressing ethical and pedagogical challenges, which demand responsible and deliberate responses. Without safeguards, algorithmic bias, misinformation, and opacity in AI-supported content risk reinforcing educational inequities, misrepresenting science, or undermining student and teacher agency. Ethical integration therefore requires transparency measures such as documentation of AI outputs, explainability features, and opportunities for students to critique, validate, and reflect on AI-supported insights. Privacy protections are equally crucial, as adaptive learning environments depend on sensitive learner data that must be anonymized, secured, and governed by clear consent protocols. Equally important is ensuring scientific integrity: AI-supported simulations and multimodal resources must be grounded in validated models and pedagogical coherence, requiring collaboration among educators, domain experts, and technologists. Accessibility features must be designed with cultural sensitivity and inclusivity in mind, ensuring that AI tools not only translate dominant scientific discourse but also affirm diverse epistemologies and linguistic identities. Moreover, the role of AI as a co-author in curriculum development must remain transparent and accountable, with human educators retaining oversight to ensure alignment with disciplinary knowledge, instructional goals, and ethical norms. By embedding such responsible practices, AI can serve as a collaborative partner that enriches inquiry, creativity, and accessibility while upholding the foundational values of equity, critical reasoning, and scientific integrity in education.

1.5.4 Assessment

In this section, Chaps. 9 and 10 attend to AI-supported assessments from different perspectives. Chapter 9 introduces how AI is reshaping science assessment by redefining constructs, introducing new formats, and transforming classroom practices in ways that both expand opportunities and complicate traditional notions of measurement. Traditional assessments, which largely focused on content recall or procedural fluency, are increasingly insufficient in an era where generative AI can provide immediate and often sophisticated answers to many standardized tasks. Instead, AI is pushing assessment design toward performance-based models that emphasize authentic reasoning, complex problem-solving, collaborative inquiry, and human–AI partnerships. These changes include the emergence of AI-resilient assessments that require students to critique, refine, or extend AI-generated outputs, and the use of multimodal data streams—including text, diagrams, gestures, and speech—to capture complex learning processes. AI also enables new possibilities for formative assessment through real-time feedback, adaptive diagnostics, and integration with classroom simulations that mirror professional scientific practice. Importantly, these innovations make it possible to trace student learning trajectories over time, revealing misconceptions, incremental conceptual changes, and opportunities for targeted intervention. Together, these transformations move science assessment away from static measurement and toward dynamic, iterative, and context-sensitive evaluations that better reflect the nature of contemporary science learning.

The chapter further argues that responsibly and ethically addressing these changes requires principled approaches that safeguard validity, fairness, and equity, while also supporting instructional utility. Developers must ensure that AI-scoring systems are transparent, interpretable, and consistently aligned with assessment constructs, rather than producing black-box judgments. Human-in-the-loop models are essential to preserve teacher oversight, while data sampling strategies must be carefully designed to mitigate bias across diverse student populations and to ensure cultural and linguistic inclusivity. Ethical assessment design also demands cultivating AI literacy among students and educators, enabling them to critically evaluate AI outputs, identify limitations, and maintain epistemic agency in inquiry. Moreover, formative assessments supported by AI must protect student data privacy, comply with legal frameworks, and amplify learning rather than replacing instructional judgment. Professional development for teachers will be crucial to ensure they can interpret AI-supported insights, contextualize them in classroom practice, and intervene when AI feedback is misleading or incomplete. By grounding AI-supported assessments in evidence-centered design, rigorous fairness audits, and sustained human oversight, educators can ensure that technological innovations strengthen rather than compromise the validity, reliability, and inclusivity of science education.

While acknowledging the potentials to science assessment, Chap. 10 also stresses the potential pitfalls of integrating AI in science assessments, particularly in three forms: cognitive outsourcing, bias, and issues of locality. Students may outsource cognitive effort by relying on AI to generate responses, undermining authentic engagement and reducing self-regulated learning opportunities, which over time can limit their ability to develop independent scientific reasoning and metacognitive skills. Teachers, too, risk over-reliance on AI for task design, scoring, and feedback, which could erode their assessment literacy, weaken pedagogical coherence, and diminish their professional agency in making context-sensitive judgments. Additionally, sociotechnical biases in AI scoring—stemming from historical inequities, skewed training datasets, or flawed algorithms—pose serious threats to fairness, disproportionately impacting marginalized learners such as English Language Learners, students from under-resourced schools, or those using non-standard dialects. Locality concerns further complicate adoption, as AI-supported assessments must align with national or regional standards, adapt to varied linguistic and cultural contexts, and remain accessible despite disparities in infrastructure, policy environments, and digital resources. Without careful attention, these challenges risk not only misrepresenting student ability but also reinforcing systemic inequities.

To ethically navigate these challenges, Chap. 10 suggests that educators and developers must reconceptualize assessment practices with human–AI collaboration at their core, ensuring that AI supplements rather than supplants teacher judgment and student engagement. This requires designing assessments that resist over-reliance on AI, such as multimodal and interactive tasks that preserve student agency, encourage critical thinking, and demand context-rich problem solving. Addressing bias involves collecting diverse, representative data across linguistic, cultural, and socioeconomic groups; adopting robust analytic rubric designs that foreground scientific reasoning rather than surface features; and applying statistical and sociological strategies for bias detection, auditing, and mitigation. Locality-sensitive design is equally vital: AI-supported assessments must respect cultural and linguistic diversity, integrate context-specific standards, and expand access through open-source models and equitable infrastructure so that assessment benefits are distributed fairly across global contexts. Transparency, frequent calibration of both human and machine scorers, and structured stakeholder involvement from teachers, students, and policymakers are essential to building trust and ensuring accountability. By embedding these practices—emphasizing fairness, explainability, data security, and ongoing human oversight—AI-supported science assessment can advance equity, authenticity, and educational value while respecting human dignity and promoting more just and inclusive science learning.

Taken together, these chapters highlight that AI is transforming science assessment into a more dynamic, responsive, and authentic practice while also surfacing new ethical and pedagogical challenges. AI enables performance-based, multimodal, and formative assessments that align more closely with the complexity of scientific reasoning and inquiry. At the same time, pitfalls such as cognitive outsourcing, sociotechnical bias, and misalignment with local contexts threaten the integrity and equity of assessment practices. Addressing these challenges requires robust frameworks grounded in validity, fairness, transparency, and human oversight. It also demands building AI literacy among educators and learners so that AI is approached not as a shortcut but as a partner in inquiry and evaluation. When guided by responsible and ethical principles, AI-supported science assessment can enhance both the rigor and inclusivity of science education, ensuring that innovations strengthen, rather than weaken, the educational enterprise.

1.5.5 Educational Outcomes

Chapter 11 highlights how AI has transformed the desired outcomes of science education by broadening its focus from factual knowledge and inquiry skills to include AI literacy, critical reflection, and ethical reasoning. Contemporary frameworks emphasize that students should not only understand scientific concepts but also know how AI functions within scientific inquiry, apply AI tools in authentic contexts, and critically evaluate AI-supported outputs. In this sense, science education now aims to cultivate competencies that align with the realities of an AI-driven world: collaboration with AI systems, recognition of algorithmic bias, and reflective judgment about the social, environmental, and epistemological implications of AI. These outcomes go beyond technical proficiency, preparing learners to navigate complex socio-scientific challenges and to exercise agency in shaping the role of AI in their personal, professional, and civic lives. Expanding on this transformation, the outcomes also incorporate dispositions such as adaptability, resilience in the face of technological uncertainty, and the ability to integrate interdisciplinary perspectives. For example, students are increasingly expected to connect their scientific reasoning with ethical, cultural, and political considerations, thereby situating AI not merely as a technical tool but as a socio-technical system with wide-ranging consequences. These expanded expectations indicate that science education is moving toward a holistic vision where cognitive, practical, and moral capacities are developed in parallel.

Addressing these new outcomes responsibly requires embedding ethical principles and reflective practices throughout science education. Ethical engagement entails teaching students to question the fairness, transparency, and accountability of AI systems, as well as to consider the implications of data bias, misinformation, and unequal access to digital resources. Reflection must be positioned as an ongoing practice, enabling students to move beyond surface-level judgments about accuracy and toward deeper inquiries into the assumptions, purposes, and societal impacts of AI technologies. To achieve this, educators need coherent learning progressions, scaffolded opportunities for inquiry, and curricula that balance technical skills with ethical reasoning. Responsible implementation also involves professional development for teachers so they can confidently integrate AI tools while guiding students through the complexities of ethical dilemmas. Schools and policy-makers must further ensure equitable access to AI resources, reducing digital divides that could exacerbate inequalities in learning outcomes. By fostering critical awareness and a sense of responsibility, science education can prepare students not merely to use AI, but to act as thoughtful stewards of its applications and consequences.

1.6 From Here Forward: With AI in Classrooms, What Roles Remain Uniquely Human?

As the preceding sections illustrate, AI is transforming nearly every dimension of science education—goals, procedures, materials, assessment, and outcomes. These transformations highlight AI's capacity to amplify efficiency, personalization, and authenticity in learning, while also raising questions of fairness, transparency, accountability, and integrity. Yet amid this sweeping change, a critical question remains: what roles in teaching and learning cannot, and should not, be delegated to machines? Identifying these uniquely human contributions is essential, not only to safeguard the integrity of education but also to ensure that AI functions as an enabler rather than a replacement of human judgment, creativity, and empathy. Without clarity on what remains distinctly human, there is a risk of drifting into educational models that privilege automation over meaning, standardization over equity, and efficiency over human growth. Thus, foregrounding the roles that remain uniquely human allows us to envision AI-supported classrooms that advance the transformative aims of science education rather than compromise them.

First, human educators remain indispensable as moral and relational anchors in classrooms. AI can generate solutions, feedback, or explanations, but it cannot genuinely care for students' well-being, cultivate belonging, or build trust across cultural and social divides. Teachers embody empathy, mentorship, and encouragement—qualities that sustain students' motivation, resilience, and identity development in ways no algorithm can replicate. For example, while AI-supported tutors can provide personalized feedback at scale, it is the teacher who notices when a student feels discouraged, recognizes subtle nonverbal cues, and responds with compassion. Beyond individual relationships, teachers also nurture classroom communities, cultivating shared norms of respect, inclusion, and collective inquiry. This relational dimension connects directly to earlier concerns about fairness and equity: without the human touch, AI risks reproducing existing disparities rather than addressing them. Teachers ground education in compassion and ethical responsibility, ensuring that science learning supports not just cognitive growth but also social and emotional flourishing.

Second, teachers and students alike bring interpretive and ethical judgment that AI lacks. Scientific reasoning often involves uncertainty, competing perspectives, and normative considerations. While AI systems may model probabilities or detect patterns, only humans can situate those results within cultural, ethical, and civic contexts. For example, AI can simulate climate change scenarios or analyze genetic data, but decisions about sustainability, equity, and justice require human deliberation. In this respect, the teacher's role extends beyond interpreting data to modeling ethical reasoning for students—demonstrating how to weigh evidence against moral principles, community needs, and long-term consequences. Likewise, students must be empowered to question AI outputs: Whose perspectives are reflected? What assumptions underlie the model? What values should guide our response? These interpretive and ethical functions connect directly to earlier discussions of accountability and transparency, ensuring that AI-supported science education remains socially responsible and aligned with democratic values.

Third, human agency is irreplaceable in fostering creativity, imagination, and curiosity. AI excels at recombining existing knowledge, yet it currently lacks the kinds of boundary-crossing insights and divergent thinking that spark true innovation. In classrooms, teachers inspire wonder, pose provocative questions, and create opportunities for students to pursue open-ended investigations. For instance, while AI might suggest multiple possible explanations for a phenomenon, it is the student's imaginative leap—the "what if" question—that drives new lines of inquiry. Teachers amplify this spirit by designing environments that value experimentation, embrace ambiguity, and legitimize failure as part of the discovery process. This aligns with earlier sections on goals and outcomes, where creativity, resilience, and interdisciplinary problem-solving were highlighted as essential competencies for an AI-driven future. Here, the uniquely human contribution lies not in efficiency but in nurturing the dispositions and curiosity that sustain lifelong learning and scientific innovation.

Ultimately, what remains uniquely human about AI-supported science education is the ability to co-construct meaning through dialogue and community. AI may simulate conversation, but it lacks the lived experience, vulnerability, and shared identity that underpin genuine dialogue. Classroom discussions, peer collaboration, and teacher-student exchanges cultivate epistemic humility and democratic participation, reminding learners that science is not only a body of knowledge but also a collective human endeavor. In dialogic interactions, misunderstandings become opportunities for learning, diverse perspectives enrich understanding, and students practice civic reasoning by negotiating differences respectfully. These human-centered practices connect to earlier warnings against over-reliance on AI and the risk of mechanizing science education. Dialogue and community remind us that knowledge is co-created, contested, and contextualized—an essential safeguard against the simplification or distortion of scientific reasoning by AI.

In this sense, the most vital human roles in AI-supported classrooms are not diminished by technology but amplified through intentional design. Teachers become orchestrators of inquiry, ethical stewards of technology use, and mentors of student identity and agency. Students, meanwhile, are not merely consumers of AI outputs but active participants who critique, extend, and reinterpret them. Thus, the integration of AI into science education calls not for a diminished view of humanity but for a sharpened appreciation of the qualities that remain distinctly human—empathy, judgment, creativity, and community. These qualities must remain at the heart of science education if AI is to truly serve the larger goals of equity, integrity, and human flourishing. Framed in this way, the partnership between humans and AI becomes less about efficiency and more about meaning, less about automation and more about agency, ensuring that the future of science education remains transformative, ethical, and unmistakably human.

Declaration of AI Use The paper was wrritten by the authors. Parts of this manuscript were prepared with the assistance of generative AI tools for language editing or paragraphing of the authors' writing. The authors reviewed, revised, and verified all AI-supported content for accuracy, appropriateness, and originality. No confidential or sensitive data were entered into AI systems. The final responsibility for the content rests solely with the authors.

Acknowledgement This material is based upon work supported by the National Science Foundation under Grant No. 2332964 (PI Zhai). Any opinions, findings, and conclusions or recommendations expressed in this material are those of the author(s) and do not necessarily reflect the views of the National Science Foundation.

References

Akgün, S., & Krajcik, J. (2024). Artificial intelligence (AI) as the growing actor in education: Raising critical consciousness towards power and ethics of AI in K-12 STEM classrooms. In X. Zhai & J. Krajcik (Eds.), *Uses of artificial intelligence in STEM education* (pp. 494–521). Oxford University Press. https://doi.org/10.1093/oso/9780198882077.003.0022

Bewersdorff, A., Hartmann, C., Hornberger, M., Seßler, K., Bannert, M., Kasneci, E., Kasneci, G., Zhai, X., & Nerdel, C. (2025). Taking the next step with generative artificial intelligence: The transformative role of multimodal large language models in science education. *Learning and Individual Differences, 118*(102601), 1–13. https://doi.org/10.1016/j.lindif.2024.102601

Bianchini, S., Müller, M., & Pelletier, P. (2022). Artificial intelligence in science: An emerging general method of invention. *Research Policy, 51*(10), 104604. https://doi.org/10.1016/j.respol.2022.104604

Cheng, C., Wang, J., Zhai, X., & Yang, Y. (2025). AI literacy and gender equity in elementary education: A quasi-experimental study of a STEAM–PBL–AIoT course with questionnaire validation. *International Journal of STEM Education*, 1–22. https://doi.org/10.1186/s40594-025-00574-y

Crippen, K. J., Zhai, X., & Lee, O. (2026). Responsible and ethical principles for the practice of AI-supported science education. In X. Zhai & K. J. Crippen (Eds.), *Advancing AI in science education: Envisioning responsible and ethical practice (pp.* 49–80), Springer. https://doi.org/10.1007/978-3-032-16871-9_3

Eraslan, G., Avsec, Ž., Gagneur, J., & Theis, F. J. (2019). Deep learning: New computational modelling techniques for genomics. *Nature Reviews. Genetics, 20*(7), 389–403. https://doi.org/10.1038/s41576-019-0122-6

Gao, Y., & Zhang, T. (2025). In X. Zhai & G.-G. Lee (Eds.), *Artificial intelligence for STEM education research—advanced methods and applications*. Springer.

Gašević, D., Greiff, S., & Shaffer, D. W. (2022). Towards strengthening links between learning analytics and assessment: Challenges and potentials of a promising new bond. *Computers in Human Behavior*, 134. https://doi.org/10.1016/j.chb.2022.107304

Gil, Y., Greaves, M., Hendler, J., & Hirsh, H. (2014). Amplify scientific discovery with artificial intelligence. *Science, 346*(6206), 171–172.

Gobert, J. D., Li, H., Dickler, R., & Lott, C. (2024). Can AI-based scaffolding promote students' robust learning of authentic science practices? In X. Zhai & J. Krajcik (Eds.), *Uses of artificial intelligence in STEM education* (pp. 250–268). Oxford University Press.

Guo, S., Latif, E., Zhou, Y., Huang, X., & Zhai, X. (2024a). Using generative AI and multi-agents to provide automatic feedback. *arXiv*. preprint arXiv:2411.07407.

Guo, S., Zheng, Y., & Zhai, X. (2024b). Artificial intelligence in education research during 2013–2023: A review based on bibliometric analysis. *Education and Information Technologies*, 1–25. https://doi.org/10.1007/s10639-024-12491-8

Herdliska, A., & Zhai, X. (2024). Artificial Intelligence-Based Scientific Inquiry. In X. Zhai & J. Krajcik (Eds.), *Uses of artificial intelligence in STEM education* (pp. 179–197). Oxford University Press. https://doi.org/10.1093/oso/9780198882077.003.0009

Hu, E. H., & Morgan, P. L. (2024). *Explaining achievement gaps: The role of socioeconomic factors*. Thomas B. Fordham Institute.

Kitano, H. (2016). Artificial intelligence to win the nobel prize and beyond: Creating the engine for scientific discovery. *AI Magazine, 37*(1), 39–49.

Krenn, M., Pollice, R., Guo, S. Y., Aldeghi, M., Cervera-Lierta, A., Friederich, P., dos Passos Gomes, G., Häse, F., Jinich, A., Nigam, A., Yao, Z., & Aspuru-Guzik, A. (2022). On scientific understanding with artificial intelligence. *Nature Reviews Physics, 4*(12), 761–769. https://doi.org/10.1038/s42254-022-00518-3

Kubsch, M., Krist, C., & Rosenberg, J. M. (2022). Distributing epistemic functions and tasks—a framework for augmenting human analytic power with machine learning in science education research. *Journal of Research in Science Teaching*. https://doi.org/10.1002/tea.21803

Latif, E., Ma, W., & Zhai, X. (2025). Intelligent tutor for STEM learning. In X. Zhai & G. Lee (Eds.), *Artificial intelligence for STEM education research—advanced methods and applications*. Springer.

Latif, E., & Zhai, X. (2025). Privacy-preserved automated scoring using federated learning for educational research. International conference of artificial intelligence in education, Palermo, Italy.

Lee, G., Yun, M., Zhai, X., & Crippen, K. (2025). Artificial intelligence in science education research: Current states and challenges. *Journal of Science Education and Technology*, 1–18.

Lee, G., & Zhai, X. (2024). Using ChatGPT for science learning: A study on pre-service teachers' lesson planning. *IEEE Transactions on Learning Technologies, 17*, 1683–1700. https://doi.org/10.1109/TLT.2024.3401457

Lee, H. S., Gweon, G. H., & Pallant, A. (2024). Supporting simulation-mediated scientific inquiry through automated feedback. In X. Zhai & J. Krajcik (Eds.), *Uses of artificial intelligence in STEM education* (pp. 198–216). Oxford University Press. https://doi.org/10.1093/oso/9780198882077.003.0010

Li, W., Chang, H.-Y., Bradford, A., Gerard, L., & Linn, M. C. (2024). Combining natural language processing with epistemic network analysis to investigate student knowledge integration within an AI dialog. *Journal of Science Education and Technology*, 1–14.

Linn, M. C., & Gerard, L. (2024). Assessing and guiding student science learning with pedagogically informed natural language processing. In X. Zhai & J. Krajcik (Eds.), *Uses of artificial intelligence in STEM education* (pp. 58–77). Oxford University Press. https://doi.org/10.1093/oso/9780198882077.003.0004

Liu, L., Cisterna, D., Kinsey, D., Qi, Y., & Steimel, K. (2024). AI-based diagnosis of student reasoning patterns in NGSS assessments. In X. Zhai & J. Krajcik (Eds.), *Uses of artificial intelligence in STEM education* (pp. 145–178). Oxford University Press. https://doi.org/10.1093/oso/9780198882077.003.0008

Mafarja, N., Zulnaidi, H., & Mohamad, M. M. (2025). Developing critical thinking skills through AI-enhanced science education for sustainability. In *Rethinking the pedagogy of sustainable development in the AI era* (pp. 173–196). IGI Global Scientific Publishing.

National Research Council. (2012). In J. W. Pellegrino & M. L. Hilton (Eds.), *Education for life and work: Developing transferable knowledge and skills in the 21st century*. National Academies Press.

Nehm, R. H. (2024). AI in biology education assessment: How automation can drive educational transformation. In X. Zhai & J. Krajcik (Eds.), *Uses of artificial intelligence in STEM education* (pp. 39–57). Oxford University Press. https://doi.org/10.1093/oso/9780198882077.003.0003

Nyaaba, M., Zhai, X., & Faison, M. Z. (2024). Generative AI for culturally responsive assessment in science: A conceptual framework. *Education Sciences, 14*(12), 1–22.

OECD. (2017). *PISA 2015 technical report*. https://www.oecd.org/pisa/data/2015-technical-report/PISA2015_TechRep_Final.pdf.

OECD. (2021). AI and the future of skills, Volume 1: Capabilities and Assessments, Educational Research and Innovation, OECD Publishing, Paris, https://doi.org/10.1787/5ee71f34-en.

Pellegrino, J. W. (2024). A new era for STEM assessment: Considerations of assessment, technology, and artificial intelligence. In X. Zhai & J. Krajcik (Eds.), *Uses of artificial intelligence in STEM education* (pp. 17–37). Oxford University Press. https://doi.org/10.1093/oso/9780198882077.003.0002

Sajja, R., Sermet, Y., Cwiertny, D., & Demir, I. (2025). Integrating AI and learning analytics for data-driven pedagogical decisions and personalized interventions in education. *Technology, Knowledge and Learning*, 1–31.

Shi, L. (2025). Global perspectives on AI competence development: Analyzing national AI strategies in education and workforce policies. *Human Resource Development Review*, 1–30. https://doi.org/10.1177/15344843251332 36

Shi, L., & Choi, I. (2024). A systematic review on artificial intelligence in supporting teaching practice: Application types, pedagogical roles, and technological characteristics. In X. Zhai & J. Krajcik (Eds.), *Uses of artificial intelligence in STEM education* (pp. 321–347). Oxford University Press.

Sinha, R. K., Roy, A. D., Kumar, N., Mondal, H., & Sinha, R. (2023). Applicability of ChatGPT in assisting to solve higher order problems in pathology. *Cureus, 15*(2).

Su, J., & Zhong, Y. (2022). Artificial intelligence (AI) in early childhood education: Curriculum design and future directions. *Computers and Education: Artificial Intelligence, 3*, 100072.

Varsik, S., & Vosberg, L. (2024). The potential impact of artificial intelligence on equity and inclusion in education.

Vartiainen, H., Toivonen, T., Jormanainen, I., Kahila, J., Tedre, M., & Valtonen, T. (2021). Machine learning for middle schoolers: Learning through data-driven design. *International Journal of Child-Computer Interaction, 29*, 100281.

Wu, X., Zhao, H., Zhu, Y., Shi, Y., Yang, F., Hu, L., Liu, T., Zhai, X., Yao, W., & Li, J. (2024). Usable XAI: 10 strategies towards exploiting explainability in the LLM era. *arXiv*. preprint arXiv:2403.08946.

Xing, W., & Li, C. (2024). Fair artificial intelligence to support STEM education: A hitchhiker's guide. In X. Zhai & J. Krajcik (Eds.), *Uses of artificial intelligence in STEM education* (pp. 522–546). Oxford University Press. https://doi.org/10.1093/oso/9780198882077.003.0023

Zhai, X. (2023). ChatGPT and AI: The game changer for education. SSRN. https://ssrn.com/abstract=4389098.

Zhai, X., Haudek, K. C., Shi, L., Nehm, R., & Urban-Lurain, M. (2020). From substitution to redefinition: A framework of machine learning-based science assessment. *Journal of Research in Science Teaching, 57*(9), 1430–1459. https://doi.org/10.1002/tea.21658

Zhai, X., & Krajcik, J. (2024). Pseudo artificial intelligence bias. In X. Zhai & J. Krajcik (Eds.), *Uses of artificial intelligence in STEM education* (pp. 568–580). Oxford University Press. https://doi.org/10.1093/oso/9780198882077.003.0025

Zhai, X., & Pellegrino, J. (2023). Large-scale assessment in science education. In N. G. Lederman, D. L. Zeidler, & J. S. Lederman (Eds.), *Handbook of research on science education* (Vol. III, pp. 1045–1098). Foutledge.

Zhai, X. (2024). AI and machine learning for next generation science assessments. In H. Jiao & R. W. Lissitz (Eds.), *Machine learning, natural language processing and psychometrics* (pp. 201–219). Information Age Publisher.

Chapter 2
AI in Contemporary Scientific Practice: Implications for AI-Integrated Science Education

Ross H. Nehm and Marcus Kubsch

2.1 Introduction

The Framework for K–12 Science Education (National Research Council [NRC], 2012) and other policy documents (American Association for the Advancement of Science [AAAS], 2010) urge educators to engage students in "authentic" science. Today, authenticity necessarily includes grappling with artificial intelligence (AI) as a part of scientific inquiry. Yet what counts as "authentic" use of AI depends both on how we conceptualize science and on the extent to which classroom uses mirror the ways scientists actually deploy AI. In short, aligning AI in education with AI in science requires clarity about the aims, norms, and practices inherent to the contemporary scientific enterprise (cf. NRC, 2012). Against this backdrop, two challenges immediately present themselves.

One challenge is that most scientific disciplines lack broad and consistent policies and standards for AI use (and misuse), posing challenges for educators seeking to characterize appropriate theoretical and empirical applications of AI in science. Machine-learning reporting may look different in chemistry than in ecology, and quality-control benchmarks for medical imaging may not mirror those in physics. These inconsistencies show some signs of resolution, as evidenced by the growing number of AI standardization documents aimed at reducing the production of

R. H. Nehm (✉)
Department of Ecology and Evolution, Stony Brook University, Stony Brook, NY, USA

Department of Applied Mathematics and Statistics, Stony Brook University, Stony Brook, NY, USA
e-mail: ross.nehm@stonybrook.edu

M. Kubsch
Department of Physics, Freie Universität Berlin, Berlin, Germany
e-mail: m.kubsch@fu-berlin.de

X. Zhai, K. J. Crippen (eds.), *Advancing AI in Science Education*, Advances in Technology-Rich Science Education 3,
https://doi.org/10.1007/978-3-032-16871-9_2

misleading or flawed AI-based research (e.g., Kapoor et al., 2024; see Table 2.1). Despite such progress, however, standardization remains incomplete.

A second challenge is descriptive: we lack integrative reviews that map how AI is used across disciplinary contexts, from observational sciences to experimental and computational domains. Without a clearer picture of the diversity of AI roles

Table 2.1 Examples of explicit ideals and reliable processes instantiated as frameworks or standards for AI in science

Field of study	Framework or standard	Description	Internet link
Medicine: Imaging	CLAIM	Checklist for AI in medical imaging	https://www.nature.com/articles/s42256-023-00717-2
Science	REFORMS	The recommendations for machine learning-based science (REFORMS) checklist comprises 32 items designed to improve reporting practices in ML-based scientific research	https://www.science.org/doi/10.1126/sciadv.adk3452
Medicine: Medical imaging	MINIMAR	The minimum information for medical AI reporting (MINIMAR) ensures transparency in AI-driven healthcare predictions	https://doi.org/10.1093/jamia/ocaa088
Medicine: Health Research	TRIPOD-AI	An extension of the TRIPOD statement, providing guidelines for reporting AI-based clinical prediction models	https://www.equator-network.org/
Medicine: Clinical trials	CONSORT-AI	An extension of CONSORT, focusing on reporting AI interventions in clinical trials	https://www.nature.com/articles/s41591-020-1034-x
Physics	ROOT, FAIR	ROOT is the de-facto standard for analyzing large datasets, FAIR (findability, accessibility, interoperability, and reuse) is a standard for data management (increasingly) required by leading journals and funding bodies such as the EU	https://root.cern/ https://www.go-fair.org/fair-principles/
Biochemistry	BioCompute object (BCO)	The BioCompute object (BCO) framework standardizes and shares computations and analyses generated from high-throughput sequencing (HTS) data. It ensures that bioinformatics pipelines, including those utilizing AI, are reproducible and interpretable, facilitating regulatory submissions and collaborative research in computational chemistry and biology	https://biocomputeobject.org/

Note. Documents like these will be valuable for understanding the development of broader standards for AI-enabled science. N.B. Complex scientific instrumentation not based on AI likewise has standards for operation, interpretation, and reporting that may be useful points of comparison when introducing students to reliable processes relating to AI and methods in practice, it is difficult to connect disciplinary core ideas to relevant AI-enabled practices or to specify essential AI proficiencies for K–16 learners (NRC, 2012; AAAS, 2010). These descriptive gaps, coupled with normative uncertainty, make it difficult to design learning environments that are both faithful to science and instructionally productive (e.g., Levine & Tu, 2024). To address these challenges, we adopt a comparative lens from science studies and anthropology: we set descriptive accounts of current scientific practice alongside normative accounts rooted in professional standards and ideals (Kitchener, 2002; Chinn et al., 2011, 2014). Using interviews with scientists employing AI in their work along with focused literature reviews (e.g., Levine & Tu, 2024), we chart how AI is deployed across and within disciplines.

The chapter begins by situating AI within the broader scientific enterprise and identifies essential similarities and differences between AI and other instruments used by scientists (e.g., sensors, scales, PCR machines). This topic is addressed at length because a clear understanding of the social structure of science and the epistemic nature of instruments in scientific networks is foundational for appreciating what is unique about AI in science (and what is not). The distributed nature of expertise, ubiquitous in science practice, requires dependencies among humans and instruments that introduce critical considerations including epistemic dependence, vigilance, and trust. Heuristics for navigating these issues are essential to scientific knowledge production with *and* without AI. The ongoing lack of attention to these topics in K-16 education makes discussing AI even more challenging.

The chapter continues by discussing the different roles that AI plays *beyond* its use as an instrument, including as a collaborator and semi-autonomous agent and the unique considerations of these roles for scientific inquiry practices. The next sections of the chapter review AI-based scientific research on a much finer scale and identify similarities with traditional science practices but also highlight differences in epistemic goals and scales. A key finding is that AI is employed in all the science practices currently identified in education frameworks (NRC, 2012). The chapter ends by identifying a limitation with using contemporary science to model K-16 AI-integrated science education: scientists currently learn about AI in fragmented, idiosyncratic, and inefficient ways in both formal and informal contexts. How students learn about AI in science will require adopting a more principled, structured, and effective strategy.

The broadest implication of this chapter for educators is that a robust understanding of the unique affordances and limitations of AI can best be understood through the lens of distributed epistemic networks and their unique properties. That is, one must understand the ways that scientific knowledge is currently produced within large and specialized research groups dependent upon complex instrumentation and the challenges inherent to them in order to make sense of AI-enabled science. Many purported challenges of AI-enabled science discussed in the media are *not* in fact unique to AI (e.g., risks of dependence and trust in AI models mirrors the risks of dependence and trust among scientists and the tools they use). This is an actionable insight for educators because many challenges ascribed to AI (e.g., trust, reliability) can and should be introduced to learners using much less complicated inquiry tools (e.g., dependence and trust considerations in data produced by temperature sensors, or in claims made by others in collaborative inquiry). AI may be productively envisioned as a more complex extension of the epistemic features of contemporary science practice.

2.2 Key Findings and Recommendations

2.2.1 Scientists with Different Areas of Expertise Work in Collaborative Networks to Investigate and Explain the Natural World. Complex Instrumentation, Including AI, Plays Essential Roles in These Networks and Necessitates Epistemic Dependence and Vigilance in Scientific Knowledge Production

Science is a collaborative enterprise in which individual laboratories, larger research groups (e.g., NSF-Simons Research Collaborations on the Mathematical and Scientific Foundations of Deep Learning), and global scientific consortia (e.g., Human Genome Project, CERN) pursue targeted epistemic aims.[1] These aims may include explaining how a chromosomal mutation impacts disease phenotypes, developing a mathematical model integrating historical weather patterns to predict future climate change, or testing for the presence of a hypothesized subatomic particle. Most epistemic networks in science, whether lab- or field-based, require both humans (of varying levels of expertise and autonomy) and a variety of tools and instruments to achieve these aims (e.g., machines to sequence genes and genomes, purify chemical compounds, analyze interstellar objects). The collective knowledge generation network of humans and instruments can achieve scientific aims that individual humans and instruments cannot. Because different humans and instruments play differentiated roles in the network (entities in the network lack the same information, skills, and expertise), each collective may be referred to as a *distributed epistemic network* or DEN (Fig. 2.1).

DENs in science are a useful framework for exploring the roles that AI can and should play in science education. Scientists in these networks (e.g., Principal Investigators [PIs], post-docs, graduate students, lab technicians) harbor different levels of autonomy, agency, and expertise. A PI, for example, may have a broad understanding of the DEN and the agency to chart its course; a technician may be the sole expert on the use and functioning of complex instrumentation; and a statistician may be the sole expert on how to analyze the data produced by the instrument. The specialized nature and complexity of epistemic aims in science mean that

[1] These networks are embedded within broader social structures, such as institutions, government agencies, and scientific societies, but we will set aside these broader contexts in our discussion.

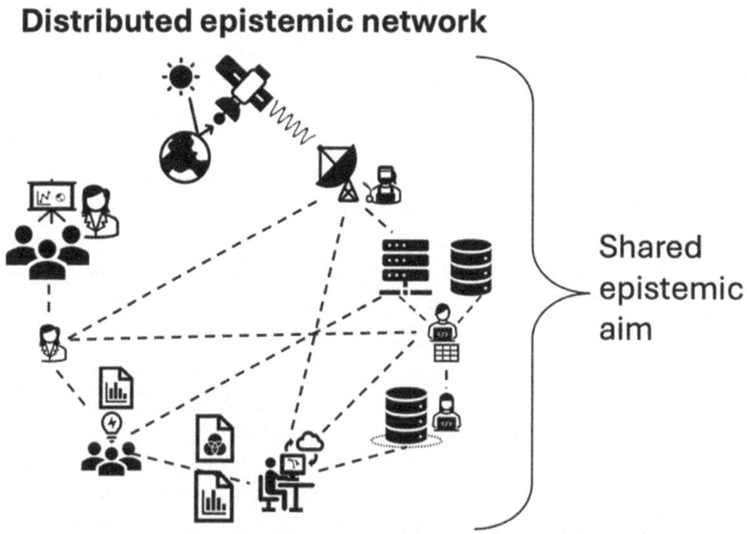

Fig. 2.1 The contemporary scientific enterprise is characterized by complex distributed networks of humans and instruments united by shared epistemic aims (research goals). Each node is characterized by many factors, including authority, autonomy, expertise, and trustworthiness. Information flows among nodes

individual actors in the DEN (e.g., the technician) typically lack a complete understanding of every component.

Necessarily, actors in a DEN must depend upon other experts and instruments; this situation is referred to as *epistemic dependence* (Alvarado 2023; Duede 2022). For example, a technician may depend upon and trust a scientific instrument to generate accurate output; a statistician may depend on and trust the technician to certify that the output is accurate; and the PI may depend upon and trust the statistician to analyze the data from the instrument appropriately. The complex and distributed nature of understanding in scientific networks results in areas of ignorance, or epistemic opaqueness, among the actors within it. Simply put, humans are dependent on one another and their instruments during scientific knowledge production. Epistemic dependence is therefore inevitable and applies to collaborators, traditional tools (e.g., thermometers, gel electrophoresis boxes, sensors) and new tools like AI.

The scale, complexity, and cost of large-scale contemporary scientific instrumentation (e.g., supercolliders, fMRI machines, automated sequencers) have produced increasing divergence from the instruments used in science classrooms. For example, scientists use scanning electron microscopes whereas students use compound microscopes. Despite these differences, deploying instruments like thermometers, balances, gel boxes, and AI in K-16 education serve as essential opportunities for teaching about the challenges of epistemic dependence and trust in DENs (e.g., groups of students, instruments, and a teacher engaged in scientific inquiry in relation to a collective epistem aim). This situation is comparable to how the use of supercolliders might create epistemic dependence and trust in large scale DENs. Epistemic dependence (i.e., relying on others for knowledge due to a personal lack of expertise) and epistemic trust (i.e., placing confidence in others as sources of knowledge) emerge as inescapable and foundational features of science inquiry in K-16 classrooms (cf. NRC, 2012) that are *not* unique to AI. Given that most of these instruments are less complex than AI, they should be used to explore epistemic dependence and trust early in school. For example, although sensors are used during photosynthesis experiments in high school and university settings, students rarely explore why we should trust the information provided by the sensor or why that particular type of sensor was employed. Experiences like these prepare students for trusting the outputs of particular AI models without reflection.

Scientists (or students and teachers) in DENs require strategies for navigating instances of epistemic dependence (Alvarado 2023; Duede 2022). For example, what approaches might a PI use after hiring a new statistician or adding a new piece of equipment to the DEN? The PI depends on these nodes in the network, but should she trust the results they generate? *Epistemic vigilance* may be viewed as a 'habit of mind' or disposition associated with epistemic cognition (Kitchener, 2002); that is, being alert to how information or understanding was generated in order to be able to evaluate its trustworthiness. Normative epistemic heuristics, 'rules of thumb,' or practical mental shortcuts need to be activated when actors (scientists, students, citizens) are faced with epistemic dependence (e.g., Chinn et al., 2014). These heuristics may include approaching information with skepticism; considering the trustworthiness of sources; gathering information from additional sources; looking for inconsistencies and commonalities across sources; seeking the evidence, justification, and reasoning behind sources of information; and examining the reliability of sources over time (Chinn et al., 2011, 2014). Epistemic vigilance is unavoidable in scientific knowledge generation and is a productive disposition in inquiry-based science education. This perspective prepares learners for the arrival of more complex tools and technologies like AI.

Science educators have already generated theoretically grounded and classroom-ready frameworks and heuristics that may be used to help learners productively enact epistemic vigilance. In the domain of epistemic cognition, Chinn et al. (2014) proposed the three-pronged AIR model (Aims and values, Ideals, and Reliable processes). In short, Aims and values relate to the *goals* that are being pursued (e.g., Aim: I want to determine if this gas pump at a gas station is accurately showing how many gallons of gas I purchased). Ideals relate to the *standards* or benchmarks that are used to evaluate an aim (e.g., I think that different instruments designed to measure the same thing [a gas pump and my car gas gauge] *should* produce consistent information. Reliable processes refer to the ways or *methods* by which data or other epistemic outputs (e.g., explanations, models) are produced (e.g., comparing the information produced by the gas pump and my car gas gauge–the pump indicates I purchased 10 gallons but my tank gauge indicates the car is empty–I do not trust the information I'm receiving). Ideals clearly relate to Reliable processes. For example, if 'authority' is used as an epistemic ideal (the gas pump is certified by the government and is therefore always accurate), then one should seek out what the authority indicates would be considered a reliable process. Explicit consideration of these issues can foster accountability.

The AIR model is applicable to a broad range of contexts and domains beyond science (e.g., news consumption, legal issues like eyewitness testimony). Having students develop and agree upon their own heuristics when confronted with epistemic dependence (Krist et al., 2019), or introducing students to heuristics commonly used in science, is essential to collaborative knowledge production and should occur prior to working with complex instrumentation (be it a sensor or AI). For example, in classroom inquiry, different students may focus on different aspects of a phenomenon (e.g., whether cans of diet or "regular" soft drinks float in a tank of water). Some students may be tasked with analyzing the ingredients listed on the cans, others may manipulate the cans by repeatedly dropping them into a tank of water, and others may gather data about the taste of what is contained within the cans. Sharing and reporting information is essential to understanding the phenomenon and as a result introduces the issues of epistemic dependence and trust to students (e.g., lacking direct knowledge of other group members' data or evidence-generating processes). Should a student trust the data another group member produced? Working with other agents (e.g., students, teachers) in a classroom DEN sets the stage for the addition of complex instrumentation (including AI) and the epistemic dependence and trust issues that emerge in knowledge production networks.

Aims and values, Ideals, and Reliable processes often remain implicit in authentic science practice, including when using AI. For example, while broad aims and values may characterize the charter of a scientific society, such values will not be explicit in a journal article or conference paper. This makes the descriptive characterization of epistemic heuristics in the context of AI challenging. Complicating the matter further, different scientific fields or disciplines have established their own Ideals and Reliable processes. For example, the CLAIM checklist of best practices for the use of AI in medical imaging has been established (Mongan et al., 2020), but it is not used in many other areas of science. A recent attempt to create Ideals and Reliable processes for one type of AI–Machine Learning (ML)—serves as an example of how the absence of general frameworks limits robust knowledge production. Establishing and modeling ideals and reliable processes is essential for science and science education.

ML is increasingly employed in scientific DENs as a tool for making scientific claims (Kapoor et al., 2024). For example, ML may be used to make predictions, gather measurements, identify patterns, and help answer scientific questions (Kapoor et al., 2024: p. 2; Jumper et al., 2021). The absence of epistemic Ideals and Reliable processes for its use in science, however, has led to irreproducible published findings and other challenges (Kapoor et al., 2024; McDermott et al., 2021). Scholars have recently worked to remedy this situation and develop explicit frameworks for ML, such as the REcommendations FOR Machine learning-based Science (REFORMS). REFORMS is a set of questions and consensus guidelines developed by computer scientists, data scientists, mathematicians, social scientists, and biomedical scientists that set standards for how to use ML to design and implement a study. It consists of eight major categories (i.e., Study goals, Computational reproducibility, Data quality, Data preprocessing, Modeling, Data leakage, Metrics and uncertainty, Generalizability and limitations) along with 32 specific benchmarks. Frameworks like REFORMS establish the ideals and reliable processes that scientists *should* use when using AI to produce new knowledge, reviewing each other's work during peer review, and for scientific journals seeking to set explicit and transparent benchmarks. Descriptively, we do not know the extent to which REFORMS is being used by scientists, but resources such as these will be essential for science educators seeking to develop guidelines for the use of AI tools in the classroom (see Table 2.1).

In summary, AI has the potential to serve as a valuable example of much broader (but rarely emphasized) aspects of authentic science and to direct attention to the roles of epistemic dependence, epistemic vigilance, and epistemic trust in knowledge production networks. Just like scientists, K-16 students working with one another and with field and laboratory equipment and instrumentation (e.g., sensors, motion detectors, microscopes) will be faced with epistemic dependence and vigilance in ways that are less complex than engaging with AI in science (as discussed below). Students should gain practical experience relying on and navigating through epistemologically opaque components of epistemic networks (e.g., outputs of students or instruments.) These opportunities provide experiences for practicing epistemic vigilance heuristics.

While valuable in their own right, epistemic heuristics serve as necessary preparation for using AI in science. Unlike other forms of scientific instrumentation, however, AI has the potential to play various roles in science that extend far beyond traditional instrumentation. For example, although AI may serve as an instrument/tool in some contexts (e.g., producing a graph based on data provided by the user), it may also serve as an agent or collaborator (e.g., brainstorm ideas for testing a hypothesis), or as an extended agent (e.g., iterative refinement of ideas based on back-and-forth interaction with humans). Each role highlights different issues relating to epistemic dependence, vigilance, and trust. The different roles that AI can play in DENs, and the unique challenges to epistemic dependence are addressed in the next section.

2.2.2 AI Plays Different Roles in Distributed Networks of Scientific Knowledge Production (e.g., Instrument, Agent) that Have Different Implications for Epistemic Dependence and Vigilance. These Roles Should Be Explicitly Discussed and Modeled in Science Learning Environments

AI has the potential to play different roles in DENs, which may be simplified into three major categories based on scientific aims (Fig. 2.2). These categories lack rigid boundaries and may be instantiated in different combinations. First, AI may be employed as an *instrument or tool.* Just as microscopes, telescopes, and sensors have been developed with the aim of helping humans see or detect invisible phenomena, AI may be used as a tool to extend human capabilities. The scientific aim is established by human agents, and the instrument or tool is used to generate a result or product that requires interpretation by human agents. In *AI as instrument* contexts, AI lacks autonomy and agency and is under the direction of human agents (e.g., scientists, students). AI-driven image classification techniques, for example, may be applied to vast datasets generated by telescopes, thereby accelerating the identification of celestial objects and phenomena by astronomers (Hezaveh et al., 2017). In climate science, machine learning models may be used to enhance simulations and improve the resolution and predictive power of climate forecasts (Reichstein et al., 2019). In particle physics, AI algorithms employed at CERN's Large Hadron Collider filter through massive volumes of collision data to search for the Higgs boson (Radovic et al., 2018). In biochemistry, AI has been used to help explain sequence–structure–property relationships in biomolecules. These situations are analogous to a student: programming an electronic sensor with the aim of using the tool to measure and report the pH of a solution over time; using a scanning electron microscope to identify and analyze patterns on the surface of a microscopic fungus; or using a machine learning model to detect tumors in a corpus of histology slides. The student is developing the scientific aim and the instrument lacks autonomy and agency.

Interviews with physicists and biologists confirmed the *AI as an instrument* role within contemporary scientific inquiry. For example, scientists discussed how they rely on AI for performing tasks such as computer code generation, pattern recognition, image processing, and data analysis (e.g., pattern recognition and classification). One physicist noted the usefulness of unsupervised learning algorithms to process vast quantities of simulation data. A biologist discussed her work with convolutional neural networks and transfer learning to identify vegetation types in satellite imagery associated with disease transmission. Another physicist explained how machine learning was used to approximate high-dimensional functions to model quantum mechanical systems. These examples illustrate the *AI as an instrument* role in contemporary science practice (Fig. 2.2a).

At the other extreme from the *AI as instrument* role is the *AI as agent* role. In such cases, AI may be provided with authority, autonomy, and agency during scientific inquiry. For example, AI may establish scientific aims (e.g., generating suitable hypotheses worthy of investigation) and have the agency to select relevant data, to interpret results, and to draw conclusions (see Fig. 2.2c). In this context, AI functions in ways that overlap with or supplant human agents (e.g., scientists or students in a DEN) by directing humans to aims worthy of study. A recent example of the *AI as agent* role is Coscientist, an AI system developed by researchers that autonomously conducts complex chemical experiments with no human oversight (Krenn et al., 2024). The system can design and execute experimental protocols, analyze results, and revise hypotheses—demonstrating full-cycle scientific reasoning and action. Similarly, NASA's rover Perseverance uses AI to autonomously navigate the Martian surface and make decisions about geological sampling based on onboard

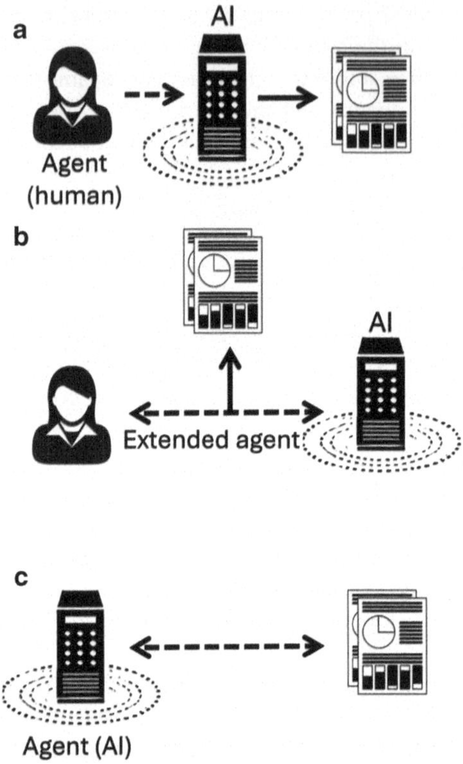

Fig. 2.2 Simplification of the many *intended uses or roles* that AI may play in scientific knowledge production. (**a**) *AI as an instrument:* a human agent (e.g., scientist) develops an aim and uses AI to perform a task to generate a product such as an analysis or visualization of a large data set. In this case AI lacks autonomy and agency. (**b**) *Extended agent*: A human agent collaborates with AI to generate a scientific product that is a result of unique contributions from both humans *and* AI. (**c**) *AI as an agent*. AI has the agency, autonomy, and authority to identify aims, make decisions, and draw conclusions during the investigative process (e.g., identify what hypotheses are worthy of investigation). Complex combinations of these scenarios are possible and likely

sensor analyses. AI allows scientific exploration in extraterrestrial environments that prohibit real-time human guidance (Gaines et al., 2016). Both examples illustrate the continuing evolution of science practice.

Although the *AI as agent* role was not as common as the *AI as instrument* role in our interviews with scientists, one physicist identified this role in current practice and spoke of fully automated laboratories where AI not only proposes new molecules but also controls robots to synthesize them. Another scientist—Dr. Electron,[2] Physicist—reflected on how neural networks serve as universal function approximators, replacing traditional analytical modeling in certain domains: "Machine learning can effectively construct models that would be extremely difficult to design

[2]All names of interviewed scientists are pseudonyms.

analytically by humans." These examples make clear that AI has begun to encroach upon roles previously occupied by human scientists.

The *AI as instrument* and *AI as agent* roles represent clear extremes along a continuum. In between these two roles are cases in which AI may lack autonomy and agency but engage in activities that extend beyond using AI as a tool or instrument. Such interactions may include brainstorming, seeking suggestions for relevant literature or methods, assisting with possible interpretations, and generating critiques. This AI role is analogous to a scientist seeking opinions, ideas, and critique from a colleague, or a student seeking ideas from a teacher for improving their research design. Importantly, in this role, the *human* agent establishes the aims, maintains authority and autonomy in the investigative process, but has extended capabilities because of AI interaction. Interviews with scientists also identified the *AI* as *extended agent* role in current practice. Dr. Proton, another physicist, described how generative models help researchers explore novel molecular structures by specifying desired properties: "You can really tell the computer: I would like to build a new molecule that has a better optical absorption activity in this range, and then it gives you a structure." Dr. Proton went on to emphasize how AI can identify latent variables in complex datasets, revealing patterns and relationships that might otherwise remain hidden. As she put it, "Before this answer would be a lot of guesswork or a lot of… real work from someone… now interacting with this system… you can [together] find the answer." In this role, AI extends the researcher's cognitive reach through iterative interactions with AI that extend beyond an instrument or tool.

These simplified roles for AI–*instrument, agent, extended agent*– are aim-dependent. That is, the AI role is identified through the *intended use of AI* rather than the entities and their interactions. It is clearly possible (and likely) that combinations of these roles occur in contemporary science, and that the role of AI in an epistemic network may shift throughout the investigative process (e.g., from *instrument* to *agent*). Although AI may play complex roles in science, it is valuable to identify possible role *types* because they motivate different considerations and concerns relating to epistemic dependence, vigilance, and trust. Understanding these roles is therefore foundational to the design of AI-enabled science learning environments.

The distinct roles that AI plays in science have implications for the design of AI-enabled science education. When AI acts as an inquiry partner, students should be encouraged to interact with it dialogically—posing questions, evaluating suggestions, and reflecting on its implications. Employing epistemic heuristics (e.g., AIR) will be essential during these interactions. As a methodology, AI introduces opportunities to evaluate model assumptions, data reliability, and computational processes. When viewed as an agent, AI challenges conventional understandings of scientific expertise. Different roles demand different forms of epistemic vigilance and trust: from critical scrutiny and technical validation to philosophical reflection on what counts as scientific knowledge and doing science. By integrating these diverse roles of AI into school science, educators can better prepare students to participate in and understand the evolving landscape of AI-augmented science. Doing so ensures not only technical literacy but also the cultivation of informed, reflective scientific reasoning in an AI and data-driven world. Without robust epistemic heuristics to scaffold actors in DENs, these interactions, like those in contemporary science, may generate flawed, false, or harmful conclusions.

2.2.3 *AI-Based Scientific Research Shares Similarities with Traditional Science Practices But Also Differs in Terms of Its Epistemic Goals and Scales. These Differences May Require Re-Evaluation of the Constructs of Science Practices in Science Education*

AI-supported scientific research shares key similarities with traditional science practices, but it also introduces significant shifts in epistemic goals, methods, and scales. These shifts may require a re-evaluation of long-standing constructs in science education and research, particularly concerning what constitutes "authentic science." The transformation spurred by AI is not merely technological; it is epistemological, reshaping how knowledge is generated, interpreted, and validated.

At a foundational level, we argue that AI-enhanced and traditional research share many characteristics. Both rely on principled data collection, preprocessing, analysis, and visualization. As Dr. Fauna, a biologist noted, "Both approaches complement each other to achieve deeper biological insights" and ultimately aim to answer domain-specific questions and validate insights using evidence. Several researchers emphasized the continuity between traditional modeling and AI-enhanced approaches. In biology, for example, Dr. Fauna explained: "…[it's] kind of fair to frame this usage of machine learning and deep learning as…the next step in the computational tools…from maybe complex statistical models to just taking these models as the next bigger, more advanced model." Similarly, Dr. Electron described neural networks as "universal function approximators" that extend the logic of regression to highly complex systems: "It's like a linear regression, but it's non-linear and high-dimensional. And this was very complicated before… we used models with a lot of intuition and previous detailed knowledge of the field to do this kind of fit. Now, the machine can learn the model for you." These examples illustrate how on the one hand AI has extended practice and on the other hand began to blur the distinction between practice and scientist.

AI-based inquiry also diverges from traditional science practice in important ways. Hypothesis-driven experimentation tied to models of phenomena is common in traditional science practice. AI, in contrast, has facilitated the expansion of hypothesis-free discovery in massive datasets. As Dr. Proton noted "So we produce a lot of data from all the simulations […and…] unsupervised learning…algorithms have helped a lot to sort through the data". In addition, in physics and materials science, AI allows researchers to bypass transparent yet laborious numeric methods by leveraging black-box machine learning methods that are speeding up the execution of tasks. Dr. Proton explained: "We do something that is called molecular dynamics… and now there is this big area of machine learning for interatomic potentials…

In the past three years, we have seen like a gain of three or four orders of magnitude in the time scale… it's a complete game changer."

AI-enabled science also repositions scientists in relation to science practices. AI systems can move beyond identifying patterns to generate hypotheses, thereby altering the researcher's role in knowledge production. As Dr. Proton reflected: "It is basically giving you a lot more powers than you had before… maybe you wouldn't be able to find the answer just by yourself, but with AI you can." Another researcher, Dr. Electron, remarked on how generative models are beginning to resemble intellectual collaborators: "You can really tell the computer: I would like to build a new molecule that has a better optical absorption activity… and it gives you a structure and composition." Here the line between agent and practice blurs. Beyond its role as a methodology, AI is increasingly becoming part of executing the science practice itself. Authentic science now includes actor-practice hybrids that do not cleanly align with traditional characterizations of practice (Table 2.2).

Table 2.2 Comparisons of AI-supported and traditional scientific work

Aspect	AI-supported methods	Traditional methods	Similarities	Differences
Data collection and manipulation	Automated collection and processing of massive, high-dimensional datasets; handles complexity characteristic of "omics" (transcriptomics, genomics)	Human collection and processing of small sets restricted by human time and effort	Both include data collection and data processing	Automated methods provide scale and consistency but may lack interactions between research object and researcher that may lead to serendipitous discoveries (e.g., discovery of penicillin)
Hypothesis testing	Often hypothesis-free, identifying patterns in large volumes of data	Primarily hypothesis-driven and anchored to conceptual models	Both seek to gather evidence using data	Sample size leads to different challenges: with traditional methods, limited sample sizes may lead to underpowered studies, in AI enabled methods, large sample sizes may lead to inflated discovery rates as it becomes harder to separate spurious from causal relationships
Prediction & Discovery	Prediction modeling using ML, neural networks, and other approaches facilitates novel pattern discovery	Relies on established theories; method dependent novel discovery is uncommon	Both are used to make predictions and discover new insights into phenomena	Traditional methods have interpretability "baked in" whereas interpretability of ML models is challenging and needs establishment through explainable ML methods

(continued)

Table 2.2 (continued)

Aspect	AI-supported methods	Traditional methods	Similarities	Differences
Speed & scalability	Automates repetitive tasks and scales to generate and analyze large datasets rapidly	Manual and time-intensive, limiting scalability	Both aim to increase research efficiency to increase the tempo of discovery	Scalability of traditional methods often depends on (expensive) training of researchers whereas scalability of AI methods depends primarily on available hardware
Interpreting results	Can be a "black box," requiring explainable AI techniques for interpretation	Results are more transparent, with clearer procedural tracking of pathways	Both require validation and contextual interpretation of findings	See "Prediction & Discovery"
Experimental design	Permits dynamic and adaptive experimental designs with automated progress (e.g., reinforcement learning)	Static experimental designs predominate	Principled design is required in both contexts	Designs that require extensive interaction with the environment (e.g., working with animals) remain out of scope for purely AI-driven methods
Applications	Ideal for large scale integrative tasks such as multi-omics integration, personalized medicine, and systems modeling	Common across disciplines in many contexts and research systems	Both have applications across diverse disciplines	Both may require different, highly specialized skill sets (e.g., setting up quantum chemistry simulations vs. microscopy)
Limitations	Prone to overfitting, bias, hallucination and interpretability challenges	Limited by human capacity for data analysis and inefficiency with large-scale data	Both are limited by data amount, quality, complexity and interpretability	Complexity of ML methods adds another challenge and thus potential sources of error to already complex traditional methods
Synergism	Guides hypotheses for experimental validation	Provides experimental validation for AI-generated models	Both approaches complement each other to achieve deeper insights	–

Note. Although what is considered 'traditional' varies across scientific disciplines, we use the term to refer to methods that are commonly encountered in undergraduate lab settings or predate the uptake of AI methods

2.2.4 AI Is Increasingly Integrated into all Science Practices, Requiring Curriculum Updates that Reflect Authentic, Contemporary Science

Over the past decade, science education frameworks have emphasized the centrality of practice in meaningful science learning (Ford, 2008; NRC, 2012). Three-dimensional learning—the integration of disciplinary core ideas, science practices, and cross-cutting concepts—integrates practices ranging from developing explanations to analyzing and interpreting data (NRC, 2012). It is therefore important to examine how practices are evolving in response to AI. Literature reviews and interviews with scientists indicate that AI pervades all science practices and has even encroached on roles previously restricted to humans (as shown in Table 2.3). This finding illustrates the increasing and widespread penetration of AI into scientific inquiry. The AI methods indicated in Table 2.3 should be integrated into K-16 science education in developmentally appropriate ways.

Table 2.3 underscores a vital point: the core epistemic *aims* of science remain, but how they are achieved are evolving rapidly. As Dr. Electron put it, "It's not that we've stopped asking scientific questions—it's that AI helps us ask better ones, and faster." In Interviews with researchers across biology, chemistry, and physics, a clear consensus emerged: AI is *not* just helping scientists do what they already did—it is enabling them to do things that were previously impossible. Theoretical physicist Dr. Electron, for example, noted: "In physics, machine learning is mostly used to approximate non-linear, high-dimensional functions... it replaces educated guesses with learned models." A chemist, Dr. Redox, stated: "In my field, AI has sped up simulations by orders of magnitude. We're simulating things we couldn't even dream of doing before." The voices of scientists and Table 2.3 motivate a key point: AI is not only involved in all science practices, but it is gaining the agency previously afforded to scientists themselves. As noted in previous sections, AI is not only used as an instrument to achieve the aims of scientists; it is now acting as an agent and scientific collaborator.

The ongoing transformation of real-world science by AI is *not* mirrored in K-16 science learning environments that continue to model and practice science as a primarily manual endeavour reliant on traditional computational tools and instruments (e.g., calculators, human produced-spreadsheets). This gap between practice and pedagogy has grown with the advent of AI, and it must be addressed if students are to understand how science is done today and how it will be done in the future. Curriculum materials must reflect the increasingly central role of AI in scientific inquiry across all science practices.

Table 2.3 AI methods have transformed how scientists engage in the core science practices in the framework for K-12 science education

Practice	AI methods	Traditional methods	Similarities
1. Constructing explanations	AI may autonomously provide data-driven explanations, identifying patterns or relationships without prior assumptions (e.g., causal inference models) which may not be apparent to humans	Explanations rely on hypothesis-driven approaches, experimental observations, and known scientific principles	Both aim to provide accurate, evidence-based explanations for scientific phenomena
2. Developing and using models	AI creates complex, dynamic models (e.g., neural networks, simulations) that can incorporate vast datasets and predict behavior in real-time, often without explicit theoretical assumptions	Models are often simplified, static, or based on theoretical frameworks (e.g., equations for population dynamics)	Both involve representing biological systems to understand, predict, or explain their behavior
3. Analyzing and interpreting data	AI excels at processing large-scale, high-dimensional data (e.g., genomics, proteomics) using machine learning and statistical methods	Traditional methods rely on manual analysis or statistical techniques applied to smaller datasets	Both focus on uncovering patterns, relationships, and insights from scientific data
4. Engaging in arguments from evidence	AI tools can identify and evaluate evidence, supporting arguments with quantitative predictions and data patterns	Arguments are built through logical reasoning, literature review, and experimental validation	Both aim to create well-supported, evidence-based arguments for scientific claims
5. Asking questions	AI can suggest new questions by identifying patterns, anomalies, or gaps in large datasets (e.g., automated hypothesis generation)	Questions arise from human curiosity, existing knowledge, and observations in the field or lab	Both aim to generate meaningful, testable questions to advance scientific understanding
6. Planning and carrying out investigations	AI automates data collection (e.g., image recognition in ecology, robotic sampling, or high-throughput sequence alignments)	Data is collected manually or through standard laboratory and field protocols	Both rely on ensuring data quality, relevance, and integrity during collection processes

(continued)

Table 2.3 (continued)

Practice	AI methods	Traditional methods	Similarities
7. Obtaining, evaluating, and communicating information	AI is used for getting an overview of the literature including automated summaries (that need to be checked by humans). AI is used to create graphs and tables from raw data, write publications, and draw illustrations	(Scientific) search engines are used to get lists of potentially relevant literature. Literature needs to be read. Writing publications, creating figures and tables, and drawing illustrations or diagrams is a manual, time-intensive task	While the core tasks remain the same, AI allows automation and thus large efficiency gains. However, depending on the type of AI used (especially relevant for LLMs), outputs of AI must be carefully checked by human experts to ensure accuracy
8. Using mathematics and computational thinking	AI makes the use of scientific and mathematical software easier (e.g., using natural language instead of dedicated programming language), AI can directly write computer code but code may be erroneous and requires checking. AI may also struggle to solve advanced problems	Use of mathematics is supported through software such as computer algebra systems; computational thinking needs to be done by humans but libraries for popular programming languages allow us to build on existing knowledge	While the core tasks remain the same, AI allows us to speed up things. However, AI generated computer code needs to be carefully checked and advanced problems still require human expertise

2.2.5 *The Fragmented and Idiosyncratic Ways that Scientists Learn about AI Highlight the Need for Principled and Structured AI Education in Science Learning*

A comprehensive review of the approaches that graduate students, post-docs, and scientists learn about AI applications in their respective disciplines is currently lacking. Interviews with scientists, reviews of online resources, and examination of graduate programs suggest that although some degree programs have integrated AI-supported courses, a wide variety of decentralized, discipline-specific approaches lacking organizational structures and standards characterize much of AI education in science. Examples of the diverse ways that scientists learn to use AI include: (1) academic courses within graduate programs, (2) workshops and bootcamps, (3) online courses, (4) research collaborations, (5) academic conferences (and workshops therein), (6) online communities and associated resource hubs, (7) industry-led programs, (8) tool and resource training, and (9) academic journal articles. Examples of resources that scientists are using to learn about AI that may be useful for science educators seeking to understand these resources, are summarized in Table 2.4.

Table 2.4 Examples of the diverse ways that scientists are learning about AI applications in their disciplines

Source of AI learning	Selected examples
Academic training graduate programs	Machine learning basics, statistical analysis for biology, applications of AI in bioinformatics, tools like python (with libraries like TensorFlow, PyTorch, scikit-learn)
Online courses	Platforms such as: Coursera (e.g., machine learning, AI for medicine), edX (AI for life sciences), Udemy and Kaggle learn (coding and AI applications)
Workshops and Bootcamps	Programs like the cold Spring Harbor laboratory's computational biology workshops, genome informatics meetings; NIH and NSF workshops (funded for teaching AI and computational tools); private sector training: Industry-led initiatives (e.g., Google's AI for scientific discovery programs), bioinformatics companies (like Illumina or Thermo fisher) offer webinars and user training sessions. Certification AI training programs: DataCamp, fast.ai
Conferences	ISMB (intelligent systems for molecular biology), RECOMB (research in computational molecular biology), symposia, (deep learning in computational biology)
Online communities and resources	GitHub repositories (e.g., AlphaFold and ColabFold for protein prediction, Scikit-bio for bioinformatics tasks), discussion forums (e.g., stack overflow for coding questions; BioStars for bioinformatics discussions; Kaggle for community learning; YouTube channels for tutorials on AI tools like TensorFlow or PyTorch
Books and online guides	*Deep learning for the life sciences* by Greenberg, et al. *bioinformatics with python cookbook* for applied bioinformatics. AI tutorials: Online series like EMBL's *machine learning for biologists* from institutions. Blogs and newsletters: Distill.Pub for machine learning research. *Nature computational science* for applications in biology
Hands-on practice with tools	Cloud platforms: Google Colab for GPU-powered AI experiments; AWS and azure offer specialized life sciences AI tools; software: Bioinformatics suites like Bioconductor or galaxy; machine learning libraries like scikit-learn and PyTorch lightning
AI journals and papers	*Bioinformatics* or *nature machine intelligence* keeps researchers updated on tools and techniques

Note. Many resources are tailored to specific scientific subdisciplines, making it challenging to characterize the scientific enterprise as a whole

Our review of current approaches to learning about AI across scientific disciplines was supported by interviews with researchers. For example, physicist Dr. Electron at Freie Universität Berlin emphasized that learning about AI was happening in a haphazard way: "how you learn is very patchy at the moment [...] because there are no textbooks really, ... you need to read a lot of papers and you need to go to conferences and schools and talk to people." The diversity of avenues for learning about AI is considerable and presents a profound challenge for science educators seeking to understand *what* needs to be learned in different disciplines and *how* such learning should be organized.

In addition to the lack of clear organizational structures for learning about AI, the *pace* of change makes it challenging for scientists to keep up with new developments and discoveries. As Dr. Electron noted: "Students go to those mini schools and conferences and are reading the literature like mad. It's very fast and the amount of papers is very high—like every day, and keeping on top of everything that goes on is very difficult." In addition, the rapid development of new AI applications introduces uncertainty regarding which methods will persist: "It's so difficult because so many people are working on [AI applications] and no one knows what will survive in the end. So, you might read a lot of stuff that … in one year from now is already not relevant anymore."

What do these findings imply about how K-16 students should learn about AI? First and foremost, the delivery of AI education must be structured more effectively and efficiently than its haphazard and idiosyncratic inroads into science practice. Centralized web-based AI resource hubs for science education are needed that: represent multiple disciplines; link AI tools to implementation, evaluation, and reporting standards; connect lessons to key principles; and foster connections among communities of practitioners.

The rapid pace of change in science also suggests that K-16 students will benefit from learning about foundational principles and methods rather than details about specific tools or techniques. Otherwise, learning about AI in science education will not be beneficial for students (what they learn will be outdated quickly). For example, discussing the limitations of a trained AI model due to the specifics of the training data provides an opportunity to learn about fundamental questions about model generalizability. Establishing core principles for AI in K-16 science education emerges as an urgent priority.

2.3 Conclusion

For half a century, complex instruments and tools have formed essential nodes in the complex epistemic networks that characterize the contemporary scientific enterprise (Fig. 2.1). Yet science education has not kept pace with the expansion of complex instrumentation, the increasing specialization and dependencies among actors and instruments, or the emergence of autonomous AI agents in these networks. Many science lessons continue to model manual, individualized, technologically outdated, and small-scale inquiry. Approaches like these misrepresent authentic contemporary science practice to students.

In cases where technological tools like sensors, satellites, and PCR machines do play a role (e.g., use of authentic datasets, participation in citizen science, engagement in networked Course-Based Research Experiences), insufficient attention has been directed at helping students navigate epistemically opaque interactions with technology and its outputs (How do I know if I can trust it?). Introducing and modelling epistemic heuristics (e.g., AIR; Chinn et al., 2011, 2014) have been promoted for navigating social media and the disinformation inherent to it, but these heuristics have immense potential for use with technology (including AI). In science lessons and labs, too often the focus is on how to *use* technology rather than tackling inevitable epistemic dependence and trust considerations. These gaps have left students (and teachers) woefully unprepared for the AI revolution in science (Zhai & Nehm, 2023).

AI is filling increasingly diverse roles within scientific networks—instrument, extended agent, autonomous agent—and reshaping the boundaries of traditional science practices (e.g., Bejnordi et al., 2017; Casalino et al., 2020; Hope et al., 2021; Rolnick et al., 2019; Schwaller et al., 2020). These changes challenge current characterizations in science education of how science is practiced (NRC, 2012) and who (or *what*) does it. They also introduce new forms and magnitudes of epistemic dependence on AI. Science educators must respond to the changing nature of science by meaningfully embedding AI into science education: not only as another tool or method, but as a subject in need of critical reflection. By foregrounding epistemic vigilance, principled reasoning, and adaptable heuristics for use in many aspects of science, we can prepare students for a future in which AI is a central actor in scientific knowledge production.

Declaration of AI Use The authors wrote the paper. Parts of it were written by the authors with the assistance of generative AI tools for language editing or paragraphing. The authors reviewed, revised, and verified all AI-supported content for accuracy, appropriateness, and originality. No confidential or sensitive data was entered into AI systems. The final responsibility for the content rests solely with the authors.

Acknowledgement This material is based upon work supported by the National Science Foundation under Grant No. 2332964. Any opinions, findings, andconclusions or recommendations expressed in this material are those of the author(s) and do not necessarily reflect the views of the National Science Foundation.

References

AAAS. (2010). Vision and change: A call to action. Washington, DC AAAS www.visionand-change.org/VC_report.pdf.

Alvarado, R. (2023). *AI as an Epistemic Technology Science and Engineering Ethics 29*(5). https://doi.org/10.1007/s11948-023-00451-3

Bejnordi, B. E., et al. (2017). Diagnostic assessment of deep learning algorithms for detection of lymph node metastases in women with breast cancer. *JAMA, 318*(22), 2199–2210.

Casalino, L., Gaieb, Z., Goldsmith, J. A., Hjorth, C. K., Dommer, A. C., Harbison, A. M., et al. (2020). Beyond shielding: the roles of glycans in the SARS-CoV-2 spike protein. *ACS Central Science, 6*(10), 1722–1734.

Chinn, C. A., Rinehart, R. W., & Buckland, L. A. (2014). Epistemic cognition and evaluating information: Applying the AIR model of epistemic cognition. Processing inaccurate information: Theoretical and applied perspectives from cognitive science and the educational sciences. (pp. 425–453).

Chinn, C. A., Buckland, L. A., & Samarapungavan, A. L. A. (2011). Expanding the dimensions of epistemic cognition: Arguments from philosophy and psychology. *Educational Psychologist, 46*(3), 141–167.

Duede, E. (2022). Instruments agents and artificial intelligence: novel epistemic categories of reliability Synthese 200(6). https://doi.org/10.1007/s11229-022-03975-6

Ford, M. (2008). 'Grasp of practice' as a reasoning resource for inquiry and nature of science understanding. *Science & Education, 17*, 147–177.

Gaines, D. M., Estlin, T., Bornstein, B., & Anderson, R. C. (2016). Machine learning for autonomous science exploration. *AI Magazine, 37*(4), 21–31.

Hope, T., Portenoy, J., Vasan, K., Borchardt, J., Horvitz, E., Weld, D. S., & West, J. (2021). SciSight: Combining faceted navigation and research group detection for COVID-19 exploratory scientific search. *Nature Machine Intelligence, 3*(6), 442–448.

Hezaveh, Y. D., Levasseur, L. P., & Marshall, P. J. (2017). Fast automated analysis of strong gravitational lenses with convolutional neural networks. *Nature, 548*(7669), 555–557. https://doi.org/10.1038/nature23463

Jumper, J., Evans, R., Pritzel, A., Green, T., Figurnov, M., Ronneberger, O., et al. (2021). Highly accurate protein structure prediction with AlphaFold. *Nature, 596*(7873), 583–589. https://doi.org/10.1038/s41586-021-03819-2

Kapoor, et al. (2024). REFORMS: Consensus-based Recommendations for Machine-learning-based Science. *Science Advances, 10*, eadk3452.

Kitchener, R. F. (2002). Folk epistemology: An introduction [Editorial]. *New Ideas in Psychology, 20*(2-3), 89–105. https://doi.org/10.1016/S0732-118X(02)00003-X

Krenn, M., Häse, F., & Aspuru-Guzik, A. (2024). Autonomous AI makes Nobel-winning chemistry discoveries. *Nature Machine Intelligence*. https://www.nature.com/articles/s42256-024-00876-y.

Krist, C., Schwarz, C. V., & Reiser, B. J. (2019). Identifying essential epistemic heuristics for guiding mechanistic reasoning in science learning. *Journal of the Learning Sciences, 28*(2), 160–205. https://doi.org/10.1080/10508406.2018.1510404

Levine, H., & Tu, Y. (2024). Machine learning meets physics: A two-way street. *Proceedings of the National Academy of Sciences, 121*(27), e2403580121. https://doi.org/10.1073/pnas.2403580121

McDermott, M. B., Wang, S., Marinsek, N., Ranganath, R., Foschini, L., & Ghassemi, M. (2021). Reproducibility in machine learning for health research: Still a ways to go. *Science Translational Medicine, 13*, eabb1655.

Mongan, J., Moy, L., & Kahn, C. E. (2020). Checklist for artificial intelligence in medical imaging (CLAIM): A guide for authors and reviewers. *Radiology: Artifical Intelligence, 2*, e200029.

National Research Council, Division of Behavioral, Board on Science Education, & Committee on a Conceptual Framework for New K-12 Science Education Standards. (2012). *A framework for K-12 science education: Practices, crosscutting concepts, and core ideas*. National Academies Press.

Radovic, A., Williams, M., Rousseau, D., Kagan, M., Bonacorsi, D., Himmel, A., et al. (2018). Machine learning at the energy and intensity frontiers of particle physics. *Nature, 560*(7716), 41–48. https://doi.org/10.1038/s41586-018-0361-2

Reichstein, M., Camps-Valls, G., Stevens, B., Jung, M., Denzler, J., Carvalhais, N., & Prabhat. (2019). Deep learning and process understanding for data-driven Earth system science. *Nature, 566*(7743), 195–204. https://doi.org/10.1038/s41586-019-0912-1

Rolnick, D., Donti, P. L., Kaack, L. H., Kochanski, K., Lacoste, A., Sankaran, K., et al. (2019). Tackling climate change with machine learning. *arXiv*. preprint arXiv:1906.05433.

Schwaller, P., Laino, T., Gaudin, T., Bolgar, P., Hunter, C. A., Bekas, C., & Lee, A. A. (2020). Molecular transformer: A model for uncertainty-calibrated chemical reaction prediction. *ACS Central Science, 5*(9), 1572–1583.

Zhai, X., & Nehm, R. H. (2023). AI and formative assessment: The train has left the station. *Journal of Research in Science Teaching, 60*(6), 1390–1398. https://doi.org/10.1002/tea.21885

Chapter 3
Responsible and Ethical Principles for the Practice of AI-Supported Science Education

Kent J. Crippen, Xiaoming Zhai, and Okhee Lee

3.1 Introduction

The landscape of science education is being reshaped in complex ways with the evolution of artificial intelligence (AI), particularly generative AI. These technologies are transforming how educators, students, and policymakers approach teaching, learning, and curriculum design, offering both significant promise and serious concerns (Jang & Choi, 2024). AI systems can simulate complex phenomena, provide real-time feedback, and adapt instructional content to individual learners' needs (Miller & Yoon, 2023; Yılmaz, 2024). Generative applications such as ChatGPT have gained widespread attention for their ability to produce human-like text dynamically, fueling experimentation and enthusiasm across educational contexts (Blonder et al., 2024; Lee et al., 2025). Within science education specifically, AI holds the potential to expand access to learning, support inquiry-based pedagogy, and personalize instruction (Heeg & Avraamidou, 2023; Jia et al., 2023; Lee et al., 2025). At the same time, the possibilities foreground pressing questions about governance, transparency, and responsibility, underscoring the need for principled approaches that guide how AI is designed, implemented, and evaluated in science education contexts.

K. J. Crippen (✉)
School of Teaching and Learning, University of Florida, Gainesville, FL, USA
e-mail: kcrippen@coe.ufl.edu

X. Zhai
AI4STEM Education Center, University of Georgia, Athens, GA, USA
e-mail: xiaoming.zhai@uga.edu

O. Lee
Steinhardt School of Culture, Education, and Human Development, New York University, New York, NY, USA

X. Zhai, K. J. Crippen (eds.), *Advancing AI in Science Education*, Advances in Technology-Rich Science Education 3,
https://doi.org/10.1007/978-3-032-16871-9_3

While the emerging and potential uses of AI are promising and increasingly practical, their integration into educational systems requires clear guidance, strategic implementation, and ongoing critical review to ensure alignment with goals and to uphold values such as fairness, privacy, and human oversight. Accordingly, many institutions are now developing formal policies to support *responsible principles* for AI use, which emphasize the *how* of educational practice. These efforts harness the benefits of AI while establishing boundaries to mitigate potential risks and unintended consequences (Eutsler et al., 2025).

Urgent concerns are also raised about AI amplifying existing inequities, reducing learners to data profiles, and eroding the agency of both teachers and students in decision-making processes (Billingsley et al., 2025; Bulathwela et al., 2024; Rivera-Novoa & Duarte Arias, 2025). These issues complement concerns about beneficial use, scientific integrity, and respect for human values (Gulumbe et al., 2025; Stokel-Walker, 2023; Thorp, 2023). In response, some institutions have adopted restrictive policies or outright bans, portraying generative AI as a threat to human intelligence and academic norms (Weissman, 2023). Overly pragmatic efforts based solely on responsible principles often overemphasize how AI should be practiced, while overlooking the equally important questions of *why* it should be used and the need for *ethical principles* for AI use. Both celebratory framings and restrictive bans risk portraying AI as either an inevitable solution or an existential threat, obscuring the nuanced middle ground.

Although the terms "responsible principles" and "ethical principles" are frequently used interchangeably, in our view, they reflect distinct dimensions of AI-supported education. Responsible principles address implementation, the methods, and safeguards that determine how AI functions in classrooms and institutions. Ethical principles address purpose, the values, and goals that determine why AI should be used in the first place. In this chapter, we offer a curated collection of eight Responsible and Ethical Principles (REP) for the practice of AI-supported science education. This collection represents a theory-based and evidence-driven framework that establishes key concepts, provides theoretical and empirical grounding, and guides interpretation and applications (Antonenko, 2015). The framework addresses both the action-oriented, implementation-focused dimension of responsible principles and the value-driven, purpose-oriented dimension of ethical principles, which are equally essential for guiding the practice of AI-supported science education.

The integration of AI in all aspects of education, including science education, raises foundational questions about the nature of knowledge and knowing. It invites us to consider what constitutes legitimate knowledge, how it is constructed, and who has the authority to define and validate it. As AI systems increasingly mediate access to information and shape learning processes, we must critically examine the assumptions these technologies embody about knowing, such as whether knowledge is fixed or evolving, individual or distributed, and the degree to which learning is a process of acquisition, construction, or participation. These epistemological questions directly influence how AI reshapes educational goals, instructional practices, and classroom experiences, underscoring the importance of integrating philosophical reflection with practical design. These questions, including the agentic potential of AI as a collaborative partner among learners and institutions, lie at the heart of how we design, implement, and evaluate AI-supported science education.

Without intentional design and critical oversight, AI integration risks narrowing the goals of science education to what can be measured, optimized, or predicted, rather than fostering more profound understanding. These tensions call for a deliberate, principled, and philosophical approach to AI integration, which should not be approached as a technical upgrade but as a strategic shift in how we conceive, design, and enact science education. This shift requires rigor and reflection to ensure that AI strengthens rather than weakens the broader aims of science education. These aims include cultivating scientific literacy, ensuring fair access to opportunities, nurturing inquiry and critical thinking, and preparing learners to apply scientific understanding fruitfully in their lives. Accordingly, we use the phrase 'AI-supported' as a modifier to the phenomenon of education, recognizing the myriad ways AI may operate and influence educational practice.

Science education practice, the specific activities educators use to advance learning goals (Windschitl et al., 2012), serves as the context in which the REP framework operates, providing both direction for action and a lens for evaluating why those actions matter. Rooted in educational theory and socio-technical critique (Kudina & van de Poel, 2024), the REP framework clarifies how educators can act responsibly and why those actions carry ethical significance. Rather than serving as a compliance checklist, the framework functions as a critical lens for examining whether AI tools and their applications (i.e., AI support) reflect the values and social responsibilities upheld by the global science education community. We propose this framework and illustrate its use to advance AI-supported science education as an accessible, inclusive, and reflective practice, contributing to the broader dialogue on how AI, depending upon our decision-making, may shape teaching, learning, and curriculum now and in the future. Accordingly, the chapter is organized into five parts:

1. The context for the ideas and presentation of the chapter
2. The REP framework
3. Integrating the REP framework into a systems model of educational transformation
4. Scenario-based vignettes translating the REP framework into classroom practice
5. Position statements that ground the REP framework

Together, these sections support the chapter's central claim: AI-supported science education must be approached responsibly and ethically (for details, see Sect. 2 of this chapter). This claim necessitates changes in how (a) educational goals are defined, (b) instructional practices are enacted, (c) learning materials are designed, (d) assessments are conducted, and (e) outcomes are envisioned (for details, see Sect. 3 of this chapter). These changes, when guided by REP, foster a systems perspective that links ethical commitments to practical action and contributes to meaningful science education for all.

3.2 Context for This Chapter

This chapter is part of a book that emerged through a process that intentionally modelled the values and commitments it proposes. The book was co-authored by a highly experienced and engaged committee of interdisciplinary scholars convened to explore the implications of AI-supported science education. The significance and consequences of the book were compounded by recent changes in Federal priorities and the political climate in the U.S. Yet, the committee's work was rooted in the practice of shared deliberation, open communication, mutual critique, and iterative refinement, which are hallmarks of collaborative, responsible, and ethical scholarship. Roles were distributed across idea generation, framework construction, case study development, and writing, with transparency maintained throughout regarding decision-making and authorship.

Building on this foundation of human collaboration, this chapter also illustrates AI-supported scholarly practice. This chapter was constructed through sustained interactions between the human authors and the AI agent ChatGPT. The human authors, who retained complete control over conceptual framing, content, direction, collaborated with ChatGPT. First, ChatGPT served as a tool for synthesizing language, organizing structure, and refining style. For example, when developing the section linking REP to educational transformation, ChatGPT helped refine the structure and coherence. Early drafts included overlapping ideas and dense phrasing; through iteration, the model supported synthesis, clarified transitions, and aligned language to strengthen conceptual clarity. Second, the human authors used ChatGPT reflexively to bring tensions to the surface, question assumptions, and improve clarity. For example, each classroom vignette includes a sketch generated by ChatGPT, drawing on our collaboratively developed key messages and framework. Rather than accepting these outputs at face value, the human authors engaged with them reflexively, again using the sketches to surface tensions, question embedded assumptions, and sharpen clarity. This iterative process strengthened the alignment of each example with the REP framework. In this sense, the collaboration with ChatGPT mirrored the same kind of AI-supported science education practice that this chapter advocates: AI was present, but always in a supportive role, guided by human oversight and ethical reflection.

This partnership between the human authors and ChatGPT exemplifies several REP in action: transparency and explainability, human oversight and autonomy, and accountability. Together, the collaborative work among human authors and the conscious use of AI as a co-creative agent offers a living example. It demonstrates how responsible principles shape how AI is implemented (e.g., oversight, safeguards, transparency). In addition, ethical principles remind us why AI is used (e.g., to promote integrity, reflexivity, and human values in scholarly practice). Rather than treating AI as a neutral assistant, this chapter frames AI and human interactions as an epistemological and pedagogical endeavour, which must be transparent, intentional, and grounded in human values and educational goals.

3.3 The REP Framework

Science education has long aspired to cultivate not only knowledge and skills but also values, such as curiosity, responsibility, and civic reasoning (Levy et al., 2021). These goals are increasingly challenged as AI-supported tools and environments become integrated into the learning process. Generative AI can produce fluent explanations, elegant code, and data-rich visualizations, of varying accuracy, often blurring the line between original student thinking and machine assistance. These capabilities risk shifting the role of learners from active constructors to passive recipients unless educators deliberately create spaces for interpretation, critique, and reflection (Darvishi et al., 2024; Zhai et al., 2024).

The context of AI-supported science education raises complex and vital questions. What responsibilities do educators and students carry in preparing for an AI-supported world? How can science education practice ensure that learners remain active, critical participants rather than passive recipients? These questions must be addressed honestly, openly, and iteratively if we are to preserve the goals of science education in an AI-supported era.

To address these challenges, we introduce the REP framework for AI-supported science education. This framework is grounded in the distinction established earlier: responsible principles guide how AI is implemented (methods, safeguards, processes), while ethical principles guide why AI is used (values, goals, purposes). Together, they provide complementary lenses for interpreting the role of AI in educational practice.

Building on this distinction and the need for both responsible and ethical guidance, the REP framework offers eight guiding principles for educators, students, designers, and researchers (see Table 3.1). These principles serve as guiding commitments for educators, designers, and researchers working to achieve AI-supported science education in thoughtful and effective ways. They represent a synthesis of ideas drawn from scholarly and international sources on REP for AI use. While many frameworks address ethics from a policy or technical standpoint, few translate those ideas into actionable considerations for teachers and students. Our contribution lies in adapting these principles to the specific context of science education, where classroom practice, disciplinary integrity, and learner agency are central.

The REP framework was developed by integrating foundational principles from the UNESCO Recommendations on the Ethics of Artificial Intelligence (2021); the IEEE Ethically Aligned Design initiative (2016); the AI4K12 Five Big Ideas (Touretzky et al., 2019); and academic literature examining fairness, bias, and socio-technical design (e.g., Buolamwini & Gebru, 2018; Selbst et al., 2019). Additional influence came from privacy guidelines (e.g., the U.S. Student Data Privacy Pledge; https://studentprivacypledge.org/); human-centered AI frameworks (Floridi et al., 2018; Holmes et al., 2019); and recent critiques of oversight, transparency, and algorithmic accountability in educational technology.

What makes the REP framework distinct is its explicit attention to the practice of AI-supported science education, the activities of teachers, learners, and researchers.

Table 3.1 Overview of the REP framework: definitions and key considerations for science education

Responsible [R] and ethical [E] principles	Definitions	Key considerations for science education
Fairness [R]	AI should optimize learning opportunities and processes for all learners	Are systems designed to be equitable for all students? How are biases identified and mitigated in practice?
Privacy & Data Security [R]	Teacher and student data should be protected, comply with legal standards, and not be misused for surveillance or other dubious purposes	How is data collected, stored, and used? Are consent, security, and context--sensitive privacy protections in place?
Human Oversight & Autonomy [R]	AI should support, not replace, human teachers and learners, fostering effective science learning and instruction	Does the system preserve the decision--making roles of teachers and students? Are users able to interpret and adjust AI recommendations?
Beneficial & Meaningful use [E]	AI should enhance science education by promoting engagement, innovation, and deeper thinking, including participation in authentic science and engineering practice	Does the AI promote active, experiential, inquiry-based learning? Is it aligned with learning goals that value curiosity, explanation, and exploration? Is the use an authentic science and engineering practice?
Robustness & Scientific Integrity [E]	AI systems should be accurate, reliable, and aligned with scientific principles to prevent misinformation	Are outputs consistent with disciplinary knowledge and practice? Are safeguards in place to ensure factual accuracy and reliability?
Respect for human values [E]	AI systems should be designed to respect learners' and teachers' dignity, autonomy, and ethical decision-making in science	Do the systems support ethical reasoning, emotional well-being, and cultural responsiveness? Are human--centered values reflected in design and use?
Transparency & Explainability [R&E]	AI processes, decisions, and recommendations should be clear, interpretable, and transparent for students and educators	Can students and educators understand how the AI system works? What data or logic is behind outputs and recommendations?
Accountability [R&E]	Educators, developers, and institutions have ownership and responsibility for AI-supported decisions and errors	Who is accountable for decisions informed by AI? Are there mechanisms for reviewing and correcting misuse or unintended impacts?

Origins are indicated as emphasizing compliance with policies, rules, and norms [Responsible; R] or moral values [Ethical; E]

Whereas many existing frameworks emphasize broad policy guidelines (e.g., UNESCO, 2021), technical reliability (IEEE, 2016), or general principles of fairness and bias in socio-technical systems (Buolamwini & Gebru, 2018; Selbst et al., 2019), they often remain abstract and detached from the day-to-day realities of classroom instruction. Likewise, human-centered AI approaches (Floridi et al., 2018; Holmes et al., 2019) tend to highlight design principles without fully addressing how teachers and students enact responsible and ethical use in science classrooms. By contrast, the REP framework is grounded in the practice of science education: for example, teachers may use AI-supported platforms to adapt laboratory investigations in real time, learners may critique AI-supported explanations of scientific phenomena to strengthen their reasoning, and researchers may analyze classroom data with AI support while ensuring transparency and privacy. More specifically, this may include students engaging in modelling by refining AI-supported visualizations of molecular interactions or constructing arguments from evidence by evaluating AI-supported claims with empirical data. The principles are not intended as a checklist but as a foundation for reflective decision-making that integrates both responsible (how) and ethical (why) perspectives. They support intentional, forward-thinking engagement with emerging technologies and promote responsible and ethical action in research, design, and the activities educators use to advance AI-supported science education. The following brief explanations further illustrate how each principle in the framework can guide educational decision-making through its application in practice.

Fairness ensures that AI is used to enhance learning opportunities and processes for all learners. As a responsible principle, it emphasizes the need for careful attention to methods and safeguards that support fairness. Educators and designers must be vigilant against systemic bias and intentional about reaching diverse learners. Fairness requires evaluating datasets and model behavior for discriminatory patterns (Buolamwini & Gebru, 2018; Memarian & Doleck, 2023). In AI-supported classrooms, this means designing inclusive and accessible learning experiences that reflect the diversity of student identities and contexts, particularly for historically minoritized groups (Roscoe et al., 2022). Science educators must therefore ensure that AI support does not inadvertently reproduce inequities but instead broadens opportunities for participation and success. Although articulated here as a distinct principle, fairness interacts with other REP commitments such as respect for human values and accountability, underscoring that the principles must be applied together rather than treated as a checklist.

Privacy and data security emphasize the protection of student and teacher information. As a responsible principle, it shapes how AI support handles data, requiring safeguards for collection, storage, and use. AI supports must not misuse data for surveillance or other dubious purposes. They must comply with legal and ethical standards, such as the Family Educational Rights and Privacy Act (FERPA) and the General Data Protection Regulation (GDPR). However, ethical privacy goes beyond compliance, requiring informed consent, data minimization, and contextual sensitivity (Floridi & Cowls, 2019; UNESCO, 2021). In the practice of science education, this principle demands vigilance about how student data is used in AI-supported dashboards, assessments, or personalized platforms, ensuring that participation in science learning never comes at the cost of security or dignity. Privacy also interacts with accountability and fairness, reinforcing the collective responsibility to protect learners.

Human oversight and autonomy affirm that educators and learners retain decision-making power. As a responsible principle, it directs AI support so that human judgment is preserved rather than displaced. AI should support, not override or replace, human judgment and educational agency. Appropriate humans must always be in the loop (Mosqueira-Rey et al., 2022). This principle is aligned with guidance from the OECD and the Partnership on AI, emphasizing that algorithmic tools should not displace the professional expertise of teachers (Holmes et al., 2019; OECD, 2015). In AI-supported science education, students must retain meaningful roles in inquiry and reflection, while teachers remain responsible for pedagogical choices.

Beneficial and meaningful use remind us that AI should enhance educational experiences by promoting deeper learning, intellectual engagement, and authentic forms of participation, rather than merely focusing on efficiency or automation. As an ethical principle, it highlights why AI should be used in science education: to enrich rather than diminish human learning and experience. This principle aligns with calls for pedagogically sound integration, where AI promotes curiosity, inquiry, and student-centered learning (Holmes et al., 2019; Zhai, 2022). In AI-supported science classrooms, this involves using tools to empower learners to explore phenomena, practice science authentically, and generate insights that extend beyond efficiency or automation. Beneficial use interacts with robustness and respect for human values, reinforcing the importance of aligning AI support with the deeper purposes of science education.

Robustness and scientific integrity require that AI support be reliable, accurate, and scientifically sound. As an ethical principle, it defines why AI must be held to disciplinary standards: to safeguard accuracy and scientific ways of knowing. Outputs must align with accepted disciplinary knowledge and avoid misinformation or oversimplification. This principle is fundamental in AI-supported science education, where scientific accuracy and evidence-based reasoning are foundational (Osborne, 2013). Developers and educators must ensure that systems reflect and support the integrity of science as a way of knowing. This principle connects directly to fairness (e.g., preventing biased or inaccurate outputs) and accountability (i.e., ensuring that failures can be corrected), underscoring that scientific integrity cannot stand alone but depends on other REP commitments.

Respect for human values guides designers and educators to ensure that AI honors student dignity, cultural identity, and ethical reasoning. As an ethical principle, it clarifies why AI should serve education: to reinforce democratic, cultural, and moral dimensions of learning. This principle draws from human-centered AI frameworks and emphasizes the cultural, emotional, and moral dimensions of learning (Levy et al., 2021; UNESCO, 2021). In AI-supported science education, this may involve surfacing the ethical implications of scientific inquiry, designing culturally responsive tools, and ensuring that AI technologies reinforce, rather than erode, the human dimensions of teaching and learning. Respect for human values interacts with fairness and beneficial use, highlighting the importance of designing AI support that strengthens the humanity of education.

Transparency and explainability imply that AI supports and its processes are understandable to those using them. As both a responsible and ethical principle, it bridges how and why: systems must be designed to make their operations clear (how) to promote informed, reflective engagement (why). Teachers and students should know how outputs are generated and how systems make decisions (Holmes et al., 2019; Selbst et al., 2019). Admittedly, achieving this level of transparency is extremely difficult, as many AI systems operate with complex, opaque models that resist straightforward explanation (von Eschenbach, 2021). However, this challenge underscores the importance of the REP framework. All users should rightly expect transparency and interpretability from any AI support, which must therefore be designed to enable and reinforce meaningful understanding. In AI-supported science classrooms, transparency fosters data literacy by helping learners interpret algorithms, models, and assumptions (UNESCO, 2021). This principle supports human oversight and autonomy, ensuring that AI support is open to critique and reflection rather than hidden from view.

Accountability places responsibility for AI's outcomes on those who design, implement, and oversee it. As both a responsible and ethical principle, it addresses how mechanisms for oversight are built and why human actors must own responsibility. Harm or error must be met with precise mechanisms for review and rectification. Accountability also includes the obligation to evaluate long-term impacts and unintended consequences (IEEE, 2016). In the practice of AI-supported science education, accountability requires schools, districts, and software developers to create governance structures that evaluate both immediate outcomes and broader impacts on learning (Kudina & van de Poel, 2024). Accountability works in tandem with fairness, privacy, and oversight to ensure that responsibility is never shifted onto the learner or the technology itself.

While the REP framework draws on precedents from existing frameworks, this chapter contributes by adapting and contextualizing them within the distinctive context and practice of AI-supported science education. We aspire to move beyond abstract guidelines and examine how REP are enacted in practice through the tools school leaders and teachers choose and the ways they design and facilitate learning. In the following section, we apply the REP framework to key components of educational transformation to highlight its practical implications for both the nature and scale of potential change. This orientation offers a conceptually grounded, user-oriented, and widely applicable framework for diverse educational contexts. Ultimately, this product reflects our belief that REP for the practice of AI-supported science education is not a fixed outcome, but an ongoing process of reflection, negotiation, and adaptation embedded in the everyday work of teaching and learning.

3.4 Integrating the REP Framework with an Educational Transformation Model

The REP framework is most potent when paired with a clear systems model of how meaningful educational change occurs. In this section, we set our focus on the educational ecosystem and apply the REP framework primarily at the mesosystem level, the scale where interactions between classroom practice and institutional structure shape implementation (Rosa & Tudge, 2013). Building on this perspective, we integrate the REP framework into the bidirectional model of educational transformation, making visible how both responsible (how) and ethical (why) principles operate within systemic change.

The bidirectional model of educational transformation recognizes five core areas of science education: goals, teaching practice, learning materials, assessment, and outcomes (Zhai, 2023) (see Chap. 1 of this book for full details on the model). In each of these areas, AI-supported science education can be examined through REP, ensuring that both purpose and implementation are aligned. These five areas also serve as the primary organizing structure for the entirety of this book, with subsequent chapters dedicated to exploring each area in depth, defining the nature of change brought about by AI, illustrating how AI is being used in response to those changes, and using the REP framework to identify implications. This approach is consistent with broader calls for systemic, embedded strategies to guide AI innovation in education, ones that recognize the interdependence of classroom, institutional, and technological layers (Sharples, 2025). In the next section, we bring this integration to life through a series of classroom-based vignettes that illustrate how the REP framework can guide the practices of AI-supported science education in everyday teaching and learning.

3.4.1 Educational Goals

AI is reshaping educational goals by making it essential for learners to understand and critically leverage emerging technologies to thrive in their lives and meet future job market needs. This transformation must be guided by principles of fairness and respect for human values (see Chaps. 4 and 5 of this book). Education fosters not only knowledge and skills but also adaptability, ethical awareness, and human-centered qualities, such as creativity and moral reasoning. As AI takes on more complex tasks, educators must emphasize uniquely human capabilities to ensure meaningful learning. The educational goals we set and the technologies we use to pursue them must advance learning by enabling all learners to engage in inquiry and authentically participate in scientific practice.

To ensure that AI supports rather than undermines learning, educational goals must emphasize cultivating creativity, critical thinking, and logical judgment. These goals align closely with the principles of beneficial and meaningful use, human oversight and autonomy, and robustness and scientific integrity. As this illustrates, REP guide both how educational goals are implemented responsibly (e.g., fairness, oversight) and why they are pursued ethically (e.g., meaningful use, scientific integrity). These principles encourage students to use AI to enhance, not replace, their understanding, agency, and scientific reasoning. For example, as outlined in Sustainable Development Goal 4 (SDG 4), this includes eliminating gender disparities and ensuring inclusive, quality education for all. For example, persistent gaps remain in access to STEM pathways for girls and women, including lower participation in advanced science and mathematics courses, limited exposure to role models in STEM fields, and reduced opportunities for research experiences (Unterhalter, 2019; Rizal et al., 2021; UNESCO, 2021). These disparities underscore the importance of aligning educational goals with reform-focused initiatives. Without careful design, AI support could unintentionally reinforce and widen these gaps.

3.4.2 Teaching Practice

Adoption of classroom practice aligning closely with the principles of fairness and respect for human values can help ensure that AI-supported educational procedures accommodate the unique learning trajectories of all students (see Chaps. 6 and 7 of this book). Teachers can use AI to automate routine tasks, such as grading or responding to frequently asked questions, allowing them to focus on guiding inquiry, facilitating discussion, and mentoring students (Zhai et al., 2020). AI-supported dashboards and learning analytics tools can aid in classroom management and offer insights into student progress, enabling educators to intervene more effectively (Gobert et al., 2023). However, these benefits come with responsibilities. Procedures that rely heavily on student data raise concerns about privacy and data security, and must adhere to legal standards in protecting sensitive information.

Moreover, learners exhibit diverse strategies and preferences (Chang et al., 2011). AI supports must be developed with this variability in mind, ensuring that personalization does not become homogenization. The principles of human oversight and autonomy, and transparency and explainability are critical here: teachers must remain in control of pedagogical choices, and students should be aware of how and why AI-supported recommendations are made. Taken together, these examples show that REP inform both how educational procedures are implemented responsibly (e.g., oversight, transparency) and why they should promote dignity ethically (e.g., beneficial and meaningful use, respect for human values).

While AI can support project-based learning and other forms of inquiry-driven learning, an over-reliance on automated systems risks narrowing the instructional approach and neglecting the relational and dialogic elements of science teaching. Generative AI should complement, not replace, educator expertise and student initiative. Effective AI-supported science education depends on developing human-AI partnerships in which tools augment professional judgment rather than supplant it. Ultimately, science educators must design AI-supported procedures that not only promote knowledge acquisition but also engage students in the practice of reasoning, modeling, and problem-solving. These aspirations reflect the principles of beneficial and meaningful use and accountability. Ongoing research and collaboration are needed to ensure that procedural innovations with AI enhance, rather than hinder, responsible and ethical teaching and learning.

3.4.3 Learning Materials

While the transformation of education needs the reconsideration of appropriate and novel learning materials, these changes are not equally distributed (see Chap. 8 of this book). Students from historically minoritized groups, including first-generation college students, underrepresented racial or ethnic minorities, and female-identifying learners, frequently encounter persistent barriers to accessing high-quality digital and AI support (Roscoe et al., 2022; Zhai & Krajcik, 2025). These disparities reflect more profound structural inequities in socioeconomics, education, and technology infrastructure. If left unaddressed, AI may amplify existing gaps in opportunity, participation, and outcomes.

To align with the principle of fairness, science educators and developers must ensure that AI-supported learning materials are inclusive, accessible, and culturally responsive. This includes designing for diverse languages, dialects, and cognitive preferences, and ensuring that content is free from bias or stereotypes in both language and representation. The language and visual processing capabilities of AI can also be harnessed to promote cross-cultural understanding, such as through multilingual resources or content that reflects students' lived experiences (Davoodi, 2024; Saddhono et al., 2024).

Transparency is yet another essential consideration. Many AI-supported materials are opaque, making it difficult for students and teachers to know how content is selected, what data was used to train the system, or how recommendations are made. The principle of transparency and explainability calls for clear communication about how AI systems function and the epistemic assumptions underlying the materials they produce. Without this clarity, students may passively accept flawed or biased outputs.

Finally, developers and educators must uphold robustness and scientific integrity by ensuring that AI-supported materials align with accepted scientific knowledge, accurately represent disciplinary content, and afford authentic forms of practice, especially as AI evolves in its influence on the practice of science disciplines. Tools should be rigorously tested to prevent misinformation and reflect current understandings in science education. Equipping students to question, validate, and critique these tools is also essential to fostering scientific agency and trust. In this way, REP shape both how learning materials are designed responsibly (e.g., fairness, transparency) and why they are used ethically (e.g., scientific integrity, respect for human values).

To fulfil the promise of AI in transforming science learning, investment is needed in underserved communities to ensure that all students have access to high-quality, AI-supported learning materials and other forms of support. Doing so will help ensure that the integration of AI supports, not undermines, meaningful learning for all.

3.4.4 Assessment

While assessment practice offers the most potential for transformation, AI challenges long-standing assumptions that undergird this practice (see Chaps. 9 and 10 of this book). For example, generative AI can produce essays or responses that blur the line between student work and AI output (Stokel-Walker, 2023). This raises concerns about authenticity, academic integrity, and the evolving role of core skills and literary competencies, such as writing. To align with the principles of accountability, transparency, and explainability, as well as human oversight and autonomy, educators must adapt assessments to emphasize creativity, critical thinking, and logical reasoning in ways that complement the current and evolving skills and science practice.

Those principles, as part of assessment practice, can also exacerbate inequities. When access to advanced tools varies by demographic group, standardized tests may reflect access more than ability (González-Calatayud et al., 2021). REP-aligned assessment must ensure fairness by questioning and testing how training data and tool design may introduce bias, especially for minoritized learners (Memarian & Doleck, 2023). Moving beyond efficiency, assessment should preserve student agency, recognize diverse demonstrations of understanding, and reflect the goals of science education. Alternative approaches, such as portfolios, performance tasks, or reflective prompts, may better align with REP commitments and the evolving needs of science learners. In short, REP guide both how assessment is responsibly structured (e.g., oversight, accountability, fairness) and why it should be used ethically (e.g., to support integrity and meaningful learning).

3.4.5 Educational Outcomes

In an era increasingly shaped by AI, educational systems must evolve to prepare students not merely to function in an AI-supported world but to thrive within it. Such changes demand competencies, such as understanding AI, applying AI, and evaluating and reflecting on AI in scientific contexts (see Chap. 11 of this book). These domains are designed not just as academic goals but as long-term competencies that support lifelong learning, adaptive expertise, and ethical reasoning. Educational outcomes, therefore, should encompass students' ability to understand how AI support operates within scientific inquiry; leverage AI support for problem-solving; and critically assess the reliability, impact, and fairness of AI support. Importantly, these outcomes must align with the principles of fairness, transparency, accountability, and human oversight and autonomy, ensuring that the next generation is equipped to make informed decisions in personal, professional, and civic domains.

For example, students should be prepared with the knowledge and skills to use AI support to generate hypotheses from scientific datasets, recognizing that these hypotheses require testing and are often shaped by underlying algorithmic assumptions. A student might apply a supervised learning model to classify climate patterns and critically evaluate the model's limitations by considering data biases or the interpretability of outputs, engaging with the principles of robustness and scientific integrity (Herdiska & Zhai, 2024). Similarly, developing competencies for evaluating AI-supported scientific content requires learners to assess the reliability of outputs, compare them with empirical evidence, and consider the responsible and ethical implications of using such data and outputs in decision-making. For instance, a student might assess whether an AI support used for medical simulations inadvertently embeds bias that could affect minoritized populations, thereby enacting the principle of fairness.

Ultimately, these educational outcomes transcend specific technologies or classroom tools; they represent a foundational shift in what it means to be scientifically literate in the twenty-first century. By emphasizing interdependent competencies that support knowledge acquisition, technical engagement, and reflective judgment, the model anticipates a world where AI is not just a tool but a partner in scientific reasoning and societal decision-making. Preparing students with these competencies ensures that they are capable not only of adapting to technological change but also of guiding such change responsibly and ethically. Embedding the REP framework into science education, therefore, has profound implications: it cultivates future professionals and citizens who are responsible in their use, morally aware, scientifically informed, and prepared to address the complex socio-technical challenges posed by an AI-supported world. Thus, REP clarify both how outcomes are responsibly shaped (e.g., fairness, oversight, accountability) and why they are ethically essential (e.g., to foster integrity and scientific literacy).

3.5 Scenario-Based Vignettes: Translating the REP Framework into Classroom Practice

The following scenario-based vignettes illustrate an application of the REP framework at the scale of science classroom practice across the five core areas of educational transformation. Each vignette is grounded in science classroom instruction to illustrate how REP support students while engaging with the Next Generation Science Standards (NGSS; NGSS Lead States, 2013). These examples demonstrate how REP can influence classroom decisions, maintaining visibility of both the responsible and ethical practices in practice.

At the outset of the vignettes, we state disclaimers. The vignettes are not intended to illustrate the grand notion of NGSS-aligned instruction where students engage in science and engineering practices, disciplinary core ideas, and crosscutting concepts (i.e., three-dimensional learning) to make sense of phenomena and problems (i.e., phenomenon-based learning). Instead, through these short vignettes, we illustrate how AI can be used to support NGSS-aligned instruction. Specifically, we demonstrate how the REP framework applies to real-world decisions educators make every day regarding what to teach, how to teach it, and how to support all students in ways that are relevant, inclusive, and responsive to current educational contexts. Each vignette also underscores that REP are interdependent; no single principle operates in isolation, even if one or two are highlighted for emphasis.

3.5.1 Vignette 1: Redefining Explanations in a ChatGPT-Era Biology Class

Core Area: Educational Goals

Use a model to illustrate that cellular respiration is a chemical process whereby the bonds of food molecules and oxygen molecules are broken and the bonds in new compounds are formed, resulting in a net transfer of energy (NGSS HS-LS1-7).

NGSS Alignment:

- Science and Engineering Practice: Constructing explanations and designing solutions
- Disciplinary Core Idea: LS1.C—Organization for Matter and Energy Flow in Organisms
- Crosscutting Concept: Energy and matter: Flows, cycles, and conservation

REP Framework: *Transparency & Explainability, Human Oversight & Autonomy, Beneficial & Meaningful Use.*

Scenario In a high school biology classroom, after a series of science investigations and class discussions about cellular respiration, Ms. Jenkins assigns a task to students: construct an explanation of cellular respiration in writing. Her students turn to ChatGPT to help craft their explanations. The explanations they bring to class are polished and fluent, but something feels off.

As class discussions unfold, it becomes clear that beneath the sleek phrasing lies a shallow grasp of the process. Students often struggle to explain why specific steps are crucial or how ATP fuels cellular activity. Ms. Jenkins notices a pattern: While AI can generate quick answers, it doesn't automatically cultivate understanding.

Fig. 3.1 Sketch of redefining explanations in a ChatGPT-Era biology class (credit to GPT-4o)

Rather than banning the technology, she shifts her teaching strategy. The next day, she brings in AI-generated explanations herself and asks students to critique them. "Where is this explanation correct? Where is it too vague or misleading? How could we make it better?" she prompts. Students now work in small groups, dissecting outputs, debating their scientific accuracy, and revising responses with a deeper understanding of the disciplinary core ideas at play. In doing so, they engage in NGSS-aligned practices such as analyzing and interpreting data, constructing explanations, and arguing from evidence, while applying crosscutting concepts such as patterns to identify consistencies across data, and cause and effect to reason about the validity of explanations. Diagrams are sketched, analogies are tested, and scientific language is reworked with greater precision.

In this evolving classroom, explanation becomes an active, iterative process of interpreting, questioning, and improving. Ms. Jenkins fosters a culture where student reasoning is visible, agency is central, and learning is grounded in curiosity and critique (Fig. 3.1).

Reflection This scenario illustrates how educational goals need to evolve to reflect the presence of generative AI. Prioritizing conceptual interpretation and critique helps ensure that students remain active thinkers with meaningful agency. Interpreting the AI output in this way challenges the false assumption that AI is a source of correct answers and instead positions AI as a learning partner that, like any human, is prone to errors, false assumptions, and misinterpretations. The vignette aligns with the principles of transparency and explainability (students must understand and assess AI outputs, especially for scientific accuracy), human oversight and autonomy (teachers and students remain decision-makers), and beneficial and meaningful use (AI supports conceptual learning). Together, these principles demonstrate not only how AI can be responsibly implemented in science education practice but also why it should be ethically integrated to preserve inquiry, agency, and deeper learning. They also underscore that REP are interdependent, working collectively rather than in isolation.

3.5.2 Vignette 2: Personalizing with Caution in a Physics Classroom

Core Area: Educational Procedures

Plan and conduct an investigation to provide evidence that the transfer of thermal energy when two components of different temperatures are combined within a closed system results in a more uniform energy distribution among the components (second law of thermodynamics) (HS-PS3-4).

NGSS Alignment:

- SEP: Planning and carrying out investigations
- DCI: PS3.B—Conservation of Energy and Energy Transfer
- CCC: Systems and system models

REP Framework: *Fairness, Privacy & Data Security, Human Oversight & Autonomy.*

Scenario In Mr. Patel's physics lab, students are preparing to investigate thermal energy transfer when hot and cold substances interact within closed systems. The learning performance is to plan and conduct an investigation of how energy redistributes and moves toward equilibrium, in line with the second law of thermodynamics. To support student investigation, Mr. Patel integrates an AI-supported learning platform that dynamically adjusts tasks based on student responses. At first, the tool seems promising; students receive tailored guidance and scaffolded support. But as the semester unfolds, he notices a troubling pattern: Students who struggle early on are repeatedly offered simplified experimental designs with fewer variables and less room for design choices.

The AI, designed to personalize, is quietly narrowing possibilities. Some students are no longer planning investigations; instead, they're following overly

Fig. 3.2 Sketch of personalizing with caution in a physics classroom (credit to GPT-4o)

constrained recipes. Mr. Patel grows concerned that the platform is unintentionally limiting challenge and reinforcing disparities.

Determined to re-center student agency, he digs into the platform's algorithmic decision rules and consults with colleagues. He learns how student data is driving the platform's choices, and how those choices may be shaping opportunity. In response, he rebalances his approach. All students are now invited to design their own experiments: choosing materials, controlling variables, and drawing evidence-based conclusions. The AI remains in use, but under thoughtful and critical oversight, with flexibility for the teacher to adapt prompts, scaffold inquiry, or override outputs when needed, and with transparency in making visible how recommendations are generated and what data are drawn upon. In doing so, the teacher models NGSS practices such as developing and using models and engaging in argument from evidence, while highlighting the crosscutting concept of systems and system models to help students see the AI itself as a system with inputs, processes, and outputs.

The classroom shifts. Diverse groups of students lean in over bubbling mixtures, debating how best to measure equilibrium or minimize heat loss. They sketch hypotheses, troubleshoot flawed designs, and compare temperature data. Across the room, every student is engaged in planning, testing, and reflecting, challenged less by what the AI has allowed but by what their thinking has made possible (Fig. 3.2).

Reflection This vignette highlights the importance of using AI to support, rather than constrain, student agency in science instruction. AI-supported procedures must be monitored for unintended bias and must protect student data. The scenario aligns with the principles of fairness (equitable access to challenge and opportunity), privacy and data security (awareness of how student data informs decision-making), and human oversight and autonomy (the teacher's role to promote inclusive pedagogy). Taken together, these principles illustrate how AI support can be responsibly designed and monitored and why it should be ethically integrated to protect learner agency. They also reinforce that REP are interdependent, requiring alignment across fairness, privacy, and oversight to function effectively.

3.5.3 *Vignette 3: Choosing Culturally Responsive AI Materials in Chemistry*

Core Area: Learning Materials

Communicate scientific and technical information about why the molecular-level structure is important in the functioning of designed materials (HS-PS2-6).

NGSS Alignment:

- SEP: Developing and using models
- DCI: PS2.B—PS1.A—Structure and Properties of Matter
- CCC: Structure and function

REP Framework: *Fairness, Respect for Human Values, Robustness & Scientific Integrity.*

Scenario In Mrs. Alvarez's chemistry classroom, students are creating explanations of how designed materials function based on their molecular structures. They use an AI-supported platform that generates visualizations. The tool also presents real-world scenarios to help students understand the everyday context of the materials (e.g., how they are used and why they are interesting and important). On the surface, it seems effective.

But something does not sit right. As she walks from table to table, Mrs. Alvarez notices puzzled expressions. One student scrolls past a scenario comparing acid strength using common sports drinks in the U.S. and expresses that she's never heard of the beverage brands being referenced. Another hesitates during a simulation about thermal decomposition in cookie baking and says, "We don't have an oven at home."

The problem is not with the chemistry; it is with the context. The AI-generated examples, though scientifically sound, are steeped in Western norms. The examples assume shared experiences that don't reflect the diverse lives of students. Slowly, Mrs. Alvarez realizes that if students can't see themselves in the science, their connection to the examples falters.

Refusing to let student engagement be a casualty of default settings, Mrs. Alvarez takes action. She starts curating new examples drawn from her students' own community. She swaps commercial drink brands for tamarind juice and vinegar-based marinades in a locally relevant acid-base lesson. Instead of cookie recipes, she introduces molecular changes in nixtamalization, a traditional maize preparation. She also invites students to explore biodegradable materials inspired by banana leaves or cactus fibers used in sustainable packaging.

With these new examples, her classroom comes alive. Students light up when they see chemistry unfolding in foods from their own kitchens or environmental innovations linked to their regions. The AI platform remains a resource, but now, its support is better anchored in an inclusive, human-centered pedagogy.

In Mrs. Alvarez's room, molecular structure still matters. But just as importantly, so do students' identities, languages, and lived experiences. Through her thoughtful adaptation, science becomes not just accurate, but accessible, affirming, and empowering (Fig. 3.3).

Fig. 3.3 Sketch for choosing culturally responsive AI materials in chemistry (credit to GPT 4o)

Reflection This vignette underscores the need to critically examine AI-supported learning materials for cultural relevance and accessibility as well as scientific accuracy. Ensuring that materials reflect students' diverse identities and uphold academic standards supports both engagement and learning. The scenario aligns with the principles of fairness (ensuring inclusive design), respect for human values (affirming student dignity and context), and robustness and scientific integrity (maintaining scientific accuracy). Together, these principles show why AI must be ethically used to affirm students' identities and how it must be responsibly designed to safeguard integrity. They also demonstrate the interdependence of REP, since fairness, values, and integrity reinforce one another in practice.

3.5.4 Vignette 4: Rethinking Writing in an AI-Supported Earth Science Class

Core Area: Assessment

Evaluate or refine a technological solution that reduces impacts of human activities on natural systems (HS-ESS3-4).

NGSS Alignment:

- SEP: Engaging in argument from evidence
- DCI: ESS3.C—Human Impacts on Earth Systems
- CCC: Stability and change

REP Framework: *Accountability, Human Oversight & Autonomy, Transparency & Explainability.*

Scenario In Ms. Brown's Earth science class, for a unit on climate solutions, students have been developing written proposals that argue for specific technological solutions to reduce human impacts on Earth systems, drawing on evidence they've collected to support their claims. As the class has just wrapped up, students are submitting their final proposals, many crafted with help from generative AI tools. The writing is polished, the tone confident. But as Ms. Brown reads through the submissions, something feels off.

Despite their fluency, the essays lack substantiated argumentation. Proposed technologies to combat climate change, like carbon capture systems or sustainable agriculture, are often asserted without relevant data or detailed justification.

Evidence is mentioned only in passing, with weak or scattered connections to claims. The persuasive lustre of AI-generated prose is masking conceptual gaps, particularly how students are expected to build arguments from evidence.

Ms. Brown doesn't want to penalize students for using AI, but she also doesn't want it to replace their thinking, either. So, she reimagines the assessment. The following week, instead of submitting a final draft alone, students are asked to include annotated versions of their work: highlighting where they used AI, what parts they edited, and where they added evidence-based reasoning. They also prepare short oral defenses of their proposals and write reflections on how AI helped or complicated their understanding of climate technologies.

In class, the atmosphere changes. Students gather in small groups to practice defending their proposals, debating the feasibility of solar roadways, or refining arguments for reforestation technology. One student explains how she revised an AI-generated claim about desalination after checking the evidence on its energy footprint. Another reworks his conclusion after realizing his solution didn't account for long-term trade-offs. By making the reasoning and evidence behind the writing visible, Ms. Brown fosters a culture of intellectual ownership. Students are no longer just handing in persuasive texts; they're grappling with science, evaluating technological solutions, and interrogating the tools that shaped their arguments.

In this classroom, assessment isn't about punishing AI use; it is about illuminating learning. With the principles of transparency, autonomy, and evidence-based argument at the center, Ms. Brown empowers students to engage in climate science as thinkers, not just writers (Fig. 3.4).

Reflection This vignette illustrates the need for assessment practice that values transparency, student agency, and authentic intellectual work. Because the task required students to construct proposals grounded in evidence, the integration of AI challenged the integrity of engaging in argument from evidence. The scenario aligns with the principles of accountability (ensuring responsible assessment design), human oversight and autonomy (preserving educator judgment and student voice), and transparency and explainability (making AI's role in student work visible and

Fig. 3.4 Sketch for rethinking writing in an AI-supported earth science class (credit to GPT 4o)

subject to scrutiny). Together, these principles highlight both how AI must be responsibly integrated into assessment procedures and why it must be ethically framed to promote integrity and authentic engagement with scientific argumentation. They also underscore the interdependence of REP commitments, as accountability, oversight, and transparency reinforce one another in shaping an assessment that truly reflects student reasoning.

3.5.5 Vignette 5: Forecasting Future Readiness with AI Dashboards

Core Area: Educational Outcomes

Analyze data from tests to determine similarities and differences among several design solutions to identify the best characteristics of each that can be combined into a new solution to better meet the criteria for success (MS-ETS1-3).

NGSS Alignment:

- SEP: Analyzing and interpreting data
- ETS1.B—Developing Possible Solutions
- CCC: Cause and effect

REP Framework: *Fairness, Robustness & Scientific Integrity, Respect for Human Values.*

Scenario In a middle school science lab, eighth graders are elbow-deep in a design challenge: building water filtration prototypes using everyday materials, sand, charcoal, mesh, and recycled bottles. They are meeting the learning performance: clean the dirtiest water sample in the most efficient way possible. Laptops hum as students record their test results and upload photos to an AI-supported dashboard designed to track their STEM readiness over time, a recent concept that has been defined by the school district as important for all students as a culmination of multiple STEM courses.

The dashboard analyzes multiple data sources, including test scores, attendance, participation, and engagement trends, to generate STEM readiness profiles, which summarize patterns and highlight strengths and areas for growth in practices such as analyzing and interpreting evidence. As the project wraps up, some students begin to feel uneasy. Jamal, who's been experimenting relentlessly with filter materials, frowns at his screen. "It says I'm unlikely to succeed in engineering," he mutters. A few seats over, Sofia sighs, "I got a low score, too... even though our filter worked best in class."

Teachers across the district begin noticing a pattern: Students from underrepresented backgrounds, multilingual learners, or those who struggle on traditional tests are disproportionately flagged as 'off-track,' even when their real-world problem-solving is strong. The readiness profile, meant to represent data-driven analysis, is instead narrowing interpretation to a single predictive judgment rather than supporting the NGSS practice of analyzing and interpreting data in multiple ways.

Alarmed, the district pivots. Instead of displaying singular scores, the dashboard now highlights learning behaviors, design strengths, and iterative progress. Students are prompted to reflect on data from multiple design trials. They compare filtration results, identify variables that influenced success, and interpret trends across their group's data, engaging directly in the practice of analyzing and interpreting data. They also consider how their performance connects to broader STEM competencies, building readiness not as a fixed score but as a profile of growth and inquiry.

In Ms. Nguyen's classroom, the shift is tangible. "You're not your score," she reminds her class. "You're a designer. And every prototype, even one that fails, teaches us something." Students now use the dashboard to set personal learning goals and map strengths, not to judge future careers but to build toward them.

By centering student voice, data transparency, and engineering practice, the revised approach empowers learners to analyze outcomes, identify promising design traits, and innovate forward. The AI platform remains a guide for growth, not a gatekeeper of potential (Fig. 3.5).

Fig. 3.5 Sketch for forecasting future readiness with AI dashboards (credit to GPT 4oT)

Reflection This vignette highlights how AI supports that shape educational outcomes must be implemented intentionally and cautiously. When used without context or reflection, AI supports risk-narrowing opportunities and reinforces disparities. Aligning with the principles of fairness (avoiding biased predictions), robustness and scientific integrity (ensuring data accuracy and validity), and respect for human values (supporting student dignity and growth), the revised approach emphasizes AI as a guide, not a gatekeeper. By reframing STEM readiness profiles as opportunities for students to compare results, identify variables, and interpret patterns across design trials, the dashboard now supports the NGSS practice of analyzing and interpreting data. This shift shows how REP can responsibly structure outcome measures and why ethical commitments to agency are essential in shaping readiness as growth rather than prediction. The reflection also underscores the interdependence of REP, since fairness, integrity, and human values must operate together to sustain meaningful use of data in science education.

The five vignettes, taken together, demonstrate that REP in science education are not a one-time decision, but a dynamic process of reflection, adaptation, and thoughtful stewardship. Whether rethinking instructional goals, choosing learning tools, or interpreting assessment data, educators must engage with AI critically and intentionally. By foregrounding the REP framework, these vignettes demonstrate how science education can harness the promise of AI while safeguarding its human, disciplinary, and democratic purposes. More comprehensive examples of REP for the practices of AI-supported science education are provided in the individual chapters of this book.

3.6 Position Statements in Proposing the REP Framework

The following position statements define the purpose and values of the REP framework in science education. They are intended to clarify the framework's foundational assumptions, emphasize its relevance in the context of AI integration, and express the responsible and ethical commitments that inform and guide its development. Together, these statements distil the broader arguments of the chapter into concise commitments that reflect both how REP are implemented responsibly and why they are ethically essential. Like the principles themselves, they are interdependent and should be read as a collective orientation rather than as isolated directives.

3.6.1 AI Must Serve Educational Goals, Not Replace Them

We reject deterministic narratives that portray AI as either a threat or a solution. Instead, we advocate for an educational approach in which AI is shaped by, and reinforces, the goals of science education. AI should extend students' engagement with scientific ideas, not merely automate instruction or assessment. For example,

AI-supported lab simulations can be deployed with a feedback architecture that seeks to promote an understanding of molecular phenomena by responding in a visual or multimodal way, prioritizing the visibility of invisible molecular processes and thus making them more accessible to learners and teachers (Mroz et al., 2025). This stance reflects the ethical purpose (why) of beneficial and meaningful use and the responsible implementation (how) of scientific integrity, showing the complementarity of REP commitments in shaping practice.

3.6.2 Transparency and Understanding Are Essential for REP

To use AI responsibly and ethically, teachers and students must understand how these systems function and influence learning. Responsibility for this outcome begins and ends with technology developers. Transparency and explainability are prerequisites for meaningful interpretation, critique, and reflection. For instance, a teacher may guide students in exploring how a recommendation algorithm selects specific science experiments over others. Still, students must be informed about the rules and conditions under which these recommendations are made. Such classroom practice aligns with the principles of transparency and explainability as well as human oversight and autonomy, reinforcing interpretive agency and algorithmic literacy. This position shows how REP require transparent implementation and why it is ethically essential for preserving agency and interpretive practice. It also highlights interdependence, since transparency cannot be achieved without oversight and accountability.

3.6.3 REP Involves all Educational Stakeholders

REP are not limited to developers or administrators. Students, teachers, researchers, and designers all share responsibility for ensuring that AI supports are used in ways that are pedagogically sound, socially responsible, and ethical. For example, a classroom debate about whether AI-generated lab reports constitute authentic student work illustrates this shared responsibility between students and teachers. On the one hand, students are expected to engage honestly in the learning process, demonstrating their understanding through original work. On the other hand, educators must clearly define what constitutes acceptable use of AI supports and guide ethical boundaries. This shared responsibility requires open dialogue about the role of AI in learning, as well as agreed-upon norms that support integrity, accountability, and meaningful engagement from all parties involved. This message reflects the principles of accountability and respect for human values. This demonstrates both how shared responsibility is enacted through oversight and accountability and why it is ethically grounded in transparency, human values, and integrity. It reinforces that REP depend on collaboration across all stakeholders rather than isolated compliance.

3.6.4 Bias and Exclusion Must Be Actively Addressed in REP

AI systems trained on biased or non-representative data risk reproducing historical injustices. Responsible and ethical science education must include efforts by all responsible parties to audit systems for bias regularly and to design supports that prioritize fairness. For example, an AI tutor might be continuously evaluated for how its feedback varies across student demographics. Doing so reflects the principles of fairness as well as privacy and data security, with the aspiration of equitable access and protection against discriminatory practice. This position highlights how REP require responsible implementation through systematic auditing and why it is ethically imperative to pursue meaningful science learning. It also demonstrates interdependence, since fairness, privacy, and human values reinforce one another in addressing bias.

3.6.5 REP Require Ongoing Oversight

Educational contexts are dynamic and require continual review and adjustment of the technical aspects of AI supports. Human educators play a critical role in monitoring and guiding the functions of these software applications. For instance, a science department may review an AI-supported formative assessment system after noticing that students are consistently flagged as struggling due to misconceptions about specific scientific terminology, even though their conceptual understanding is sound. Teachers require access to revise the system's tagging algorithms and supplement them with human-reviewed explanations. This scenario illustrates the principles of human oversight and autonomy as well as accountability, emphasizing adaptive, human-centered governance in support of accurate and fair evaluation. This position highlights the importance of maintaining oversight structures responsibly and why ethical commitments to fairness and accountability necessitate ongoing adjustments. It reinforces interdependence by showing that oversight, autonomy, and accountability must operate together.

3.6.6 REP Are a Form of Literacy

Just as scientific and data literacy are essential for understanding and engaging with the natural and technological world, responsible and ethical literacy about AI must become a core competence for both students and teachers. This literacy must address their dual roles as *consumers* and *creators* of AI-supported content and decisions. Understanding how AI systems are designed, what data they use, and how they influence learning and decision-making is critical for participating in an increasingly algorithmic society. For example, students may prompt generative AI tools to simulate molecular interactions or visualize data, while teachers may use AI to create custom assessments or adapt instruction. Both roles require an understanding of how the tools function and the associated responsible and ethical considerations that come with their use. Integrating AI ethics into science teacher education, including professional development, can help cultivate this form of literacy and support the principles of transparency and explainability, respect for human values, and beneficial and meaningful use (Ding et al., 2024). This position illustrates how literacy is built responsibly through transparency and interpretability and why it is ethically essential for democratic participation and human values. It highlights the interdependence of literacy, fairness, and agency in preparing educators and students for an AI-driven world.

3.7 Concluding Remarks

The rapid emergence of AI-supported science education brings both exciting possibilities and urgent challenges. As a community committed to science education as a human endeavour, rooted in collective responsibility and shared ethics, we must navigate a path that balances innovation with integrity. As this chapter has argued, integrating AI into science teaching and learning is not a neutral act; it shapes what becomes possible, what is prioritized, and ultimately, what science education stands for. In this evolving landscape, responsible and ethical reflection must be central and embedded in every stage of AI use, from the design of technologies to the policies shaping their availability and ultimately to classroom implementation. REP are not optional; it is a shared imperative among all participants.

Through the development and application of the REP framework, this chapter responds to the questions raised by AI's expanding role in science education. It emphasizes both how AI should be implemented responsibly and why it should be pursued ethically, ensuring that practice is guided by safeguards as well as values. By anchoring AI use in principles of transparency, professional responsibility, and human agency, and by demonstrating AI applications across the chapters of this book, we provide researchers, educators, and policymakers with a coherent path forward for integrating AI in ways that support responsible and ethical educational transformation.

The REP framework, coupled with the core areas of educational transformation, offers a structured yet adaptable foundation for guiding this work. It underscores the interdependence of REP, which must operate together to shape inclusive and human-centered change. Together, they show that REP for the practice of AI-supported science education requires more than technical safeguards or policy checklists. It calls for intentional choices grounded in human values, inclusive practice that reflects the diversity of learners, and sustained dialogue across the educational community. By demonstrating how the REP framework can be implemented across educational goals, procedures, materials, assessments, and outcomes, this chapter provides a practical and principled roadmap for aligning AI with the system of science education.

This roadmap is strengthened by the collaborative process bridging systemic design with the lived values of interdisciplinary scholarship. This vision emerged from the collective effort of a multidisciplinary committee comprising highly recognized scholars who engaged in critical reflection and shared inquiry. Our collaboration for the book was guided by the very commitments that we advocate: transparency, ethical, mutual responsibility, and scholarly discourse. The insights we've presented in the book reflect diverse perspectives and values shaped through ongoing, often challenging, conversations about what science education can and should become in an increasingly complex world. Ultimately, our goal in this book is to ensure that the future of science education is not driven by convenience or commercial agendas, but by responsible and ethical commitment to inclusive, transformative learning for all learners.

Declaration of AI Use Parts of this manuscript were prepared with the assistance of generative AI tools. The AI tools were used for language editing of the authors' writing. The authors reviewed, revised, and verified all AI-supported content for accuracy, appropriateness, and originality. No confidential or sensitive data were entered into AI systems. The final responsibility for the content rests solely with the authors.

Acknowledgement This material is based upon work supported by the National Science Foundation under Grant No. 2332964 (PI Zhai). Any opinions, findings, and conclusions or recommendations expressed in this material are those of the author(s) and do not necessarily reflect the views of the National Science Foundation.

References

Antonenko, P. D. (2015). The instrumental value of conceptual frameworks in educational technology research. *Educational Technology Research and Development, 63*(1), 53–71. https://doi.org/10.1007/s11423-014-9363-4

Billingsley, B., Grzes, M., & Annetta, L. (2025). The future of knowledge: Conversations about artificial intelligence and epistemic insight. *Science & Education, 34*(2), 645–647. https://doi.org/10.1007/s11191-025-00636-1

Blonder, R., Feldman-Maggor, Y., & Rap, S. (2024). Are they ready to teach? Generative AI as a means to uncover pre-service science teachers' PCK and enhance their preparation program. *Journal of Science Education and Technology*. https://doi.org/10.1007/s10956-024-10180-2

Bulathwela, S., Pérez-Ortiz, M., Holloway, C., Cukurova, M., & Shawe-Taylor, J. (2024). Artificial intelligence alone will not democratise education: On educational inequality, techno-solutionism and inclusive tools. *Sustainability, 16*(2), 781. https://doi.org/10.3390/su16020781

Buolamwini, J., & Gebru, T. (2018). Gender shades: Intersectional accuracy disparities in commercial gender classification.

Chang, C.-Y., Hsiao, C.-H., & Chang, Y.-H. (2011). Science learning outcomes in alignment with learning environment preferences. *Journal of Science Education and Technology, 20*(2), 136–145. https://doi.org/10.1007/s10956-010-9240-9

Darvishi, A., Khosravi, H., Sadiq, S., Gašević, D., & Siemens, G. (2024). Impact of AI assistance on student agency. *Computers & Education, 210*, 104967. https://doi.org/10.1016/j.compedu.2023.104967

Davoodi, A. (2024). EQUAL AI: A framework for enhancing equity, quality, understanding and accessibility in Liberal arts through AI for multilingual learners. *Language, Technology, and Social Media, 2*(2), 178–203. https://doi.org/10.70211/ltsm.v2i2.139

Ding, A.-C. E., Shi, L., Yang, H., & Choi, I. (2024). Enhancing teacher AI literacy and integration through different types of cases in teacher professional development. *Computers and Education Open, 6*, 100178. https://doi.org/10.1016/j.caeo.2024.100178

Eutsler, L., Rivera, B., Barnes, M., & Cummings, J. (2025). Artificial intelligence policies in K-12 school districts in the United States: A content analysis shaping education policy. *Journal of Research on Technology in Education*, 1–25. https://doi.org/10.1080/15391523.2025.2476589

Floridi, L., Cowls, J., Beltrametti, M., Chatila, R., Chazerand, P., Dignum, V., Luetge, C., Madelin, R., Pagallo, U., Rossi, F., Schafer, B., Valcke, P., & Vayena, E. (2018). AI4People—an ethical framework for a good AI society: Opportunities, risks, principles, and recommendations. *Minds and Machines, 28*(4), 689–707. https://doi.org/10.1007/s11023-018-9482-5

Floridi, L., & Cowls, J. (2019). A unified framework of five principles for AI in society. *Harvard Data Science Review*. https://doi.org/10.1162/99608f92.8cd550d1

Gobert, J. D., Sao Pedro, M. A., & Betts, C. G. (2023). An AI-based teacher dashboard to support students' inquiry. In N. G. Lederman, D. L. Zeidler, & J. S. Lederman (Eds.), *Handbook of research on science education* (1st ed.). Routledge.

González-Calatayud, V., Prendes-Espinosa, P., & Roig-Vila, R. (2021). Artificial intelligence for student assessment: A systematic review. *Applied Sciences, 11*(12), 5467. https://doi.org/10.3390/app11125467

Gulumbe, B. H., Audu, S. M., & Hashim, A. M. (2025). Balancing AI and academic integrity: What are the positions of academic publishers and universities? *AI & Society, 40*(3), 1775–1784. https://doi.org/10.1007/s00146-024-01946-8

Heeg, D. M., & Avraamidou, L. (2023). The use of artificial intelligence in school science: A systematic literature review. *Educational Media International, 60*(2), 125–150. https://doi.org/10.1080/09523987.2023.2264990

Herdiska, A., & Zhai, X. (2024). Artificial intelligence-based scientific inquiry. In X. Zhai & J. Krajcik (Eds.), *Uses of artificial intelligence in STEM education* (pp. 179–197). Oxford University Press.

Holmes, W., Bialik, M., & Fadel, C. (2019). *Artificial intelligence in education: Promises and implications for teaching and learning* (p. 242). Center for Curriculum Redesign.

IEEE. (2016). *Ethically aligned design: A vision for prioritizing human well-being with artificial intelligence and autonomous systems*. IEEE.

Jang, H., & Choi, H. (2024). A double-edged sword: Physics educators' perspectives on utilizing ChatGPT and its future in classrooms. *Journal of Science Education and Technology*. https://doi.org/10.1007/s10956-024-10173-1

Jia, F., Sun, D., & Looi, C. (2023). Artificial intelligence in science education (2013–2023): Research trends in ten years. *Journal of Science Education and Technology*. https://doi.org/10.1007/s10956-023-10077-6

Kudina, O., & van de Poel, I. (2024). A socio-technical system perspective on AI. *Minds and Machines, 34*(3), 21. https://doi.org/10.1007/s11023-024-09680-2

Lee, G., Yun, M., Zhai, X., & Crippen, K. (2025). Artificial intelligence in science education research: Current states and challenges. *Journal of Science Education and Technology*. https://doi.org/10.1007/s10956-025-10239-8

Levy, B. L. M., Oliveira, A. W., & Harris, C. B. (2021). The potential of "civic science education": Theory, research, practice, and uncertainties. *Science Education, 105*(6), 1053–1075. https://doi.org/10.1002/sce.21678

Memarian, B., & Doleck, T. (2023). Fairness, accountability, transparency, and ethics (FATE) in artificial intelligence (AI) and higher education: A systematic review. *Computers and Education: Artificial Intelligence, 5*, 100152. https://doi.org/10.1016/j.caeai.2023.100152

Miller, K. M., & Yoon, S. A. (2023). Teaching complexity in biology through agent-based simulations: The relationship between students' knowledge of complex systems and metamodeling knowledge. *Frontiers in Education, 8*. https://doi.org/10.3389/feduc.2023.1198307

Mosqueira-Rey, E., Hernández-Pereira, E., Alonso-Ríos, D., Bobes-Bascarán, J., & Fernández-Leal, Á. (2022). Human-in-the-loop machine learning: A state of the art. *Artificial Intelligence Review*. https://doi.org/10.1007/s10462-022-10246-w

Mroz, A. M., Basford, A. R., Hastedt, F., Jayasekera, I. S., Mosquera-Lois, I., Sedgwick, R., Ballester, P. J., Bocarsly, J. D., Del Río, A., Chanona, E., Evans, M. L., Frost, J. M., Ganose, A. M., Greenaway, R. L., Kuok Mimi Hii, K., Li, Y., Misener, R., Walsh, A., Zhang, D., & Jelfs, K. E. (2025). Cross-disciplinary perspectives on the potential for artificial intelligence across chemistry. *Chemical Society Reviews*. https://doi.org/10.1039/d5cs00146c

NGSS Lead States. (2013). Next generation science standards: For states, By States. The National Academies Press. https://www.nextgenscience.org/.

Organisation for Economic Co-operation and Development [OECD]. (2015). Future of Education and Skills 2030/2040. https://www.oecd.org/en/about/projects/future-of-education-and-skills-2030.html.

Osborne, J. (2013). The 21st century challenge for science education: Assessing scientific reasoning. *Thinking Skills and Creativity, 10*, 265–279. https://doi.org/10.1016/j.tsc.2013.07.006

Rivera-Novoa, A., & Duarte Arias, D. A. (2025). Generative artificial intelligence and extended cognition in science learning contexts. *Science & Education*. https://doi.org/10.1007/s11191-025-00660-1

Rizal, R., Rusdiana, D., Setiawan, W., Siahaan, P., & Ridwan, I. M. (2021). Gender differences in digital literacy among prospective physics teachers. *Journal of Physics: Conference Series, 1806*(1), 012004. https://doi.org/10.1088/1742-6596/1806/1/012004

Rosa, E. M., & Tudge, J. (2013). Urie Bronfenbrenner's theory of human development: Its evolution from ecology to bioecology. *Journal of Family Theory & Review, 5*(4), 243–258. https://doi.org/10.1111/jftr.12022

Roscoe, R. D., Salehi, S., Nixon, N., Worsley, M., Piech, C., & Luckin, R. (2022). Inclusion and equity as a paradigm shift for artificial intelligence in education. In F. Ouyang, P. Jiao, B. M. McLaren, & A. H. Alavi (Eds.), *Artificial intelligence in STEM education: The paradigmatic shifts in research, education, and technology* (1st ed.). CRC Press.

Saddhono, K., Rohmadi, M., Lestari, T. A., Simanungkalit, K. E., & Sukmono, I. K. (2024). The role of AI in facilitating multilingual literacy and cross-cultural understanding. 2024 international conference on IoT, communication and automation technology (ICICAT). (pp. 366–371). https://doi.org/10.1109/ICICAT62666.2024.10923254.

Selbst, A. D., Boyd, D., Friedler, S. A., Venkatasubramanian, S., & Vertesi, J. (2019). Fairness and abstraction in socio-technical systems. *Proceedings of the conference on fairness, accountability, and transparency—FAT* '19*. (pp. 59–68). https://doi.org/10.1145/3287560.3287598.

Sharples, M. (2025). A systems approach to AI and education in a post-digital world. *Theory Into Practice*, 1–9. https://doi.org/10.1080/00405841.2025.2528549

Stokel-Walker, C. (2023). ChatGPT listed as author on research papers: Many scientists disapprove. *Nature, 613*(7945), 620–621. https://doi.org/10.1038/d41586-023-00107-z

Thorp, H. H. (2023). ChatGPT is fun, but not an author. *Science, 379*(6630), 313. https://doi.org/10.1126/science.adg7879

Touretzky, D., Gardner-McCune, C., Martin, F., & Seehorn, D. (2019). Envisioning AI for K-12: What should every child know about AI? *Proceedings of the AAAI Conference on Artificial Intelligence, 33*, 9795–9799. https://doi.org/10.1609/aaai.v33i01.33019795

United Nations Educational, Scientific and Cultural Organization [UNESCO]. (2021). Recommendation on the ethics of artificial intelligence. UNESCO. https://unesdoc.unesco.org/ark:/48223/pf0000381137.

Unterhalter, E. (2019). The many meanings of quality education: Politics of targets and indicators inSDG 4. *Global Policy, 10*(S1), 39–51. https://doi.org/10.1111/1758-5899.12591

von Eschenbach, W. J. (2021). Transparency and the black box problem: Why we do not trust AI. *Philosophy & Technology, 34*(4), 1607–1622. https://doi.org/10.1007/s13347-021-00477-0

Weissman, J. (2023, February 8). ChatGPT is a plague upon education (opinion). Inside Higher Education. https://www.insidehighered.com/opinion/views/2023/02/08/chatgpt-plague-upon-education-opinion.

Windschitl, M., Thompson, J., Braaten, M., & Stroupe, D. (2012). Proposing a core set of instructional practices and tools for teachers of science. *Science Education, 96*(5), 878–903. https://doi.org/10.1002/sce.21027

Yılmaz, Ö. (2024). Personalised learning and artificial intelligence in science education: Current state and future perspectives. *Educational Technology Quarterly, 2024*(3), 255–274. https://doi.org/10.55056/etq.744

Zhai, C., Wibowo, S., & Li, L. D. (2024). The effects of over-reliance on AI dialogue systems on students' cognitive abilities: A systematic review. *Smart Learning Environments, 11*(1), 28. https://doi.org/10.1186/s40561-024-00316-7

Zhai, X., Yin, Y., Pellegrino, J. W., Haudek, K. C., & Shi, L. (2020). Applying machine learning in science assessment: A systematic review. *Studies in Science Education, 56*(1), 111–151. https://doi.org/10.1080/03057267.2020.1735757

Zhai, X., & Krajcik, J. (2025). *Uses of artificial intelligence in STEM education* (p. 624). Oxford University Press.

Zhai, X., Neumann, K., & Krajcik, J. (2023). Editorial: AI for tackling STEM education challenges. *Frontiers in Education, 8*. https://doi.org/10.3389/feduc.2023.1183030

Zhai, X. (2022). ChatGPT user experience: Implications for education. *SSRN Electronic Journal*. https://doi.org/10.2139/ssrn.4312418

Part II
Educational Goals

Chapter 4
AI-Enhanced Science Literacy: Science Literacy Plus

Xiufeng Liu and Jennifer N. Tripp

AI, particularly generative AI, is rapidly transforming science and society; AI will become an integral part of everyday life for all citizens. This chapter focuses on the implications of AI on goals of science education. It will first review notions of science literacy, a commonly accepted goal of science education around the world. It will then examine how AI is enabling science, scientific problem-solving, as well as addressing complex social and environmental issues, which calls for promoting AI literacy for everyone in K-12. This chapter concludes by arguing that science literacy plus, which is the intersection of science literacy and AI literacy, be an important goal of K-12 science education in the twenty-first century.

4.1 Science Literacy as a Goal of Science Education

Since the conception of science literacy was first proposed by Hurd (1958) and McCurdy (1958), science literacy has often been described as a desirable goal for K-12 science education in the United States and around the world. There have been a few comprehensive literature reviews on science literacy (e.g., DeBoer, 2000; Laugksch, 2000; Miller, 1983; Osbome, 2023; Roberts, 2007). Most of these definitions stem from what Roberts (2007) classifies as two types of science literacy: Vision I and Vision II. Vision I science literacy is based on what is required to be

X. Liu (✉)
University of Macau, Macao SAR, China
e-mail: xiufengliu@um.edu.mo

J. N. Tripp
State University of New York (SUNY) Geneseo, Geneseo, NY, USA
e-mail: jtripp@geneseo.edu

X. Zhai, K. J. Crippen (eds.), *Advancing AI in Science Education*, Advances in Technology-Rich Science Education 3,
https://doi.org/10.1007/978-3-032-16871-9_4

competent in science fields and careers, while Vision II science literacy is based on what people need to know and can do with science in relation to everyday life and society, regardless of whether they are working in science fields. Vision I science literacy is about mastery of science knowledge and processes, and Vision II science literacy is the everyday, contextual applications of science in society. Vision I and Vision II can be thought of as two ends of a science literacy continuum (Roberts, 2007). Accordingly, learning expectations for K-12 science curricular standards, such as the *National Science Education Standards* (NRC, 1996) and the *Next Generation Science Standards* (NGSS Lead States, 2013), are often presented as a balance between Vision I and Vision II.

Emphasizing personal agency in science, researchers have also proposed a participatory notion of science literacy (Liu, 2013; Roth, 2002; Roth & Calabrese, 2004; Roth & Lee, 2004a, 2004b). This notion of science literacy, also called Vision III science literacy, promotes the use of science knowledge for actions, particularly in relation to social, cultural, political, and environmental issues (Bencze et al., 2020; Hodson, 2009; Sjöström, 2024). Vision III science literacy aims at transforming society through science. For example, the PISA 2025 science assessment includes a construct as part of science literacy called *environmental awareness, concern and agency* (OECD, 2023, p. 32). This construct measures students' competencies in taking a critical, evidence-informed perspective on personally and socially relevant environmental issues, as well as a sense of personal agency, which is informed by scientific and environmental understanding.

While the literature on science literacy has focused almost exclusively on individuals (i.e., science literacy as a personal attribute), a National Research Council committee on science literacy proposed that science literacy could be demonstrated at three aggregation levels: individual, community, and society (NRC, 2016). Community and societal science literacy—while including individual science literacy as their foundations—assume that science literacy is distributed among members and agents of community and society. The necessity of science literacy for democracy, civic engagement, culture, and economic development is more dependent on science literacy at the community and societal levels than at the individual level. Because addressing complex social, cultural, political and environmental issues requires people's agency and active engagement, this collective aspect of science literacy can be considered as part of Vision III science literacy.

Synthesizing the above science literacy perspectives, and consistent with Guerrero and Torres-Olave (2022), we represent science literacy as leveled visions with increasing sophistication in Fig. 4.1. Vision III encompasses or subsumes Vision II, which also subsumes Vision I. Across the visions, there is progression from lower levels to higher levels, from fundamental knowing (e.g., reading, writing, and numeracy), to using (e.g., explaining, evaluating, reasoning with, applying), to acting (e.g., engaging in citizen science and socioscientific issues).

Based on the above conceptualization of layered science literacy, the following sections will examine how AI is impacting each of the three visions of science literacy.

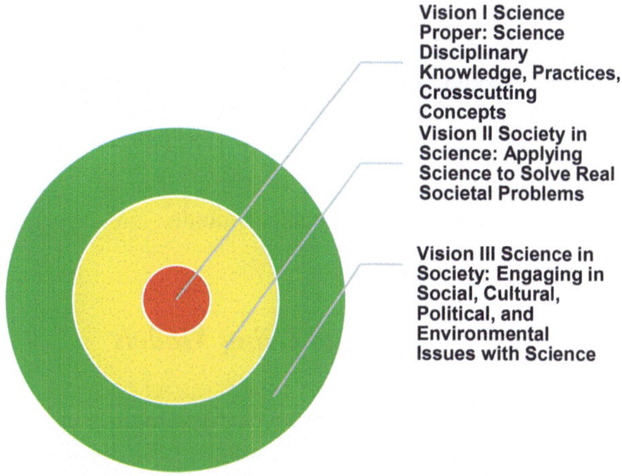

Fig. 4.1 Science literacy as layered competencies at the individual, community and societal levels

4.2 AI-Enabled Science: Vision I Science Literacy Plus

There is increasing convergence between AI and the sciences (NRC, 2025). In a comprehensive review published in *Nature*, Wang et al. (2023) describe how "artificial intelligence (AI) is being increasingly integrated into scientific discovery to augment and accelerate research, helping scientists to generate hypotheses, design experiments, collect and interpret large datasets, and gain insights that might not have been possible using traditional scientific methods alone" (p. 47). Eight science practices are identified as components of science literacy: (1) asking questions and defining problems; (2) developing and using models; (3) planning and carrying out investigations; (4) analyzing and interpreting data; (5) using mathematics and computational thinking; (6) constructing explanations and designing solutions; (7) engaging in argument from evidence; and (8) obtaining, evaluating, and communicating information (NRC, 2012). Within this section, we describe the ways in which AI is facilitating and informing each of these practices within science disciplines, thus modifying how scientists do science. This facilitation of science by AI is what is called "AI as tools" in Chap. 2.

4.2.1 Asking Questions and Defining Problems

Scientists have used AI to help with devising scientific hypotheses, including generating symbolic mathematical expressions and proposing structures of molecules in more time and cost-efficient ways (Wang et al., 2023). For example, AlphaFold2 can predict the 3D atom coordinates of proteins from amino acid sequences, which has facilitated the prediction of protein folding structures, a grand challenge in the field of molecular biology for over 50 years (Jumper et al., 2021). In pharmaceutical research, designing a new drug typically involves hundreds or thousands of possibilities; however, AI has been used to identify the most viable paths based on specific criteria or reward values (Zhavoronkov et al., 2019). BrainBench, a large language model (LLM), has surpassed human experts in predicting neuroscience experimental outcomes by a considerable margin with high confidence (Luo et al., 2024). This margin was amplified through additional LLM training in underlying patterns in neuroscience. The resulting BrainGPT generative model can be used to form hypotheses and design experiments.

4.2.2 Developing and Using Models

Scientists have also leveraged AI in developing and using models, from predicting weather patterns to diagnosing medical conditions. For example, GraphCast is a machine learning-based method for weather forecasting that leverages deep learning hardware (Lam et al., 2023). Trained with historical reanalysis atmospheric measurement data, such as observations and analysis data, GraphCast out-performed conventional weather forecasting approaches, or numerical weather predictions that primarily relied on solving equations based on physics laws with supercomputers (Lam et al., 2023). GraphCast has produced fairly accurate 10-day forecasts in under 1 min, including next weather states with 6-hour lead times. GraphCast capabilities also entail more accurate, efficient forecasts of tropical cyclone tracks, atmospheric rivers (responsible for heavy precipitation), and extreme temperatures. These models have been refined; for example, the machine learning-based weather prediction model, GenCast, has produced more accurate 15-day weather forecasts in less than 10 min, as compared with the 10-day forecasts based on traditional models (Price et al., 2025). In epidemiology, researchers have examined the synergistic possibility of AI and mechanistic epidemiological modeling to better understand disease dynamics, including how and why diseases unfold the way that they do, with implications for public health decision and policy making (Ye et al., 2025).

4.2.3 Planning and Carrying Out Investigations

AI has also been harnessed in scientific measurement by increasing precision and reducing errors in scientific investigations. AI can help visualize regions of black holes, capture a physics particle collision, improve the resolution of live-cell images, and better detect cell types across biological contexts, all of which have implications for how scientists plan and carry out investigations (Wang et al., 2023). These features extend human capabilities and make certain investigations possible. The use of drone technology and deep learning algorithms have been used to generate reliable population counts and predict potential poacher behaviors, which can be used alongside longstanding, Indigenous environmental monitoring and conservation efforts in several African countries (Petso & Jamisola, 2023). In these ways, AI systems can potentially contribute to environmental monitoring in the sea, land, and sky, which can assist conservation-based investigations and efforts (Dodhia, 2024).

4.2.4 Analyzing and Interpreting Data

AI is also facilitating data analysis and interpretation. Supervised learning requires large labeled or annotated data sets. For large unlabeled data, AI self-supervised learning techniques can allow for the development of computer models to learn the general features of a dataset without relying on explicit labels. Self-supervised learning can be used to pre-process large, unlabeled datasets for fine-tuning models on small, labeled datasets to perform subsequent data analysis and interpretation tasks (Wang et al., 2023). In particle physics, typically over 99.9% of raw instrument data represents background events and must be detected in real-time to increase data transmission efficiency and storage capacity. Deep-learning methods have been used to detect these background events and identify outlying signals or rare, potentially significant phenomena (Karagiorgi et al., 2022). In protein biology, less than 1% of sequenced proteins are annotated with biological functions. Trained AI systems are used to label unlabeled samples and use them to make predictions (Barkas et al., 2019).

4.2.5 Using Mathematics and Computational Thinking

AI-enhanced mathematics and computing are now routinely used in scientific research. In computational biology, AI tools have uncovered gene-regulatory interactions; combined with other cellular information and tools, they can analyze disease-relevant cell types which are difficult to collect, such as brain or heart cells (Brożek & Theodoris, 2025). Beyond cell type, AI-enhanced machine learning has been leveraged for computational protein design towards creating cost-effective, stable mini-binders and phospholipases in snake antivenoms (Callaway, 2025). The importance of AI-enhanced mathematics and computing is also evidenced by the 2024 Nobel Prize in Chemistry, which was awarded to David Baker, of the University of Washington and Howard Hughes Medical Institute, and Demis Hassabis and John M. Jumper of Google DeepMind for their work leading to AlphaFold3, and the 2024 Nobel Prize in Physics that was awarded to John J. Hopfield and Geoffrey Hinton for work contributing to machine learning with artificial neural networks.

4.2.6 Constructing Explanations and Designing Solutions

AI has also assisted with developing explanations of patterns in phenomena, such as in the fields of animal behavior and ecology conservation. For instance, AI-enhanced machine learning methods have been used to detect patterns in communication among sperm whales, elephants, marmoset monkeys, elephant seals, and crows (Savage, 2024). From analyzing spectrograms, or graphical representations of whale sound recordings, to visualizing sound along volume and frequency dimensions, researchers have come to explain these differences as whales communicating in their own dialects. These dialects vary according to clan and differ based on tempo and pauses. Typically, analyzing these recordings is quite time-consuming, in that 1 min of sound recording can take a team member about 10 min to analyze. However, machine-learning algorithms have drastically reduced processing times and have helped to differentiate specific whale sounds. As such, researchers have uncovered what they describe as a whale phonetic alphabet, used for sharing complex information (Savage, 2024). In addition to new research questions, uncovering wildlife communication patterns can lead to increased empathy among humans for other forms of life, while bolstering ecological conservation efforts.

4.2.7 Engaging in Argumentation from Evidence

AI has contributed to developing arguments from evidence. In medicine and public health, AI has facilitated medical image analysis, drug discovery, disease and cancer diagnostics, clinical outcome prediction, and treatment optimization (Christiansen et al., 2025; Eisemann et al., 2025; Moon et al., 2025; Ye et al., 2025). For instance, MedFound is a generalist medical language model that is pre-trained on a large corpus of diverse medical texts and clinical records; it has outperformed other models and LLMs for assisting clinicians in diagnosing medical conditions, including rare diseases (Liu et al., 2025). Although MedFound focused on language interaction, the researchers highlighted the potential for training LLMs with multimodal data, such as vision-language models that have been utilized within fields of pathology, radiology, and echocardiography for a more comprehensive approach to medical diagnoses. Likewise, Moon et al. (2025) used AI to classify renal histological types, such as normal tissue, benign tumors, and malignant tumors. Their study pointed to the potential of AI in efficient and precise diagnoses, particularly in situations with limited pathological expertise to facilitate quality control and diagnostic oversight.

4.2.8 *Obtaining, Evaluating, and Communicating*

AI is now routinely used in scientific research to retrieve, summarize, evaluate, and disseminate information. Today, large amounts of information can be summarized and explained quickly in simplified language through generative AI tools, such as LLMs and chatbots that can also rapidly generate text, images, video, and code with a few prompts. These, in turn, can be incorporated into meeting discussions, conference presentations, and journal publications. With the exponentially increasing scientific literature, possibly "disruptive findings go unnoticed in the deluge of articles," and processing and synthesizing "the myriad of relevant findings may already surpass humans' abilities" (Luo et al., 2024, p. 1). LLMs trained on vast datasets of general text and scientific literature have been able to efficiently process and integrate immense amounts of information from decades of research. AI is extending scientific writing in areas of idea generation, outlining, synthesizing literature, managing and organizing data, and assisting with revisions and editing. AI has also been leveraged in translating across named languages, which can facilitate science interactions, collaborations, and contributions across the globe.

4.2.9 *Implications for K-12 Science Education*

In this section, we have described some of the ways scientists have leveraged AI in their work across science fields, from ecology to particle physics. As scientists have harnessed AI in each of the abovementioned science practices, there is a need to rethink K-12 science education goals in alignment with Vision I science literacy. AI is altering how scientists engage in science investigations. No longer is mastery of traditional science content sufficient. At the K-12 level in science education, AI has primarily been used as a personalized tutor or to streamline formative assessment, often embedded in existing educational software programs that span subject areas. Here, AI is used as a partner to learn science content, but rarely do students have opportunities to engage in these AI-enhanced science practices as scientists. As Herdliska and Zhai (2024) ask, "How could future citizens be able to survive and succeed in their careers if they are not able to use AI to investigate problems and figure out solutions?" (p. 179). As we revisit the goals of science and AI in education, science educators must consider how they are providing opportunities for students to engage in AI-enhanced science practices where they can develop understandings of science content and careers through this engagement.

We recommend increased, intentional collaboration among K-12 educators, counselors, leaders, teacher organizations, developers, and community, industry, workforce, and university partners in science and AI fields. Such collaborations should manifest in professional development opportunities; field trips, out-of-school programs, and research symposia for students; and the co-construction of new courses, AI-science tools, and curricula for use across the lifespan. An example of such a collaboration and its outcome is an educational software product that allows for accessible molecular design through a web browser (Xie & Ding, 2025). High school students and science teachers can create novel molecules using generative AI as an assistant, such as creating molecules that bind to proteins on the tongue that engender a sweetness response. This tool aligns with engagement in authentic science practices and problems, as scientists have been using similar models for their work in structure-based drug discovery (Das, 2025). These opportunities for youth to engage in authentic performances and competencies, where they can explore their possible future selves through career exploration, are promising for fostering STEM professional identity (Tripp & Liu, 2024).

In the next section, we explore AI's application in science education for investigating and solving problems. This type of facilitation by AI is what is called "AI as an extended human agent."

4.3 AI-Facilitated Applications of Science: Vision II Science Literacy Plus

With the rapid advancement of AI, particularly generative AI since 2022 with the release of ChatGPT, AI is accessible to virtually everyone and everywhere. More and more K-12 students are using AI in their everyday lives. As discussed in Chap. 2, AI is now not only a tool to perform specific tasks such as summarizing texts, generating images, and conducting statistical analyses, but it also has been leveraged as an extended agent or partner in solving real-world problems. For example, Herdliska and Zhai (2024) engaged eighth grade students in scientific inquiry using Google's Teachable Machine, an AI platform that allows users to train their own machine-learning models using self-collected data. Students worked in groups to identify real-world problems that can be investigated and addressed with Google's Teachable Machine. They uploaded sample data, developed an algorithmic model, and then used new data to ascertain the accuracy of the model. Students presented their work to their peers. Herdliska and Zhai (2024) found that an overwhelming majority of students had a good understanding of AI fundamentals and applications to solve real-world problems. However, it is important to note that while there was general understanding of AI and its potential applications—including the significance of large sample sizes in AI training—there was a need for more in-depth and comprehensive AI understandings, from the importance of data quality to AI's limitations and ethical considerations for its use.

As another example, AI can support students in engaging in engineering design problems. The following prompt was given to ChatGPT 3.5 as part of a wind-powered car design task: *I need to design a wind-powered car. I have four plastic wheels, two axles, many popsicle sticks, a motor, a plastic fan, and hot gun glue. Please suggest a design.* Figure 4.2 presents ChatGPT's responses.

Designing a wind-powered car can be a fun and educational project! Here's a simple design utilizing the materials you have:
Materials Needed:
- **4 Plastic Wheels**
- **2 Axles**
- **Many Popsicle Sticks**
- **1 Motor**
- **1 Plastic Fan**
- **Hot Glue Gun**
Steps to Design Your Wind-Powered Car:
1. **Base Structure**:
 - Use Popsicle sticks to create a rectangular base for your car. Connect several stick pieces parallel to each other and glue them to two long Popsicle sticks to form a sturdy structure.
- Make sure the base is wide enough to support the wheels and the motor.
2. **Attaching Axles**:
 - Use the two axles for the wheels. Drill or create small holes on the sides of your base (without breaking it) where the axles can be inserted.
 - Secure each axle with glue to prevent movement, leaving enough space for the wheels to spin freely.
3. **Mounting the Wheels**:
 - Attach a wheel to each end of the axles. Ensure they are firmly glued so they don't come off during use.
4. **Mounting the Motor**:
 - Position the motor on top of the base. Use hot glue to secure it in place. Make sure the motor's shaft is accessible and aligned with the fan.
5. **Attaching the Fan**:
 - Secure the plastic fan to the shaft of the motor. The fan should be positioned at the back of the car to push air and propel the car forward when the motor is activated.
6. **Wind Activation**:
 - You can set up a simple manual mechanism like a hand crank, or if you have a battery or wind generator, connect the motor to it to allow the fan to turn.
7. **Testing**:
 - Once everything is assembled, test your car in a windy area to see how effectively it moves. Adjust the fan's angle or secure loose pieces if necessary.
Additional Tips:
- **Aerodynamics**: To improve performance, consider shaping the front of the car with more Popsicle sticks to reduce drag.
- **Weight Balance**: Ensure that the car is balanced; if it's too heavy on one end, it may not move well.
- **Experimenting**: Try adjusting the size and angle of the fan to see how it affects the car's speed.

Fig. 4.2 ChatGPT suggestions for wind-powered car design

From Fig. 4.2, we can see that the ChatGPT-suggested design is reasonable upon first glance; it can be a good starting point for students to design their cars. Note that the prompt did not mention batteries to power the motor, but ChatGPT suggested using a battery or a hand crank. However, a hand crank might not be feasible given the small size of the car. For someone who is new to designing a car like this, what ChatGPT suggests can be of great help in initiating the brainstorming process.

Through subsequent iterative designing, constructing, testing, and improving processes, ChatGPT can further support identifying aspects of the design to improve. Through the entire problem-solving process, ChatGPT acts as a knowledgeable partner.

Although AI has the capability to facilitate problem-solving as described above, there is a potential danger that AI could limit student creativity. Reducing the engineering design process to one that is no longer messy, uncertain, and cyclical to one that is linear, procedural, and prescriptive as it appears in the AI-generated solution depicted in Fig. 4.2, can do a disservice in promoting student learning, i.e., Vision II science literacy. The output does not propose additional questions or other scaffolds to develop students' science understandings, such as what aerodynamics is and how it applies to the wind-powered car or to other situations.

4.3.1 Implications for K-12 Science Education

Given the above potential of AI as a partner in scientific problem-solving, metacognitive criticality and reflective practices are components of AI literacy that students should develop in conjunction with AI-facilitated applications of science. That is, students need to learn to become aware of when to use and when not to use AI, such as at what stage of the engineering design process and in what context of specific tasks. Students also need to learn to critique some of the suggestions provided by AI chatbots and be able to decide why some of the proposed solutions are inadequate or unfeasible for a given situation or context. Science educators can help nurture this process through reflective discussions with students, which can be facilitated by debrief prompts and checklists. Science educators should also engage in reflective practices themselves, while modeling this for their students. Additionally, to use AI to apply science principles to solve problems, students need to learn basic skills of prompt engineering. Examples of prompt engineering skills that science educators can practice with students are clarity and precision of requests, providing contextual information, iterative refinement of prompts, providing guidance, and giving feedback. In the next section, we further examine how AI can be used to tackle complex social and environmental issues, which is called "AI as Agent" in Chap. 2.

4.4 Responsible and Ethical Uses of AI for Science Engagement: Vision III Science Literacy Plus

Someone who is literate in science should actively and collectively participate in understanding and addressing vexing social, cultural, political, and environmental issues. These issues cannot be solved by simply applying science or engineering practices or AI; they require engagement with a variety of stakeholders, building consensus, and jointly taking actions that grapple with the complexity of these issues interacting with other systems, including AI in society. Vision III science literacy envisions learners as change agents who are empowered by science to transform society. On the one hand, AI can transform society through information gathering, trade-off reasoning, and decision-making. For these purposes, specialized AI systems can be created to perform these tasks. On the other hand, involving AI in this process introduces ethical and social problems. Unlike conventional tools that we mostly trust (e.g., a temperature probe is calibrated, and the readings are considered valid and reliable, see Chap. 2 for more information), AI itself can be biased because the data that was used to train AI models were limited in scope and representation. Also, the validity and reliability of AI-provided advice and suggestions are unknown, and there is no auditory trail to trace and evaluate the AI reasoning process. Thus, AI might intentionally or unintentionally do harm. Guidelines informing AI usage are needed.

Some professional scientific societies have developed guidelines regarding AI use in scientific research. For example, the American Geophysics Union published a document entitled, Ethical and Responsible Use of AI/ML in the Earth, Space, and Environmental Sciences: Modules for Understanding and Capability (Stall et al., 2022); in the document, four principles for researchers are described: (a) transparency, documenting, and reporting, (b) intentionality, interpretability, expandability reproducibility, and replicability, (c) risk, bias, and effects, and (d) participatory methods. A National Research Council committee on assessing and navigating biosecurity concerns and benefits of AI use in the life sciences released a report entitled, The Age of AI in the Life Sciences: Benefits and Biosecurity Considerations (NRC, 2025). This report makes several recommendations regarding the biosecurity risks and uplifted capacities of AI-enabled biological tools. There is also a need for a similar guide for using AI in science education. What follows are factors that limit AI for science engagement in line with Vision III science literacy, along with recommendations for addressing them through the responsible and ethical practices (REP) guidelines proposed in Chap. 3.

Despite the capabilities and potential of AI in science engagement for personal, community, and societal good, it is important to be mindful of the ethnical and responsible uses of AI in science education. To achieve this goal, we need to acknowledge and remain resolute about the limitations of AI systems and potential for harm that co-exist. We also need to ask such questions as how problems are defined (including who and what are considered problems), whose problems are considered, and who is involved with developing and executing solutions to these problems to dismantle, extend, or transform systems. As Waight et al. (2022) remark, "Without an in-depth understanding of the problem, context, and people, technologies simply reinforce and replicate existing, unjust systems" (p. 1509). Nature of science and technology literature foregrounds and demands criticality around the role of culture and values; insatiable demands for technological efficiency; science and technology as value-laden systems, as part of systems, and interacting with other systems; definitions of expertise; and problematization of technologies as fix, where technologies do little to solve problems whose origins are non-technical in nature, and in fact engender new problems (Waight & Abd-El-Khalick, 2012; Waight et al., 2022).

With this lens, considering the impact of these technologies, inclusion in training and testing of AI systems, and proactively addressing the "bad borg" dangers of intrusion, uncontrollability, overload, alienation, narrowing, deceit, degradation, and disembodiment are imperative (Waight et al., 2022, p. 1504). Without such attention and care, AI-enhanced science literacies will perpetuate problems as opposed to upending them. In the section that follows, we offer an overview of some of the threats to responsible and ethical uses of AI, which include unequal access to AI and the potential for bias in AI-generated information and solutions to problems (Barelli et al., 2024). We then close with some suggestions for an asset-based approach that attends to these issues, towards responsible, ethical AI-science systems.

4.4.1 Unequal AI Access and Participation

Although AI is becoming increasingly embedded in the fabric of daily life, disparities in access remain, which reflect and amplify long-standing societal issues. There are several dimensions of unequal access that must be confronted. First is the digital divide, where access to modern computing devices; high-speed, reliable Internet; and paid subscription services with the most up-to-date AI features remain a considerable issue across geographic areas around the world (Boateng, 2024; Cardona et al., 2023). Such divisions persist due to legacies of inequality, which manifest differently across contexts. Disparities in benefiting from the full potential of AI may also result from differential investments and infrastructure allocated to building and operating AI data centers and systems. Moreover, discrepancies in opportunities for learning and developing AI systems across the lifespan also pose potential barriers to accessing AI (Regmi, 2024). As AI systems inevitably evolve so too must education systems. AI systems must be inexpensive and friendly for beginners or learners developing familiarity with them.

Consistent with the Chap. 1 REP fairness guideline, it is imperative that learners, who have historically and currently been excluded from rigorous, affirming opportunities for AI-enhanced learning and participation, are prioritized in efforts to restructure and broaden AI-science opportunities. While AI systems have been touted for their potential to promote accessibility, through voice assistants, personalized scaffolding features, and language translation and interpretation, it is important that science educators remain vigilant towards potential ruptures preventing full accessibility and inclusion (Heidt, 2024). Without ongoing care to multi-layered issues of access, problematic power hierarchies could be reified and magnified. Consider, for example, AI systems that fail to recognize the diverse languages of users in their training and content generation (Flores, 2023). Science educators must recognize these limitations and can use them as a learning opportunity by explicitly highlighting these limitations and countering them through their words and pedagogical practices.

4.4.2 Disinformation, Misinformation, and Bias in AI-Generated Information and Solutions

In a world where AI systems are increasingly present, it becomes easier to spread false information and lies, in accordance with individuals' interests and predispositions, which become more difficult to parse from reality. Chatbots can provide dangerous information, such as erroneous medical recommendations and diagnoses and promote climate change denial and pseudo-scientific beliefs (Dodhia, 2024). In such cases, misconceptions are reinforced rather than targeted. Furthermore, Saumure et al. (2025) highlight the ways in which prompts to ChatGPT to make generated images "funnier" led to images that were biased against older, heavier, and visually impaired people, highlighting the ways in which algorithmic bias according to age, body weight, and disability status—reflecting societal biases—have been neglected. LLMs are prone to hallucinations, making up sources, reshuffling patterns, and spreading false information. When overloaded with a barrage of misinformation, it is challenging and time consuming to sift through this information to evaluate what is true and what is not, resulting in more work for users. Since chatbots can display emotions through language output and images generated—although they cannot feel these emotions—the potential for exploiting emotions to spread science disinformation and misinformation is enhanced. As Ovsyannikova et al. (2025) found, third party evaluators perceive AI to be more compassionate and convey greater understanding and care as compared with expert human responses. AI-generated responses were preferred over human-generated responses, particularly when humans were unaware that the responses were produced by AI.

Training data can also be problematic and steeped in biases, especially when they are further amplified and internalized by humans (Glickman & Sharot, 2024). In 2020, a study published in *Nature Medicine* revealed that an AI system trained to detect skin cancer exhibited racialized bias, detecting better among lighter-skinned individuals than darker-skinned individuals (Wiens et al., 2019). This bias in AI-generated information and solutions are products of coded data that feed algorithms and train models (Barelli et al., 2024). Thus, bias can be embedded in the detection and automation of patterns and result in algorithmic discrimination based on incomplete, unrepresentative training data and problematic assumptions when coding. An example is voice recognition software that does not detect regional dialects when training occurs in standardized English, such that AI systems fail to recognize a plurality of pronunciations and language varieties (Boateng, 2024). This bias also occurs when medical diagnostic models are trained on data of white, healthier, and wealthier patients. With biased data, AI learns and replicates the patterns that reinforce stereotypes and bias. As Dodhia (2024) warned, "AIs learn from data created by society, and this data is permeated with society's biases often veiled so they are not immediately obvious…AIs inherit society's biases and will perpetuate those same biases, deepening divisions among different groups" (p. 171). Mitigating bias requires a multipronged approach, from data collection to model development and training, to model deployment and monitoring, to interpretability and accountability. Considering the intended uses of these AI systems, while also considering the potential misuses, evolution of uses, and ensuring regular feedback loops that can ameliorate these biases are also critical.

In science education, models are often referred to as approximations of reality, incomplete, and having limits; AI models are no different in this regard. Surely, hallucinations generated could result in opportunities to move out of normal science paradigms, producing anomalies that lead researchers to think and ask questions differently, to embark on revolutionary paradigms (Kuhn, 1970). Nevertheless, perspicacious recognition of their limitations is key, towards the REP guideline of robustness and scientific integrity. Lam et al. (2023) indicate how their AI-enhanced weather forecasting models have limitations in how uncertainty is handled, especially considering the nonlinearity and dynamicity of weather events. AI-enhanced weather prediction systems are not meant to be a replacement to traditional weather forecasting approaches "which have been developed for decades, rigorously tested in many real-world contexts, and offer many features" (p. 6). AlphaFold3 also has limitations in how the models address stereochemistry; there are chirality limitations in hallucinations that affect rare ligands and amino acids, and they favor static structures and interactions, not necessarily dynamic behaviors of these biomolecular interactions in different contexts or solutions (Abramson et al., 2024).

Thus, it is important to ensure representation amongst multiple stakeholders. To paraphrase Waight et al. (2022), "When we champion [AI] in science classrooms, whose voices and presence do we privilege and evoke?" (p. 1508). Diversity demands attention to the strengths and assets of learners, not only their needs and weaknesses. It entails nuances of cultures that shift, including science cultures, of multiple ways of knowing, being, thinking, engaging and identifying with science and AI systems. This entails ensuring that Indigenous knowledge is protected and incorporated into decision making; where Indigenous peoples, lands, and knowledge systems are respected, where AI protocols "evaluate clearly what kind of relationships we are materializing into the world" (Lewis, 2020, p. 8). Thus, close attention must be paid to interactions with AI systems, avoiding exploitation, naivete, and overreliance.

High quality field work and interpretations that mitigate bias are still necessary in studying wildlife communication, where human intuition and the understanding that humans have about the world and animal behavior are things machines cannot do (Savage, 2024). AI-predictive modeling for disease transmission is also limited because it "is a complex process influenced by a confluence of epidemiological, biological, and socio-behavioral factors" (Ye et al., 2025, p. 8), and medical LLMs must be continuously trained on feedback from physicians in clinical settings (Liu et al., 2025). Aligning with the REP guideline human oversight and autonomy, humans need to be in the loop, for data collection and decision and policy making

(Cardona et al., 2023). Human judgment is necessary to interpret subtleties of AI outputs and for verification, although as Vaccaro et al. (2024) point out, human-AI collaborations do not always result in better results than humans or AI alone, and communication barriers, trust issues, and ethical concerns must also be considered for more effective human-AI coordination and collaboration. Science educators and students must be aware of these challenges posed by AI systems, to ensure opportunities for students to not only use these tools but to critically reflect and evaluate them and their evolving science understandings. These discussions can entail contemplating what and whose ideas and values are overlooked in AI-generated responses, tools, and agents that deserve further attention. Do they agree with the outputs? How do they know what they know? What evidence was provided or not provided? How could the AI models be trained differently? Deep science disciplinary knowledge remains vital for identifying inaccuracies. Science educators must be aware of alternative conceptions in science and how to address them.

4.4.3 Costs to Privacy, Data Security, Ownership, and the Environment

Although AI tools have potential for increased contextualization and personalization, there is a cost to privacy, as "one's privacy serves as a tradeoff that allows outside agencies to surveil and exploit this collected information" (Waight et al., 2022, p. 1505). Consistent with the transparency and explainability, accountability, privacy and data security, and respect for human values guidelines, for responsible and ethical AI tools that build on learners' strengths and assets, learners' demographic information, preferences, interests, experiences, and relationships might be collected; however, the collection of this data could have problematic consequences for users if used for nefarious purposes that further marginalization. What happens to this data, how it is being used, and how it is deleted are key concerns (Schmarzo, 2023). In what ways are youth's data and privacy protected, and for young children and youth, are their parents or guardians appropriately informed of this use? To what extent do they have control over this data? Who benefits, and who is harmed? Which groups are subjected to beta testing? Essentially, how is data "solicited, collected, analysed, and operationalized" and data sovereignty respected? (Lewis, 2020, p. 22).

Ownership is another area of concern, as who owns this data and is responsible for the ideas generated with AI-tools becomes blurred. Existing generative AI tools like ChatGPT are not transparent about the sources of their training materials and methods (Nam & Bai, 2023). Immediate advantages and desires for convenience might lead to overlooking risks, tradeoffs, morals, ethics, and long-term consequences, including social impact. Indeed, through surveys deployed in Germany and Spain, Vesely and Kim (2024) found public support for AI safety oversight, including that which pertains to misinformation, privacy invasion, and criminal misuse. The REP guideline of transparency and explainability is fundamental to providing connections between recognizing patterns and judgments, reasoning, decision making and other actions to override faulty or biased decisions that AI systems might conjure.

Other costs include demands on the environment, for the extraction and exploitation of land and other resources for AI hardware and infrastructure. Time-efficient computations require a significant amount of energy and freshwater, which serves as a cooling mechanism in an already precarious climate reality (Gillani et al., 2023). Who and what will be displaced and/or harmed when AI data centers are constructed, including through pollution and material extraction? How are harms offset, to repair relations to a healthy state (Lewis, 2020)? As noted in the 2020 "Indigenous Protocol and Artificial Intelligence" position paper, from remarks by linguist and te reo Māori specialist Dr. Hēmi Whaanga (Ngāti Kahungunu, Ngāi Tahu, Ngāti Mamoe, Waitaha), there are serious risks for AI systems to extend "colonial practices of exploitation, extraction and control, particularly those that displace a peoples' understanding of themselves with a worldview that favors the colonizer" (Lewis, 2020, p. 12). These reflections should be part of discussions in science classrooms, particularly as it relates to climate science and sustainability, highlighting respect for the land and all relations, with careful consideration and questioning of whose data is collected, how it is collected and used, and must it be collected in the first place. Care must be taken to respectfully honor and credit Indigenous traditional knowledge and customary practices rather than engage in appropriation.

4.4.4 Implications for K-12 Science Education

AI-enhanced science literacy through AI agents should enhance our potential and nurture community and sustainability. As team composition affects the design, development, and deployment of AI systems, there needs to be responsible design and accountability mechanisms with many different stakeholders across disciplines and sectors of society towards real-world impact, rooted in asset-based values and a moral compass pointed towards responsible, ethical AI use in science education. As Lewis (2020) asserts, "Community-specific values of relationality, kinship, reciprocity and care should be built into the fundamental protocols governing how such practices are implemented" (p. 17). AI systems in science education should be designed in partnership with a variety of stakeholders and communities; they should be responsible and accountable to communities; this includes students, families, and science educators.

AI nor science alone will resolve all problems, particularly those that have deep roots that are non-technical in nature. AI-enhanced science literacy goals should promote healthy skepticism; flexibility; problem solving; self-regulation; collaboration; critical, creative, and imaginative thinking; empathy; and action. Science educators can pose the following questions along with engagement in AI-enhanced science: Who benefits? Who is harmed? Whose needs are met? Whose needs are neglected? How can there be mutualistic symbiotic relationships among all involved? Are we enabling and augmenting strengths and opening up varied forms of use in creative, expansive ways? How are a multiplicity of perspectives sustained in asset-based ways that promote complexity rather than oversimplification, that further diversify ways of knowing, being, thinking, and doing in both science and AI?

Moving past deficit notions of AI and science literacy requires that we critically contend with the barriers and conditions engendering unfair access and disparities in participation in the first place. Towards uncovering and redressing these obstacles, "AI system designers need to be aware of their own cultural frameworks, socially dominant concepts and normative ideals; be wary of the biases that come with them; and develop strategies for accommodating other cultural and social frameworks" (Lewis, 2020, p. 22). Science educators should also be compelled to reflect deeply on these questions as well. Further, how are we promoting agency in the selection and use of AI in science education? Do science educators, learners, and their families have seats at the decision-making tables from design to use?

Critical awareness recognizes when AI is and is not necessary; just because we can engage AI in scientific endeavors does not mean we should (Gillani et al., 2023). Societal issues must be addressed so that they do not get woven deeply into AI systems. Science education should instil values of integrity, empathy, compassion, generosity, and solidarity that serve as protective, preventative factors dissuading against deception, exploitation, theft, and greed. As Browne et al. (2024) contend, "AI ethics is necessarily a long-term and complex endeavour" and "cannot simply be achieved through one-off 'bias checks' in the development process" (p. 8). Each one of us has a responsibility to advocate for an asset-based approach to AI-enhanced science literacy, towards responsible, ethical imagination, development, deployment, use, and discard of AI-science tools.

4.5 Enhancing Science Literacy with AI: Science Literacy Plus

The above sections have examined how AI is enhancing all three visions of science literacy. In this final section, we conclude this chapter by arguing for a new form of science literary, science literacy plus, that should be promoted in K-12 science education.

Common to the above AI-enhanced science literacies is the requirement for a set of AI knowledge, skills, and dispositions, or AI literacy. AI literacy, like computer literacy, technology literacy, and other literacy frameworks, refers to a core set of competencies that all students should develop and master to be engaged citizens in a future AI-permeated world. UNESCO specifies this set of core competencies in a two-dimensional matrix of 12 blocks (see Table 4.1).

Table 4.1 AI competency framework for students

	Progression levels		
Competency aspects	Understand	Apply	Create
Human-centered mindset	Human agency	Human accountability	Citizenship in the era of AI
Ethics of AI	Embodied ethics	Safe and responsible use	Ethics by design
AI techniques and applications	AI foundations	Application skills	Creating AI tools
AI system design	Problem scoping	Architecture design	Iteration and feedback loops

Note. Recreated from UNESCO (2024). AI competence framework for students. Paris: UNESCO

According to the UNESCO AI framework, students in K-12 education should develop competencies in human-centered mindset, the ethics of AI, AI techniques and applications, and AI system design. Human-centered mindset refers to:

> Students' values, beliefs and critical thinking skills, applied to the examination of whether AI is fit for purpose, whether its use is justified, how humans should interact with it, and what responsibilities individuals and institutions should take on to contribute to the building of safe, inclusive and just AI societies. (UNESCO, 2024, p. 22)

Human-centered mindset includes human agency, human accountability, and citizenship in the AI era. Ethics of AI refers to ethical value judgements, embodied reflections, and social and emotional skills that students should develop to navigate, understand, practice, and contribute to the adaptation of principles and regulatory rules that are related to the life cycle of AI systems. Ethics of AI includes embodied ethics (i.e., understanding how AI could impact human rights and climate change within specific contexts and personal lives, human determination in human-AI collaboration, transparency and explainability, safe and responsible use, and ethics by design). Fundamental ethical principles include "do no harm," proportionality, non-discrimination, and sustainability. AI techniques and applications refer to conceptual knowledge and associated operational skills in connection with concrete AI tools or authentic tasks. Students should develop an AI foundation with respect to understanding data and algorithms, be able to critically evaluate and leverage free and/or open-source AI tools, programming libraries and datasets, and customize existing AI toolkits to create task-based AI tools. Finally, AI system design refers to systematic design thinking and comprehensive engineering skills required for problem solving, designing, building, training, testing, and optimization of AI systems.

The above four aspects of AI literacy, along with their progression from understanding to applying to creating in K-12 education are aligned with the three visions of science literacy reviewed above. Specifically, the AI techniques and applications and AI system design aspects of AI literacy are aligned with Vision I science literacy, while human-centered mindset and ethics of AI are aligned with Vision II science literacy. The three progression levels of AI literacy from understanding to applying to creating are aligned with Vision III science literacy. Therefore, by

Fig. 4.3 Science literacy plus

incorporating AI literacy into science literacy, science literacy will be enhanced (i.e., science literacy plus). Figure 4.3 presents a model of science literacy plus.

Figure 4.3 shows how *science literacy plus* is the intersection between science literacy and AI literacy. As there are currently well-defined science literacy and AI literacy frameworks as well as curriculum standards (e.g., NRC, 2012; UNESCO, 2024), we have not yet developed a similar framework for science literacy plus, although Chap. 11 is good start toward this goal. *Science literacy plus* must integrate competencies from both science literacy and AI literacy and map out a progression from K through 12 for all students. *Science literacy plus* is consistent with the categorization of AI as tools, AI as extended human agent, and AI as agent in Chap. 2, as well as the responsible and ethical practice principles and guidelines in Chap. 3. *Science literacy plus* extends and enhances science literacy, becoming a highly desirable goal of K-12 science education in the twenty-first century.

Acknowledgement This material is based upon work supported by the National Science Foundation under Grant No. 2332964. Any opinions, findings, and conclusions or recommendations expressed in this material are those of the author(s) and do not necessarily reflect the views of the National Science Foundation.

References

Abramson, J., Adler, J., Dunger, J., Evans, R., Green, T., Pritzel, A., Ronneberger, O., Willmore, L., Ballard, A. J., Bambrick, J., Bodenstein, S. W., Evans, D. A., Hung, C.-C., O'Neill, M., Reiman, D., Tunyasuvunakool, K., Wu, Z., Žemgulytė, A., Arvaniti, E., Beattie, C., Bertolli, O., Bridgland, A., Cherepanov, A., Congreve, M., Cowen-Rivers, A. A., Cowie, A., Figurnov, M., Fuchs, F. B., Gladman, H., Jain, R., Khan, Y. A., Low, C. M. R., Perlin, K., Potapenko, A., Savy, P., Singh, S., Stecula, A., Thillaisundaram, A., Tong, C., Yakneen, S., Zhong, E. D., Zielinski, M., Žídek, A., Bapst, V., Kohli, P., Jaderberg, M., Hassabis, D., & Jumper, J. M. (2024). Accurate structure prediction of biomolecular interactions with AlphaFold 3. *Nature, 630*(8016), 493–500. https://doi.org/10.1038/s41586-024-07487-w

Barelli, E., Lodi, M., Branchetti, L., & Levrini, O. (2024). Epistemic insights as design principles for a teaching-learning module on artificial intelligence. *Science & Education*. https://doi.org/10.1007/s11191-024-00504-4

Barkas, N., Petukhov, V., Nikolaeva, D., Lozinsky, Y., Demharter, S., Khodosevich, K., & Kharchenko, P. V. (2019). Joint analysis of heterogeneous single-cell RNA-seq dataset collections. *Nature Methods, 16*(8), 695–698. https://doi.org/10.1038/s41592-019-0466-z

Browne, J., Drage, E., & McInerney, K. (2024). Tech workers' perspectives on ethical issues in AI development: Foregrounding feminist approaches. *Big Data & Society, 11*(1). https://doi.org/10.1177/20539517231221780

Bencze, L., Pouliot, C., Pedretti, E., Simonneaux, L., Simonneaux, J., & Zeidler, D. (2020). SAQ, SSI and STSE education: Defending and extending "science-in-context". *Cultural Studies of Science Education, 1–27.* https://doi.org/10.1007/s11422-019-09962-7

Boateng, G. (2024). Leveraging AI to advance science and computing education across Afirca: Challenges, progress and opportunities. In G. Boateng, R. Wegerif, and I. Casebourne (eds.), Artificial intelligence in education: The intersection of technology and pedagogy (chapter 10, pp. 205–222). Zurich, Switzerland: Springer Nature.

Brożek, A., & Theodoris, C. V. (2025). AI learns from chromatin data to uncover gene interactions. *Nature, 637*(8047), 799–800. https://doi.org/10.1038/d41586-024-04107-5

Callaway, E. (2025). AI-designed proteins tackle century-old problem-making snake antivenoms: Machin learning has supercharged the field of computational protein design. *Nature, 637*, 776. https://doi.org/10.1038/d41586-025-00133-z

Cardona, M., Rodríguez, R., & Ishmael, K. (2023). Artificial intelligence and the future of teaching and learning: Insights and recommendations. *Office of Educational Technology*. Retrieved from https://coilink.org/20.500.12592/rh21zz

Christiansen, F., Konuk, E., Ganeshan, A. R., Welch, R., Palés Huix, J., Czekierdowski, A., Leone, F. P. G., Haak, L. A., Fruscio, R., Gaurilcikas, A., Franchi, D., Fischerova, D., Mor, E., Savelli, L., Pascual, M. À., Kudla, M. J., Guerriero, S., Buonomo, F., Liuba, K., Montik, N., Alcázar, J. L., Domali, E., Pangilinan, N. C. P., Carella, C., Munaretto, M., Saskova, P., Verri, D., Visenzi, C., Herman, P., Smith, K., & Epstein, E. (2025). International multicenter validation of AI-driven ultrasound detection of ovarian cancer. *Nature Medicine, 31*(1), 189–196. https://doi.org/10.1038/s41591-024-03329-4

Das, U. (2025). Generative AI for drug discovery and protein design: The next frontier in AI-driven molecular science. *Medicine in Drug Discovery, 27*, 100213. https://doi.org/10.1016/j.medidd.2025.100213

DeBoer, G. E. (2000). Scientific literacy: Another look at its historical and contemporary meanings and its relationship to science education reform. *Journal of Research in Science Teaching, 37*(6), 582–601.

Dodhia, R. (2024). *AI for social good: Using artificial intelligence to save the world.* Wiley.

Eisemann, N., Bunk, S., Mukama, T., Baltus, H., Elsner, S. A., Gomille, T., Hecht, G., Heywang-Köbrunner, S., Rathmann, R., Siegmann-Luz, K., Töllner, T., Vomweg, T. W., Leibig, C., & Katalinic, A. (2025). Nationwide real-world implementation of AI for cancer detection in population-based mammography screening. *Nature Medicine, 31*(3), 917–924. https://doi.org/10.1038/s41591-024-03408-6

Flores, N. (2023). Toward a raciolinguistic perspective on translation and interpretation. *Language in Society, 52*(5), 893–901. https://doi.org/10.1017/S0047404523000660

Gillani, N., Eynon, R., Chiabaut, C., & Finkel, K. (2023). Unpacking the "Black Box" of AI in education. *Educational Technology & Society, 26*(1), 99–111. https://www.jstor.org/stable/48707970

Glickman, M., & Sharot, T. (2024). How human–AI feedback loops alter human perceptual, emotional and social judgements. *Nature Human Behaviour*. https://doi.org/10.1038/s41562-024-02077-2

Guerrero, G. R., & Torres-Olave, B. (2022). Scientific literacy and agency within the Chilean science curriculum: A critical discourse analysis. *The Curriculum Journal, 33*(3), 410–426. https://doi.org/10.1002/curj.141

Heidt, A. (2024). 'Without these tools, I'd be lost': How generative AI aids in accessibility. *Nature, 628*(8007), 462–463. https://doi.org/10.1038/d41586-024-01003-w

Herdliska, A., & Zhai, X. (2024). Artificial intelligence-based scientific inquiry. In X. Zhai & J. Krajcik (Eds.), *Uses of artificial intelligence in STEM education*. Oxford University Press. https://doi.org/10.2139/ssrn.4591628

Hodson, D. (2009). Putting your money where your mouth is: Towards an action-oriented science curriculum. *Journal of Activist Science and Technology Education, 1*(1), 1–15. www.wepaste.org/jaste1.1.html

Hurd, P. (1958). Science literacy: Its meaning for American schools. *Educational Leadership, 16*, 13–16.

Karagiorgi, G., Kasiesczka, G., Kravitz, S., Nachman, B., & Shih, D. (2022). Machine learning in the search for new fundamental physics. *Nature Reviews Physics, 4*, 399–412.

Kuhn, T. (1970). *The structure of scientific revolutions* (2nd ed.). University of Chicago Press.

Jumper, J., Evans, R., Pritzel, A., Green, T., Figurnov, M., Ronneberger, O., Tunyasuvunakool, K., Bates, R., Žídek, A., Potapenko, A., Bridgland, A., Meyer, C., Kohl, S. A. A., Ballard, A. J., Cowie, A., Romera-Paredes, B., Nikolov, S., Jain, R., Adler, J., Back, T., Petersen, S., Reiman, D., Clancy, E., Zielinski, M., Steinegger, M., Pacholska, M., Berghammer, T., Bodenstein, S., Silver, D., Vinyals, O., Senior, A. W., Kavukcuoglu, K., Kohli, P., & Hassabis, D. (2021). Highly accurate protein structure prediction with AlphaFold. *Nature, 596*(7873), 583–589. https://doi.org/10.1038/s41586-021-03819-2

Lam, R., Sanchez-Gonzalez, A., Willson, M., Wirnsberger, P., Fortunato, M., Alet, F., Ravuri, S., Ewalds, T., Eaton-Rosen, Z., Hu, W., Merose, A., Hoyer, S., Holland, G., Vinyals, O., Stott, J., Pritzel, A., Mohamed, S., & Battaglia, P. (2023). Learning skillful medium-range global weather forecasting. *Science, 382*(6677), 1416–1421. https://doi.org/10.1126/science.adi2336

Laugksch, R. C. (2000). Scientific literacy: A conceptual overview. *Science Education, 84*(1), 71–94.

Lewis, J. E. (Ed.). (2020). *Indigenous protocol and artificial intelligence (AI) position paper*. Indigenous Protocol and Artificial Intelligence Working Group.

Liu, X. (2013). Expanding notions of scientific literacy: A reconceptualization of aims of science education in the knowledge society. In N. Mansour & R. Wegerif (Eds.), *Science education for diversity—theory and practice* (pp. 23–39). Springer.

Liu, X., Liu, H., Yang, G., Jiang, Z., Cui, S., Zhang, Z., Wang, H., Tao, L., Sun, Y., Song, Z., Hong, T., Yang, J., Gao, T., Zhang, J., Li, X., Zhang, J., Sang, Y., Yang, Z., Xue, K., Wu, S., Zhang, P., Yang, J., Song, C., & Wang, G. (2025). A generalist medical language model for disease diagnosis assistance. *Nature Medicine*. https://doi.org/10.1038/s41591-024-03416-6

Luo, X., Rechardt, A., Sun, G., Nejad, K. K., Yáñez, F., Yilmaz, B., Lee, K., Cohen, A. O., Borghesani, V., Pashkov, A., Marinazzo, D., Nicholas, J., Salatiello, A., Sucholutsky, I., Minervini, P., Razavi, S., Rocca, R., Yusifov, E., Okalova, T., Gu, N., Ferianc, M., Khona, M., Patil, K. R., Lee, P.-S., Mata, R., Myers, N. E., Bizley, J. K., Musslick, S., Bilgin, I. P., Niso, G., Ales, J. M., Gaebler, M., Murty, N. A. R., Loued-Khenissi, L. L., Behler, A., Hall, C. M., Dafflon, J., Bao, S. D., & Love, B. C. (2024). Large language models surpass human experts in predicting neuroscience results. *Nature Human Behaviour*. https://doi.org/10.1038/s41562-024-02046-9

McCurdy, R. (1958). Toward a population literate in science. *The Science Teacher, 25*, 366–368.

Miller, J. D. (1983). Scientific literacy: A conceptual and empirical review. *Daedalus, 112*(2), 29–48.

Moon, S. W., Kim, J., Kim, Y. J., Kim, S. H., An, C. S., Kim, K. G., & Jung, C. K. (2025). Leveraging explainable AI and large-scale datasets for comprehensive classification of renal histologic types. *Scientific Reports, 15*(1), 1745. https://doi.org/10.1038/s41598-025-85857-8

Nam, B. H., & Bai, Q. (2023). ChatGPT and its ethical implications for STEM research and higher education: A media discourse analysis. *International Journal of STEM Education, 10*(1), 66–24. https://doi.org/10.1186/s40594-023-00452-5

NRC. (1996). *National science education standard*. National Academies Press.

NRC. (2012). *A framework for K-12 science education: Practices, crosscutting concepts, and core ideas*. National Academies Press.

NRC. (2016). *Science literacy: Concepts, contexts, and consequences*. National Academies Press.

NRC. (2025). *The age of AI in the life sciences: Benefits and biosecurity considerations*. The National Academies Press. https://doi.org/10.17226/28868

OECD. (2023). PISA 2025 science framework (draft). https://pisa-framework.oecd.org/science-2025/assets/docs/PISA_2025_Science_Framework.pdf

Osborne, J. (2023). Science, scientific literacy, and science education. In N. G. Lederman, D. L. Zeidler, & J. S. Lederman (Eds.), *Handbook of research on science education* (Vol. III.) (chapter 25, pp. 785–816). Routledge.

Ovsyannikova, D., de Mello, V. O., & Inzlicht, M. (2025). Third-party evaluators perceive AI as more compassionate than expert humans. *Communications Psychology, 3*(1), 4. https://doi.org/10.1038/s44271-024-00182-6

Petso, T., & Jamisola, R. S. (2023). Wildlife conservation using drones and artificial intelligence in Africa. *Science Robotics, 8*(85). https://doi.org/10.1126/scirobotics.adm7008

Price, I., Sanchez-Gonzalez, A., Alet, F., Andersson, T. R., El-Kadi, A., Masters, D., Ewalds, T., Stott, J., Mohamed, S., Battaglia, P., Lam, R., & Willson, M. (2025). Probabilistic weather forecasting with machine learning. *Nature, 637*(8044), 84–90. https://doi.org/10.1038/s41586-024-08252-9

Regmi, K. D. (2024). The rise of learning technology in an unequal world: Potentials and limitations in enhancing lifelong learning. *International Review of Education, 70*(3), 433–452. https://doi.org/10.1007/s11159-023-10058-2

Roth, W. M. (2002). Taking science education beyond schooling. *Canadian Journal of Science, Mathematics and Technology Education, 2*(1), 37–48.

Roth, W.-M., & Calabrese, A. (2004). *Rethinking scientific literacy*. Routledge-Falmer.

Roth, W.-M., & Lee, S. (2004a). Science education as/for participation in the community. *Science Education, 88*(2), 2630291.

Roth, W. M., & Lee, S. (2004b). Science education as/for participation in the community. *Science Education, 88*(2), 263–291. https://doi.org/10.1002/sce.10113

Roberts, D. A. (2007). Scientific literacy/science literacy. In S. K. Abell & N. G. Lederman (Eds.), *Handbook of research on science education*. Lawrence Erlbaum Associates, Publishers.

Saumure, R., De Freitas, J., & Puntoni, S. (2025). Humor as a window into generative AI bias. *Scientific Reports, 15*(1), 1326. https://doi.org/10.1038/s41598-024-83384-6

Savage, N. (2024). AI decodes the calls of the wild. *Nature*. https://doi.org/10.1038/d41586-024-04050-5

Schmarzo, B. (2023). *AI and data literacy: Empowering citizens of data science* (1st ed.). Packt Publishing Ltd.

Sjöström, J. (2024). Vision III of scientific literacy and science education: An alternative vision for science education emphasising the ethico-socio-political and relational-existential. *Studies in Science Education*. https://doi.org/10.1080/03057267.2024.2405229

Stall, S., Gershenfeld, J. E. C., & Coward, C. (2022). Ethical and responsible use of AI/ML in the earth, space, and environmental sciences modules for understanding and capability A community report facilitated by the American Geophysical Union. *Brandeis University*. https://doi.org/10.22541/essoar.167080673.37483484/v1

Tripp, J. N., & Liu, X. (2024). Towards defining STEM professional identity: A qualitative survey study. *Journal for STEM Education Research*. https://doi.org/10.1007/s41979-024-00131-2

UNESCO. (2024). *AI competence framework for students*. UNESCO.

Vaccaro, M., Almaatouq, A., & Malone, T. (2024). When combinations of humans and AI are useful: A systematic review and meta-analysis. *Nature Human Behaviour, 8*(12), 2293–2303. https://doi.org/10.1038/s41562-024-02024-1

Vesely, S., & Kim, B. (2024). Survey evidence on public support for AI safety oversight. *Scientific Reports, 14*(1), 31491. https://doi.org/10.1038/s41598-024-82977-5

Waight, N., & Abd-El-Khalick, F. (2012). Nature of technology: Implications for design, development, and enactment of technological tools in school science classrooms. *International Journal of Science Education, 34*, 2875–2905.

Waight, N., Kayumova, S., Tripp, J., & Achilova, F. (2022). Towards equitable, social justice criticality: Re-constructing the "Black" box and making it transparent for the future of science and technology in science education. *Science & Education, 31*(6), 1493–1515. https://doi.org/10.1007/s11191-022-00328-0

Wang, et al. (2023). Scientific discovery in the age of artificial intelligence. *Nature, 620*(3), 47–60. https://doi.org/10.1038/s41586-023-06221-2

Wiens, J., Saria, S., Sendak, M., Ghassemi, M., Liu, V. X., Doshi-Velez, F., Jung, K., Heller, K., Kale, D., Saeed, M., Ossorio, P. N., Thadaney-Israni, S., & Goldenberg, A. (2019). Do no harm: A roadmap for responsible machine learning for health care. *Nature Medicine, 25*(9), 1337–1340. https://doi.org/10.1038/s41591-019-0548-6

Xie, C., & Ding, X. (2025). *Using GenAI as an assistant for molecular design*. Institute for Future Artificial Intelligence. https://intofuture.org/aims-ai-design-assistant.html

Ye, Y., Pandey, A., Bawden, C., Sumsuzzman, D. M., Rajput, R., Shoukat, A., Singer, B. H., Moghadas, S. M., & Galvani, A. P. (2025). Integrating artificial intelligence with mechanistic epidemiological modeling: A scoping review of opportunities and challenges. *Nature Communications, 16*(1), 581. https://doi.org/10.1038/s41467-024-55461-x

Zhavoronkov, A., Ivanenkov, Y. A., Aliper, A., Veselov, M. S., Aladinskiy, V. A., Aladinskaya, A. V., Terentiev, V. A., Polykovskiy, D. A., Kuznetsov, M. D., Asadulaev, A., Volkov, Y., Zholus, A., Shayakhmetov, R. R., Zhebrak, A., Minaeva, L. I., Zagribelnyy, B. A., Lee, L. H., Soll, R., Madge, D., Xing, L., Guo, T., & Aspuru-Guzik, A. (2019). Deep learning enables rapid identification of potent DDR1 kinase inhibitors. *Nature Biotechnology, 37*, 1038–1040.

Chapter 5
AI Transformations of Science Education Goals to Prepare Students as Lifelong Investigators

Marcia C. Linn, Libby Gerard, and Allison Bradford

5.1 Introduction

To respond to the pressing need to transform the goals of science instruction to meet the needs of today's students, designers can leverage Artificial Intelligence (AI) to advance teaching and learning so students can grapple with contemporary dilemmas and prepare for a rapidly changing workplace. By crafting instruction that guides students to discover new ideas and distinguish them from their prior knowledge, informed by Knowledge Integration (KI) pedagogy, designers can motivate and support learners to take on increasingly complex investigations while also refining their understanding of scientific concepts (Linn & Eylon, 2011). By integrating AI advances with KI pedagogy, we can transform science classes from passive memorization or formulaic experimentation to innovative design laboratories. We can prepare students as creative designers who take initiative to attack science challenges that impact their lives. This paper identifies promising directions for achieving this transformation.

Both students and teachers are increasingly using AI, yet many teachers think AI is doing more harm than good (L. Lin, 2024). Most teachers wish for more professional development to help them use AI effectively (e.g. Langreo, 2024). In 2023

This material is based upon work supported by the National Science Foundation under Grant No. 2101669. Any opinions, findings, and conclusions or recommendations expressed in this material are those of the author(s) and do not necessarily reflect the views of the National Science Foundation.

M. C. Linn (✉) · L. Gerard · A. Bradford
University of California, Berkeley, Berkeley, CA, USA
e-mail: mclinn@berkeley.edu; libbygerard@berkeley.edu; allison_bradford@berkeley.edu

X. Zhai, K. J. Crippen (eds.), *Advancing AI in Science Education*, Advances in Technology-Rich Science Education 3,
https://doi.org/10.1007/978-3-032-16871-9_5

about 20% of teachers reported using AI, while some surveys place that figure at 50% in 2024 (L. Lin, 2024). The most common uses of AI are for productivity such as lesson plan ideas, draft letters of recommendation, or email edits (Tyson et al., 2021; L. Lin, 2024). In the classroom, some teachers use AI to adapt written materials to the reading level of their students and others are using AI for giving feedback in writing intensive courses (e.g. Randazzo, 2024). Over 50% of students use generative AI but few use it regularly. Of students who use AI, about 20% use it for writing drafts. Many report using AI to get answers to questions or as a brainstorming partner. In summary, AI is being used sporadically by teachers and students.

AI is rapidly permeating the practice of science (Eger et al., 2025) and of computer science (Kersting, 2020). These changes necessitate rethinking the goals for science education. AI literacy, including the ability to critique, interpret, and employ AI has become integral to science education. Recent AI advances offer science educators new and promising ways to personalize instruction, including supporting students to carry out science investigations (Erduran, 2023). Transforming science instruction to focus on supporting student investigations aligns with the goals of the Next Generation Science Standards (Reiser et al., 2021). The availability of AI scaffolds for student investigations substantially expands opportunities for students to carry out their own investigations.

5.1.1 AI Supported Goals for Science Instruction

Advances in AI make it possible to transform the goals of science instruction to focus on investigations. By transforming science instruction to engage students in AI scaffolded investigations that address contemporary dilemmas, courses will also prepare students for a workplace where AI is rapidly being adopted. Students need to learn how to leverage AI to advance their own learning, collaborate with others, and develop workplace readiness. At the same time, we need to prepare practitioners to use AI resources as they scaffold students to tackle intriguing dilemmas. By integrating AI advances with KI pedagogy, we can transform science classes into innovative, responsive design laboratories. We can prepare students as creative designers who leverage AI to work individually and collaboratively to learn science principles and processes while exploring innovative solutions to pressing problems. We can support teachers to foster the development of students who advance science by leveraging AI.

5.1.2 Supporting a Robust Understanding of the Inquiry Cycle

To focus on investigations, science instruction can instill a deep, generative understanding of the inquiry cycle. We no longer think of the scientific method as a set of superficial steps. Today, students need to form an epistemological view of inquiry as a process for establishing validated insights using a wide range of methodologies and warrants for their argument. Rather than learning a series of steps that only work under certain circumstances, AI resources can support students to internalize a rich view of inquiry that encompasses multiple genres of investigations including controlling variables to explore mechanical systems; examining patterns of volcanoes and earthquakes to understand geologic transformations; and exploring biological mechanisms to understand everything from genetic inheritance to patterns of population growth and decline. Considerable work has gone into the design of inquiry cycles for scientific work in K-12 education (Gerard et al., 2010; Pedaste et al., 2015; White & Frederiksen, 2000). Web-based materials are beginning to support a robust view of the inquiry cycle for student investigations (e.g. Donnelly et al., 2014). Yet, many students remain hesitant when asked to design an investigation, rather than appreciating the strengths and limitations of their approach.

To reimagine science classrooms as design laboratories, we need sequences of activities that build on each other. KI pedagogy is based on years of research determining how to design activities that promote coherent, generative understanding. Students can start with on-ramp activities where they investigate familiar challenges such as determining which cup material is best at keeping hot things hot and cold things cold (Chen & Linn, 2019), building up to more complex challenges such as designing and testing a solar oven (McBride et al., 2018). Ideally science courses would build on these investigations to support exploration of emerging societal challenges such as reducing the dangers from urban heat islands or mitigating the impact of particulate matter pollution on the incidence of asthma (Gerard, Bradford, et al., 2022).

To ensure that students make judicious use of AI, science classrooms can guide students to develop criteria for evaluating the impact of AI tools used in their own investigations. A series of activities that engages students in experimenting with AI requests can help students develop a feel for both promising and unproductive uses of generative AI tools such as Gemini or CoPilot (two AI tools currently approved for classroom use by many school districts). To refine their scientific understanding, students can design and tune their AI requests and validate the resulting outcomes as well as the AI sources used. For example, classroom science activities might guide students to write a prompt asking AI to summarize an article. The student might then critique the summary by analyzing the article themselves. They might then refine their prompt to improve the AI response and reflect on ways to design better prompts. In another example, students can ask AI to identify models to help them understand a topic such as photosynthesis, heat transfer, or plate tectonics. They can then compare the AI suggestions to open-source interactive activities available in their classrooms or narrative accounts in textbooks. Activities where students compare scientific models, tune their AI request, and reflect on the strengths and limitations of AI not only prepare them to conduct more sophisticated investigations but also develop AI literacy.

Combining AI and KI can enable teachers to support their middle school students (often 150 students in 5 sections of 30 students) to conduct their own inquiry investigations. Supporting each student to participate in an inquiry project personalized to their needs and interests can prepare lifelong learners to solve the problems of the future. Enabling teachers to personalize AI tools to support their students as they conduct their own investigations while also critiquing the effectiveness of the AI supports offers promise.

To be sure, there are both opportunities and risks in incorporating AI to transform the goals of science education. However, to prepare students for the workplace, learning about AI is crucial. We can optimize uses of AI while addressing the known and emerging risks. These include emphasizing the importance of protecting privacy, validating outcomes, and testing for bias. Students who learn to detect hallucinations, gaps in arguments, and logical failures are poised to refine their understanding of AI. By transforming the goals of science education to prioritize investigations, we can identify where AI increases instructional effectiveness, when AI helps some students but not others, which aspects of the curriculum need reformulation, and which long term impacts of AI are most valuable.

In the next sections we synthesize the KI research that supports design of instruction to promote a deep, rich understanding of the inquiry cycle. We illustrate how a combination of AI and KI resulted in improved guidance for student scientific explanations. We then summarize aspects of the inquiry cycle that AI can enhance and illustrate how this might work in a curricular sequence.

5.2 Knowledge Integration (KI) Pedagogy

The Knowledge Integration (KI) pedagogy advances educational theory, originating from extensive empirical studies conducted over decades that reveal ways to build cumulative understanding of the inquiry cycle, especially as reflected in ability to interpret scientific representations such as graphs (e.g. Boda et al., 2021). These investigations include in-depth analyses of students' current insights, design-based research aimed at refining student ability to conduct investigations, and longitudinal studies that track student learning trajectories over extended periods, spanning weeks, months, and even years. A consistent finding from these foundational studies is that learners frequently grapple with a multitude of ideas concerning scientific phenomena, which are often conflicting, incomplete, and confusing. Further, the success of KI instruction depends on supporting students to deepen their understanding and connect one topic to another (e.g. Clark & Linn, 2003).

Central to the KI framework is the concept of a "repertoire of ideas," which refers to the diverse, and sometimes contradictory, views that students hold about the same phenomena. These ideas are not solely derived from formal instruction but also emerge from personal experiences, social interactions, and observations. They can manifest in various forms, including fleeting or enduring, consistent or seemingly contradictory, and can be expressed as causal, observational, visual, analogical, mathematical, abstract, or experiential notions.

Knowledge integration involves the process of adding, distinguishing, organizing, and evaluating accounts of phenomena, situations, and abstractions. This process encompasses both the development of a rich and varied repertoire of ideas and the crucial task of sorting out the intricate knowledge webs that connect these ideas. The KI perspective synthesizes these investigations into a set of inquiry patterns aimed at promoting coherent and cohesive understanding of the inquiry cycle, alongside design principles that guide the customization of these patterns in diverse educational contexts (Linn & Eylon, 2011).

A fundamental re-evaluation inherent in the Knowledge Integration perspective concerns the nature of student "misconceptions." Traditional pedagogical approaches often treat these as errors to be corrected or replaced. However, the KI framework views these "nonnormative ideas" as reflections of genuine intellectual effort and keen observation on the part of the student. This reframing shifts the pedagogical imperative from simply eliminating errors to actively building upon and refining existing cognitive structures. By acknowledging and working with students' diverse ideas, even those that are scientifically non-normative, KI aims to guide learners in distinguishing between scientifically accepted views and their own intuitive understandings, ultimately fostering a more cohesive and robust knowledge structure.

Furthermore, the KI pedagogy emerged as a direct response to the pervasive problem of fragmented and disconnected knowledge acquisition often observed in typical educational settings. This fragmentation is exacerbated by curricula that cover numerous topics superficially, leading students to view science as a collection of isolated facts rather than an interconnected body of knowledge. The core objective of KI—to promote coherent and cohesive understanding of science concepts and the science inquiry cycle—directly addresses this issue by emphasizing the importance of the relationships and interconnections among ideas, moving beyond accumulation of ideas to foster genuine understanding.

Research documents that learners tend to accumulate new ideas without necessarily generating deep connections to existing ones (Linn, Davis, & Eylon, 2004). This may narrow the contexts where students apply their science class insights, possibly limited to classroom activities rather than their lived experiences. To avoid a superficial sense of coherence, KI emphasizes broadening the application of scientific ideas to everyday and consequential situations that ultimately enables the learner to build the robust view of scientific inquiry they need for future investigations.

Studies of ways to leverage AI to support students to design and carry out investigations is informed by and strengthens KI Pedagogy. In particular AI tools can help guide science learning by tracking the specific ideas that students develop and supporting them to grapple with these and other ideas suggested by their peers, teachers, personal investigations, or classroom experiences. Rather than narrowing the contexts for applications of science ideas, AI tools can help designers ensure that students generalize the application of their ideas. Similarly, AI tools can keep track of ideas students encounter in lectures (like metals are good conductors), and encourage students to distinguish them from personal experience, such as arguing that metals at room temperature feel cold so they could cool things down. Highlighting opportunities to distinguish between their idea and instructed ideas supports students to discern the evidence underlying each idea and which idea(s) to prioritize to explain the phenomena. By testing their own idea against an idea they encounter in their science class, students take a first step towards conducting their own investigation.

5.2.1 KI Mechanisms for AI Designs

Creating the conditions for students to initiate investigations is crucial. Initial efforts to design environments that promote student investigations identified four mechanisms (Linn, Davis, Eylon, 2004). These mechanisms are valuable for designing ways to use AI for guidance. Rather than simply acquiring new information, we sought ways to ensure that AI guidance focuses on how students integrate complex ideas, reconcile different models, and develop more sophisticated ways of thinking. To support students to initiate investigations, KI pedagogy emphasizes four mechanisms:

1. **Make Science Accessible (Build on Student Ideas; Provide Personally Relevant Examples):** This mechanism emphasizes that learning should start with what students already know, including their intuitive ideas, everyday experiences, and even "non-normative" conceptions. Instruction should connect new scientific concepts to these existing ideas and provide examples that are personally relevant to students' lives. The goal is to avoid abstract or incomprehensible models that might lead students to simply memorize information in isolation.
2. **Make Thinking Visible (Model Scientific Thinking; Scaffold Students to Make Their Thinking Visible; Provide Multiple Representations):** This mechanism highlights the importance of making the often-hidden processes of scientific thinking explicit. This can involve modeling how scientists compare and test ideas, providing tools or activities that allow students to articulate and represent their own evolving ideas (e.g., through drawings, explanations, or debates), and offering multiple representations of concepts (e.g., diagrams, simulations, textual descriptions). When thinking is visible, students can better reflect on their own understanding and identify areas where their ideas might be incomplete or conflicting. By making their thinking visible, students also give teachers insight into their progress. Responses to embedded activities, thus, make student ideas visible and serve as assessments of student progress.
3. **Help Students Learn from Others (Encourage Listening to Others; Design Discussions; Highlight Cultural Norms):** Learning is viewed as a social process. This mechanism promotes collaborative learning environments where students can share, compare, and critique each other's ideas. Designing structured discussions, encouraging active listening, and highlighting the cultural norms of scientific discourse (e.g., valuing evidence, constructive criticism) are key to fostering this collaborative integration of knowledge.
4. **Promote Autonomy and Lifelong Learning (Encourage Monitoring; Provide Complex Projects; Revisit and Generalize Inquiry Processes; Scaffold**

Critique): The ultimate goal of KI is to equip students to become independent, lifelong learners. This mechanism involves encouraging students to monitor their own understanding, by involving them in open-ended investigations that require sustained inquiry. It entails helping students generalize the inquiry processes they learn in one context to new situations. Scaffolding critique, both of their own ideas and those of others, helps students develop the critical evaluation skills necessary for continuous knowledge integration.

5.3 Combining AI and KI to Support the Inquiry Cycle

To help learners generalize their insights about the nature of investigation, designers have devised many variations on the inquiry cycle (Linn & Eylon, 2011). A robust understanding of the inquiry cycle involves appreciating variations on the cycle, some suited to specific types of inquiry, others valuable for preliminary or highly specialized inquiry, and others useful for science learning. Rather than learning one "scientific method", students benefit from recognizing the variety of methods available, and developing a robust, epistemological perspective on inquiry. Students who recognize that the scientific method involves trial and refinement, conjecture and failure, innovation and redesign have a foundation for lifelong scientific inquiry.

Key to the success of the investigation curriculum is the motivation students have when engaged in studying their own question. Students often need support to find a question. However, once the topic is selected, they are often able to elaborate and test their ideas, which has been shown to enhance learning in numerous studies (Cazden, 1988; Fassinger, 1995; Howard et al., 1996; Wang & Eccles, 2012).

5.3.1 Inquiry Cycle Activities

To transform the goals of science education to focus on investigations, we encourage designers to build on and refine the rapid advances in AI to support *inquiry cycle activities.* Inquiry cycle activities include:

- Problem finding and brainstorming. Framing an idea as a research question, even when students care about the topic, often requires multiple iterations with critique from classmates, avatars, or teachers (Billings et al., 2024). AI-supported tools can direct students to sources to explore, help frame a researchable question, and synthesize brainstorming activities (e.g. Maier et al., 2022).
- Drawing and visualizing ideas. AI tools are emerging that support creative representation of scientific ideas. Students can compare open source alternatives and draw their own visualization.
- Finding resources and fleshing out ideas. AI can help students summarize academic papers and extract key insights. AI can help teachers create resources

aligned with their students' interests, reading level, or language proficiency (e.g. Ogunleye et al., 2024). Students report that AI dialogs provide a safe space for students to flesh out their perspectives about an unfamiliar topic without judgment from a teacher or peer (Holtmann et al., 2023). Developing these tools involves guarding against bias (e.g. Baker & Hawn, 2022). Bias can arise when the tool fails to recognize a student idea, gives guidance that disparages rather than building on student ideas, or offers guidance that is misleading for the learner.

- Offering personalized, adaptive guidance. AI-supported adaptive learning platforms (e.g., web-based activities; intelligent tutoring systems) adjust instruction based on student performance, helping students build on their ideas at their own pace (Bradford et al., 2023). Bias can arise if student ideas are missed or misinterpreted.
- Supporting virtual experimentation. AI-supported models and simulation tools can scaffold students to make predictions, design tests of their conjectures, and iterate on their ideas in virtual experiments. Tools can support students to analyze large datasets fostering scientific practices while also testing scientific conjectures (e.g. de Jong et al., 2013; Donnelly et al., 2014; F. Lin & Puntambekar, 2025). AI-supported simulations make vast, fast, complex, or hazardous experiments more accessible and safe.
- Reviewing and commenting on student work. AI tools can guide construction of oral and written reports, posters, public service announcements, and final papers. A wide range of AI tools can help teachers review or give feedback on lab reports, coding exercises, and scientific essays. Guarding against bias remains a challenge since AI tools may ignore student ideas or disparage promising reasoning (e.g. Baker & Hawn, 2022).
- Enhancing collaboration and communication. AI-supported chatbots, collaborative tools, and natural language processing applications support scientific discussions, helping students refine their scientific writing and reasoning (e.g. Billings et al., 2024). Classroom interactions often falter because students have difficulty interpreting their peers' ideas. Offering students AI generated prompts can advance peer to peer conversations.
- Developing STEM Skills. AI-based platforms can help students learn to design code, analyze data, and use other AI tools, preparing them for future careers in science and technology (e.g. Yim & Su, 2025). AI tools can generate fictional student work with specific gaps or inaccurate ideas, to support student critique. These activities need careful review and refinement to ensure that they align with student expertise.
- Increasing accessibility of educational materials. AI tools can help bridge educational gaps by offering language translation, speech-to-text services, and personalized support for diverse learning needs (e.g. Mintz et al., 2023). While valuable, these tools can also introduce bias by misinterpreting the original source.

5.4 Refining an Inquiry Cycle Activity: Offering Personalized, Adaptive Guidance

To illustrate how to combine AI and KI, we developed support for the inquiry cycle activity of offering personalized, adaptive guidance for questions that require integrating ideas. We conducted a systematic research program where we leveraged AI to design guidance to support students to generate and revise science explanations to questions that require KI (Flower et al., 1986).

This inquiry cycle activity is based on research showing that productive revision can deepen student science understanding (Berland et al., 2016). Expert teachers guide students to deepen their understanding by building on ideas they develop in their everyday lives, reinforcing the value of gathering evidence to clarify their ideas, and broadening the perspectives students consider (Rosebery et al., 2016). Designing guidance that motivates students to seek evidence to distinguish between their own ideas and those of other students promotes coherent understanding (Hammer, 1995). For example, when students argue that metal cups can keep beverages cold because they feel cold, an expert teacher might prompt the student to also consider how metal feels on a hot day. Building on each students' ideas is a challenge when a classroom of 35 students might have 10–25 distinct ideas about any given topic (Holtmann et al., 2023; Luna, 2018). Adaptive guidance for student written science explanations can assist teachers to support student investigations by providing each student with guidance for KI.

5.4.1 *Partnership Research*

Our guidance design research is conducted by a partnership of classroom teachers and their students from over 12 participating middle schools, software designers who have created the Web-based Inquiry Science Environment (WISE), learning science researchers, psychometricians, discipline experts, and computer science NLP experts at ETS. Partners contribute to design research to create and test guidance assigned based on NLP scores. Our partnership meets regularly at school site meetings, professional development workshops, and on-campus seminars. In partnership meetings, we analyze the results of each guidance design study and discuss ways to improve the guidance, drawing on the KI pedagogy, advances in NLP, and related learning sciences research.

5.4.2 AI and KI Holistic Guidance

In 2008 we began collaborating with ETS to use NLP to develop holistic scores of student responses to KI Explanation questions embedded in WISE activities. We used the scores to explore ways to design guidance for each score. Our first KI guidance acknowledged the students' progress and prompted them to visit a link to a relevant activity. The activity was designed to help the student move up one level in the KI rubric. Rather than relying on authorities, the guidance encouraged students to gain appreciation for relying on evidence to refine their knowledge (see Fig. 5.1). Our initial set of studies developed NLP models to assess the knowledge integration level of student written arguments and tested guidance designs to help students revise their arguments by integrating a new idea.

5.4.3 NLP Method for Holistic Guidance

To develop the NLP model for student scores, we started with human scores of 1000 student KI explanations. Two or more humans used the KI rubric to score a subset of the responses until they developed good agreement. Our collaborators at ETS then used the c-rater NLP tool to build a scoring model. They used about 700 of the responses to build the model and then tested the model on the remaining responses. Early versions of NLP used "bag of words" solutions that filtered out common words and then used the remaining words to achieve a score. Over the years the NLP tools have been updated to improve filtering, take advantage of strings of words, and eventually leverage Large Language Models (LLMs) to improve the tools. Today, our NLP models for KI items have excellent accuracy for assigning guidance. Using a metric called Quadratic Weighted Kappa, our items have scores above 0.85. This means that scores are mostly accurate, and when they are inaccurate, they are typically only off by one level. This level of accuracy is fine for designing guidance but not sufficient for high stakes testing.

5.4.4 Holistic Guidance Findings

In our four initial studies of NLP-based KI guidance for student explanations, we found that the adaptive KI guidance was more effective than other types of guidance typically used in middle school classrooms to promote KI. We compared KI guidance to simulated teacher guidance (e.g., Redo. What does increased carbon dioxide do to global temperature?), generic guidance (e.g., Go back and review the visualization to improve your answer), and specific guidance (e.g., Light energy transformed into __ kind of energy). Studies of the logged revisions and student navigation indicated that in comparison to the other forms of guidance, the KI

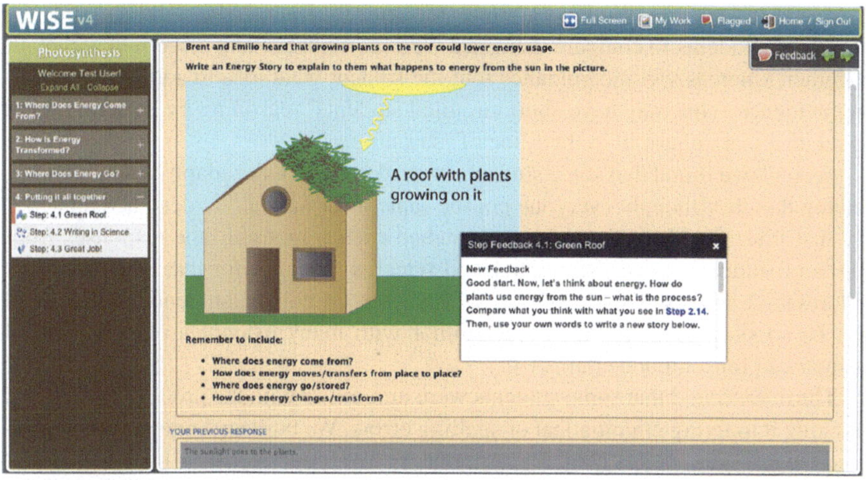

Fig. 5.1 KI guidance for science arguments using holistic NLP scoring

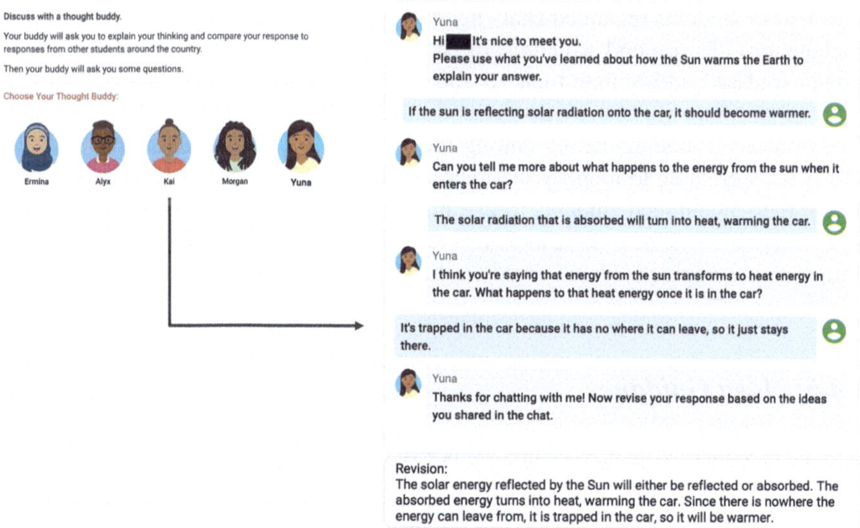

Fig. 5.2 Example online dialog for the Car on a Cold Day item

guidance was more likely to support students to revisit specific evidence in the unit when revising (Gerard et al., 2015). Students who received KI guidance were also more likely to integrate a new idea into their initial response and improve their explanations when revising. Highly specific guidance or guidance providing an accurate answer had limited or only short-term benefits.

Further, students who received KI guidance and revisited evidence were more likely to improve their explanations, than students who received specific or generic guidance and revisited evidence (Vitale, McBride, & Linn, 2016). The KI guidance supported students to compare their idea with the evidence and discern how they fit together, whereas specific guidance told the student what idea to add when revisiting evidence. This may have short circuited students' reasoning by directing them to an idea that did not fit in their line of reasoning.

Second, we found that some students were skeptical of the adaptive KI guidance, saying that they thought everyone got the same information. We designed an activity that illustrated how the software matched each response to the responses from 100's of students across the country and selected guidance for that response. This information increased students' rate of revision. Of course, students today would not be so skeptical since they are familiar with many examples of personalized responses (Tansomboon et al., 2017).

Third, we found that some students were unfamiliar with the process of revision, likening it to fixing grammatical or spelling errors. We built an Annotator that modeled the revision process (Fig. 5.2: Annotator). We found that students who had not revised in the past were more likely to revise by integrating a new idea after interacting with the Annotator (L. Gerard & Linn, 2022).

Finally, we found that in spite of gains compared to other forms of guidance, at least half of students remained challenged to add new ideas when they revised their explanations. To respond we conjectured that targeting guidance to the ideas students expressed, rather than their KI score might add value. This aligns with our study of an Annotator activity that helped students who were not sure how to revise by emphasizing distinguishing among ideas (Gerard & Linn, 2022). We decided that if we were able to identify the ideas students hold, we could create an online dialog that encouraged students to distinguish between the idea we identified and other alternatives. These insights led us to use NLP to identify student's ideas and design ways to guide students based on the ideas we identified.

5.4.5 *Idea Guidance*

Starting in 2020, we began developing NLP idea detection models to identify students' ideas within their explanations. We created online dialogs to engage students about their ideas. We investigated research questions about guidance design and science learning that emerged from our studies of holistic guidance and that were feasible due to our new capacity to detect students' specific idea-changes in responses to guidance: Will guidance aligned to a particular idea in a students' explanation encourage them to express and incorporate more of their ideas? Will a dialog that emulates a conversation with a teacher encourage students' productive revision to their arguments?

We developed a new NLP technology, idea detection, to detect each of the ideas within a student's argument. For example, for the 'Car' item in the Climate Change unit, 'how will the temperature inside a car parked in the snow feel on a sunny day?', the model can detect 25 different student ideas in their explanations, such as

Table 5.1 Car on a Cold Day Knowledge Integration Rubric

KI Prompt: On a cold winter day, Akbar is walking to their car that is parked in the sun. Their car has not been driven for one week. How will the temperature inside the car feel?" Choose (a) Colder than the outside air, (b) Warmer than the outside air, or (c) Exactly the same as the outside air. Explain your choice.

KI Level		Student Examples
1	Off-task or irrelevant	"I don't know"
2	Vague or incomplete ideas or links	"The air is warmer inside because it had no contact with cold air outside" "The car is hotter because it's been parked in the sun." *[accurate conclusion, no mechanism]*
3	Partial link Isolated normative idea OR normative and non--normative ideas linked	"Since cars absorbed heat doesn't go anywhere, it warms up the car inside rather than the outside" *[heat is trapped inside of car]* "Energy from the sun warms up the car" *[explains energy source]*
4	Full link One complete and valid connection between two ideas	"The temperature will be hotter than the outside because the heat inside the car not escape and is trapped inside, the heat is added and added until is hotter than the outside" *[heat is trapped + reabsorbed]*
5	Complex links Two or more scientifically complete and valid connections	"Energy comes from the sun and is absorbed in the car to turn to heat. The heat is stuck in the car because the windows trap it making it hotter" *[energy from sun + solar radiation to heat + heat is trapped]*

the 'sunlight turns into heat' or 'the car will feel cold like their mom's car' (see sample ideas and associated guidance in Table 5.1). The research team develops guidance for each of the identifiable ideas. The guidance is inserted in the dialog when the idea is detected. Partners design the guidance to ask students to elaborate their thinking, inspired by KI guidelines from initial studies and ways that experienced teachers guide their students (Fig. 5.2). We have implemented the partnership guidance in the online dialog in multiple units.

5.4.6 AI Method for Idea Guidance

To develop the NLP idea detection model we started with a dataset of at least 1000 individual student responses to a KI item. Two partners (researchers or teachers) analyzed a representative set of around 10% of the student responses in the dataset (selected from a range of schools and teachers) to create an idea detection rubric (see Table 5.2). An idea typically includes a phrase within a sentence. The partners identify expressions within a student's argument that they would respond to in the classroom. Thus, they identify distinct ideas in each student's explanation that they would want to build on to help the student deepen their understanding of the science topic.

Each student argument could include zero (e.g. IDK), one, or more ideas. The initial idea detection rubric lists the distinct ideas that emerge when the partners annotate the training dataset. The emerging ideas are then reviewed by other members of the partnership. Often the reviewers modify the idea list by merging overlapping ideas and distinguishing others.

Once the rubric is refined, the partners use the updated idea detection rubric to recategorize the ideas in the 10% sample. The team makes sure that the new idea categories capture all the ideas expressed in the sample. We check to be sure that the raters have high agreement. Then the partners each score about half of the remaining responses using annotation software. For a typical KI item, we end up with about 25 ideas. We create a training dataset consisting of about 80% of the responses and save the rest as a testing dataset to determine the accuracy of the model.

Table 5.2 Sample from Car Idea-Detection Rubric and Guidance

Item prompt: On a cold winter day, Akbar is walking to his car that is parked in the sun. His car has not been driven for one week. Does the air inside the car feel colder, hotter or exactly the same as the outside air? Explain.

#	Idea	KI	Assigned Adaptive Eliciting Guidance in Dialog
2	The car feels colder than outside (DI)	2	Interesting ideas! Can you tell me about how you feel if you are sitting inside a car and the sunlight shines on you through the window?
5	Heat (or cold) is conducted between the car and the surrounding area (DI)	2	You're right that heat energy is conducted between objects that touch each other. Since the sun is far away, how can its energy warm the car?
8	The sun warms the car directly (DI)	2	Interesting idea! How does the sun make the car warmer?
15	Solar radiation is transformed into heat energy (MI)	3	I think you're saying that energy from the sun transforms to heat energy in the car. What happens to that heat energy?
18	Heat or infrared energy is trapped inside the car (MI)	3	Nice progress. Can you tell me more about why heat gets trapped in the car but solar radiation doesn't?

Note. The full rubric includes 24 ideas across all KI levels and includes both descriptive and mechanistic ideas (Bradford et al., 2023)

A computer scientist then builds the NLP idea detection model using the training dataset. We test the model on the whole dataset and determine how well it works on data that was not used for training. We compute a measure of agreement between the NLP idea detection and the human scores.

From an NLP perspective, the idea detection model solves a multi-label classification task: for each word, the model makes multiple binary classification predictions about whether the word belongs to an idea category, eventually making a selection. Pre-trained transformer models are used for building the idea detection models. Transformer models use a pre-trained Large Language Model. The idea detection model is fine-tuned and validated with tenfold cross validation for hyperparameter tuning. When the NLP model achieves sufficient accuracy across the designated idea categories the idea detection model is ready for instructional use. It can be embedded in an online science unit. NLP idea detection has advanced our previous holistic NLP scoring of KI to detect specific information about students' thinking to inform guidance (e.g. Gerard et al., 2025; Schulz et al., 2019).

5.4.7 *Idea Detection Guidance Findings*

Our analysis of the student ideas detected by the NLP models in middle and high schools serving different student demographics point to the benefits of refining NLP models regularly to capture the new student ideas that emerge in new classroom contexts. Comparing idea detection rates across schools for the Car on a Cold Day item we found that the distribution of ideas in students' initial responses differed across schools. Further, some low frequency, intriguing ideas occurred in one school and not the other, including the idea that the color of the materials in the car would affect temperature. This provides empirical evidence that students often start from different places when seeking to understand science phenomena, and students' starting points vary from classroom to classroom based largely on students' prior instruction and cultural experiences. A guideline emerges to iterate the NLP model regularly to customize the detectable set of student ideas to the instructional context.

In addition, studies of the online dialogs provide detailed insight into how students revise their initial arguments to form a more coherent argument and a more integrated science understanding. We developed new methods leveraging idea detection analytics to evaluate the efficacy of idea-based guidance. This included (a) analyzing the frequency of new accurate and vague ideas that students add in response to each round of guidance during an online dialog, in comparison to the ideas in their initial argument, and (b) we compare the frequency of new ideas students add from their initial argument to their revised argument, before/after the dialog, to identify if and to what degree students integrated the new ideas they raised in the dialog into their arguments.

Across dialogs, our analyses point to the benefits of idea-detection guidance that responds to a student's vague idea and prompts them to add a more precise idea—rather than prompting the student to discard their vague idea in favor of a more precise idea. Further, student engagement with the online idea detection dialogs has resulted in improved student KI in multiple unit contexts (Bradford et al., 2023; Gerard et al., 2024; Holtmann et al., 2023; Li et al., 2023). While these gains are similar to gains with holistic guidance, the idea detection analysis reveals that the change in KI score is mainly driven by the change in the number of accurate ideas students integrated into their arguments after the dialog, as opposed to dropping of their initial vague ideas (Gerard et al., 2024).

Furthermore, the analyses of students' revision in the online dialogs and of student perceptions of idea-based guidance within the dialogs clarifies the benefits of guidance that responds to and extends a student's specific idea to expand their sense-making. The online dialogs improved the rate of student revision of their explanations compared to holistic adaptive guidance across the six online dialogs. For example, in the Car dialog embedded in the Global Climate Change unit, the majority of students revised their initial explanation after the dialog (91%) (Gerard, Bichler, et al., 2022). This is a higher rate of revision compared to prior studies reporting under 50% of students engaging in sustained revision when the holistic NLP guidance called for it (Tansomboon et al., 2017). This higher rate of revision is largely driven by students' adding of new ideas during the dialogs. Students significantly increased the number of ideas, particularly accurate ideas, that they linked in their explanations from their initial to their revised explanations, before/after the dialog (Bradford et al., 2023; Gerard et al., 2024). For example, as reported in Bradford et al. (2023) a study across four items found that each round of the idea-based guidance in the dialog elicited new student ideas. In a study of the Siblings dialog, Gerard et al. (2024) found that students who progressed to Full KI added multiple and different accurate ideas regarding genetic variation. Taken together these studies reinforce the KI perspective that building from one's ideas allows students to follow multiple pathways to a deeper scientific understanding, as opposed to there being a singular correct pathway.

We found that in spite of revision rates improving, students were still challenged to integrate ideas they raise in the dialog when revising their arguments. While the majority of the students expressed two to three new ideas during the dialog, students tended to integrate only one at most of these new ideas they raised when they revised their argument after the dialog (L. Gerard et al., 2024).

As this example illustrates, there is great potential to use AI to transform the goals of science courses. In addition, research programs are needed to refine AI tools for the investigation curriculum. Many of these research programs are illustrated in other chapters in this book.

5.5 Building an Investigation Curriculum

To transform the goals for science education for an investigation curriculum, we need to build tools for each of the inquiry cycle activities identified in this chapter:

- Problem finding and brainstorming
- Drawing and visualizing ideas
- Finding resources and fleshing out ideas
- Offering personalized, adaptive guidance
- Supporting virtual experimentation
- Reviewing and commenting on student work
- Enhancing collaboration and communication
- Developing STEM Skills
- Increasing accessibility of educational materials

We illustrated the process for offering personalized and adaptive guidance in the section above. Next steps include creating inquiry cycles for a sequence of activities and for multiple topics.

By creating a series of activities for each science topic that introduces students to the Inquiry Cycle and guides them to refine their insights we can build the curriculum. We also need a repository of resources that supports customizing the curriculum to local issues. These activities, when combined in an inquiry cycle, can support students to carry out their own investigations (Fig. 5.3). To build an inquiry curriculum for any topic, designers might:

- Design an inquiry unit that introduces the science topic such as a unit on middle school thermodynamics that engages students in using models or temperature probes to explore heat flow, thermal equilibrium, insulation and conduction and related concepts (King Chen & Linn, 2019).
- Create an on-ramp activity where students use a virtual or hands-on activity where they investigate a set question such as, "Which cup material is best for keeping hot drinks hot and cold drinks cold" or test a new design for windows that keep houses warm in winter and cool in summer.
- Design a challenge activity where students select a question or create their own. They might be encouraged to refine a question based on their interests or select from among a set of challenges. One challenge might be to design and test a solar oven.

This sequence could build a deep understanding of thermodynamics while also building up an understanding of the inquiry cycle. After repeating this approach for a few topics, students might be able to select a question and combine activities to design their own investigation by combining steps on the inquiry cycle.

Overall, research suggests that when used effectively, AI can play an important role in supporting students to build the ability to design and conduct their own investigations, enabling reimagining the goals of science education to focus on investigations. AI tools also improve accessibility to resources including text, support personalized guidance and assessments, and provide opportunities for interactive experimentation. Challenges including ethics and bias need consideration. Importantly, changing goals for science education requires thoughtful teacher collaboration and professional development.

Fig. 5.3 Inquiry cycle. Note. Students traverse the inquiry cycle in multiple ways, repeating some steps and skipping others

5.6 Building a Repository for the Investigation Curriculum

How might transforming the goals of science education to focus on investigations work? Partnerships of practitioners, researchers, designers, and computer scientists are needed to create a repository of open source activities and curriculum materials that offer on-ramp activities, sequences of the activities that promote KI, and indicators of progress that capture progress in KI. Materials need to be designed, tested and refined.

To build a repository for the inquiry curriculum, designers can draw on many open source resources, developed with funding from multiple foundations including the National Science Foundation (NSF), National Institutes of Health (NIH), Institute of Education Sciences (IES), Spencer, Gates, and Hewlett. Building on these excellent and available resources will make transforming science instruction for investigations efficient and effective. Once students and their teachers get started on using investigations for science learning, they will shape the resources in the repository and provide opportunities for others.

Using the resources, partnerships can customize inquiry cycle activities and build sequences of activities that leverage their own competencies. Thus, students will experience multiple opportunities to gain the ability to direct their own learning. This in turn will motivate them to apply these capabilities to the next science topic. Similarly, as teachers implement the transformed curriculum, they will continuously enhance it by capitalizing on the resources their students bring to the classroom. Transforming instruction to focus on investigations will continuously accelerate student learning.

5.7 Ensuring Alignment with the Responsible and Ethical Principles for AI in Science Education

Transforming the goals of science education to focus on AI-supported investigations aligns with the Responsible and Ethical Principles for AI advocated in Chap. 3 of this volume. The partnership leading the transformation represents all relevant stakeholders who will jointly ensure that the principles are implemented. First, transforming the goals of science education to focus on investigations supported by AI guidance addresses *fairness* by ensuring that guidance respects and responds to each students' ideas. Bias can arise if guidance ignores some student ideas or disparages rather than building on a students' idea. Second, the partners rely on KI pedagogy for decision-making, ensuring that the process is transparent and open. Teachers and students refine their ability to evaluate the AI analyses and teacher-led responses and ensure that then follow the KI Processes and Mechanisms. Third, the partners are accountable to the users by providing detailed information about the uses of AI and the ways that bias has bee addressed. The transformation is documented in multiple cycles of design, testing, and refinement using KI assessments and analyzing student work. Teachers and students appreciate that the investigation activities in WISE are built by a partnership that is responsive to any concern raised by users.

The partnership addresses p*rivacy and data security* by implementing innovations using the WISE system. WISE meets all requirements for privacy and security, as stipulated in its IRB. All uses of AI are monitored by *human oversight*. Teachers can monitor student progress with teacher tools. They can customize and refine all aspects of the investigation curriculum. To ensure *beneficial use, WISE* uses continuous assessment, guided by logged data and collaboration between teachers and other partners. Analyses by teachers and researchers ensure that the investigation activities benefit each learner. The WISE system is *robust*. The WISE system is open-source and has been used by 1000s of teachers and students. As AI features are incorporated, it is constantly monitored and maintained.

The partnership implements a practice of mutual respect. This includes *respect for human values.* The partnership guiding the investigation curriculum monitors how the AI systems honor students' dignity, foster ethical decision-making, and support values consistent with democratic education.

5.8 Conclusions

In conclusion, by combining advances in AI and advances in KI, we can transform the goals for science education to prepare learners who can take initiative as self-directed investigators. Focusing on the investigation curriculum will motivate both AI research and pedagogy research to build the resources needed to transform science education. Ultimately, we will prepare learners who can help develop ways to sustain and support the health of the planet.

Acknowledgement This material is based upon work supported by the National Science Foundation under Grant No. 2332964. Any opinions, findings, and conclusions or recommendations expressed in this material are those of the author(s) and do not necessarily reflect the views of the National Science Foundation.

References

Baker, R. S., & Hawn, A. (2022). Algorithmic bias in education. *International Journal of Artificial Intelligence in Education, 32*(4), 1052–1092. https://doi.org/10.1007/s40593-021-00285-9

Berland, L. K., Schwarz, C. V., Krist, C., Kenyon, L., Lo, A. S., & Reiser, B. J. (2016). Epistemologies in practice: Making scientific practices meaningful for students. *Journal of Research in Science Teaching, 53*(7), 1082–1112. https://doi.org/10.1002/tea.21257

Billings, K., Chang, H.-Y., Lim-Breitbart, J. M., & Linn, M. C. (2024). Using artificial intelligence to support peer-to-peer discussions in science classrooms. *Education Sciences, 14*(12)., Article 12. https://doi.org/10.3390/educsci14121411

Boda, P. A., Bathia, S., & Linn, M. C. (2021). Longitudinal impact of interactive science activities: Developing, implementing, and validating a graphing integration inventory. *Journal of Research in Science Teaching, 58*(2), 225–248. https://doi.org/10.1002/tea.21653

Bradford, A., Li, W., Steimel, K., Riordan, B., & Linn, M. C. (2023). Adaptive dialog to support student understanding of climate change mechanism and who is most impacted. In P. Blikstein, J. Van Aalst, R. Kizito, & K. Brennan (Eds.), *Proceedings of the 17th international conference of the learning sciences—ICLS 2023* (pp. 816–823). International Society of the Learning Sciences. https://doi.org/10.22318/icls2023.681776

Cazden, C. B. (1988). *Classroom discourse: The language of teaching and learning*. Heinemann.

Clark, D., & Linn, M. C. (2003). Designing for knowledge integration: The impact of instructional time. *The Journal of the Learning Sciences, 12*(4), 451–493.

de Jong, T., Linn, M. C., & Zacharia, Z. C. (2013). Physical and virtual laboratories in science and engineering education. *Science, 340*(6130), 305–308. https://doi.org/10.1126/science.1230579

Donnelly, D. F., Linn, M. C., & Ludvigsen, S. (2014). Impacts and characteristics of computer-based science inquiry learning environments for precollege students. *Review of Educational Research, 84*(4), 572–608. https://doi.org/10.3102/0034654314546954

Eger, S., Cao, Y., D'Souza, J., Geiger, A., Greisinger, C., Gross, S., Hou, Y., Krenn, B., Lauscher, A., Li, Y., Lin, C., Moosavi, N. S., Zhao, W., & Miller, T. (2025). *Transforming science with large language models: A survey on AI-assisted scientific discovery, experimentation, content generation, and evaluation* (arXiv:2502.05151). arXiv. https://doi.org/10.48550/arXiv.2502.05151

Erduran, S. (2023). AI is transforming how science is done. Science education must reflect this change. *Science, 382*(6677), eadm9788. https://doi.org/10.1126/science.adm9788

Fassinger, P. A. (1995). Understanding classroom interaction: Students' and professors' contributions to students' silence. *The Journal of Higher Education, 66*(1), 82–96. https://doi.org/10.2307/2943952

Flower, L., Hayes, J. R., Carey, L., Schriver, K., & Stratman, J. (1986). Detection, diagnosis, and the strategies of revision. *College Composition and Communication, 37*(1), 16–55. https://doi.org/10.2307/357381

Gerard, L., Bichler, S., Bradford, A., Linn, M. C., Steimel, K., & Riordan, B. (2022). Designing an adaptive dialogue to promote science understanding. In C. Chinn, E. Tan, C. Chan, & Y. Kali (Eds.), *Proceedings of the 16th international conference of the learning sciences—ICLS 2022* (pp. 1653–1656). International Society of the Learning Sciences.

Gerard, L., Bradford, A., DeBarger, A., Wiley, K., & Linn, M. C. (2022). Cultivating teacher efficacy for social justice in science. *Science Scope, 45*(5), 40–48.

Gerard, L., Holtman, M., Riordan, B., & Linn, M. C. (2025). Impact of an adaptive dialog that uses natural language processing to detect students' ideas and guide knowledge integration. *Journal of Educational Psychology, 117*(1), 63–87. https://doi.org/10.1037/edu0000902

Gerard, L., & Linn, M. C. (2022). Computer-based guidance to support students' revision of their science explanations. *Computers & Education, 176*, 104351. https://doi.org/10.1016/j.compedu.2021.104351

Gerard, L., Linn, M. C., & Holtmann, M. (2024). A comparison of responsive and general guidance to promote learning in an online science dialog. *Education Sciences, 14*(12)., Article 12. https://doi.org/10.3390/educsci14121383

Gerard, L., Matuk, C., McElhaney, K., & Linn, M. C. (2015). Automated, adaptive guidance for K-12 education. *Educational Research Review, 15*, 41–58. https://doi.org/10.1016/j.edurev.2015.04.001

Gerard, L. F., Spitulnik, M., & Linn, M. C. (2010). Teacher use of evidence to customize inquiry science instruction. *Journal of Research in Science Teaching, 47*(9), 1037–1063. https://doi.org/10.1002/tea.20367

Hammer, D. (1995). Student inquiry in a physics class discussion. *Cognition and Instruction, 13*(3), 401–430.

Holtmann, M., Gerard, L., Li, W., Linn, M. C., & Riordan, B. (2023). How does an adaptive dialog based on natural language processing impact students from distinct language backgrounds? In P. Blikstein, J. Van Aalst, R. Kizito, & K. Brennan (Eds.), *Proceedings of the 17th international conference of the learning sciences—ICLS 2023* (pp. 1350–1353). International Society of the Learning Sciences. https://doi.org/10.22318/icls2023.921177

Howard, J. R., Short, L. B., & Clark, S. M. (1996). Students' participation in the mixed-age college classroom. *Teaching Sociology, 24*(1), 8–24. https://doi.org/10.2307/1318894

Kersting, K. (2020). Rethinking computer science through AI. *KI - Künstliche Intelligenz, 34*(4), 435–437. https://doi.org/10.1007/s13218-020-00692-5

King Chen, J. Y., & Linn, M. C. (2019). Impact of choice on students' use of an experimentation model for investigating ideas about thermodynamics. In K. Lund, G. P. Niccolai, E. Lavoue, C. E. Hmelo-Silver, G. Gweon, & M. Baker (Eds.), *A wide lens: Combining embodied, enactive, extended, and embedded learning in collaborative settings* (Vol. 2). International Society of the Learning Sciences. https://par.nsf.gov/biblio/10106365-impact-choice-students-use-experimentation-model-investigating-ideas-about-thermodynamics

Langreo, L. (2024, March 25). Teachers desperately need AI training. How many are getting it? *Education Week*. https://www.edweek.org/technology/teachers-desperately-need-ai-training-how-many-are-getting-it/2024/03

Li, W., Lim-Breitbart, J., Bradford, A., Linn, M. C., Riordan, B., & Steimel, K. (2023). Explaining thermodynamics: Impact of an adaptive dialog based on a natural language processing idea detection model. In P. Blikstein, J. Van Aalst, R. Kizito, & K. Brennan (Eds.), *Proceedings of the 17th international conference of the learning sciences—ICLS 2023* (pp. 1306–1309). International Society of the Learning Sciences. https://doi.org/10.22318/icls2023.199424

Linn, M. C., Davis, E. A., & Eylon, B.-S. (2004). The Scaffolded Knowledge Integration Framework for Instruction. In M.C. Linn, E.A. Davis, & P. Bell (Eds.), *Internet Environments for Science Education (pp. 47–72). Mahwah, NJ: Lawrence Erlbaum Associates.*

Lin, F., & Puntambekar, S. (2025). Synergizing the scaffolds from teacher and digital text tool for progressive inquiry. *Interactive Learning Environments, 33*(10), 5904–5922. https://doi.org/10.1080/10494820.2025.2488990

Lin, L. (2024, May 15). A quarter of U.S. teachers say AI tools do more harm than good in K-12 education. *Pew Research Center.* https://www.pewresearch.org/short-reads/2024/05/15/a--quarter-of-u-s-teachers-say-ai-tools-do-more-harm-than-good-in-k-12-education/

Linn, M. C., & Eylon, B.-S. (2011). *Science learning and instruction: Taking advantage of technology to promote knowledge integration*. Routledge.

Luna, M. J. (2018). What does it mean to notice my students' ideas in science today?: An investigation of elementary teachers' practice of noticing their students' thinking in science. *Cognition and Instruction, 36*(4), 297–329. https://doi.org/10.1080/07370008.2018.1496919

Maier, T., Soria Zurita, N. F., Starkey, E., Spillane, D., McComb, C., & Menold, J. (2022). Comparing human and cognitive assistant facilitated brainstorming sessions. *Journal of Engineering Design, 33*(4), 259–283. https://doi.org/10.1080/09544828.2022.2032623

McBride, E., Vitale, J., & Linn, M. (2018). Learning design through science vs. science through design. In J. Kay & R. Luckin (Eds.), *Rethinking learning in the digital age: Making the learning sciences count, 13th international conference of the learning sciences (ICLS) 2018* (Vol. 1, pp. 17–24). International Society of the Learning Sciences. https://repository.isls.org//handle/1/474

Mintz, J., Holmes, W., Liu, L., & Perez-Ortiz, M. (2023). Artificial intelligence and K-12 education: Possibilities, pedagogies and risks. *Computers in the Schools, 40*(4), 325–333. https://doi.org/10.1080/07380569.2023.2279870

Ogunleye, B., Zakariyyah, K. I., Ajao, O., Olayinka, O., & Sharma, H. (2024). A systematic review of generative AI for teaching and learning practice. *Education Sciences, 14*(6)., Article 6. https://doi.org/10.3390/educsci14060636

Pedaste, M., Mäeots, M., Siiman, L. A., de Jong, T., van Riesen, S. A. N., Kamp, E. T., Manoli, C. C., Zacharia, Z. C., & Tsourlidaki, E. (2015). Phases of inquiry-based learning: Definitions and the inquiry cycle. *Educational Research Review, 14*, 47–61. https://doi.org/10.1016/j.edurev.2015.02.003

Randazzo, S. (2024, July 2). That essay got a B+. An AI bot graded it. *Wall Street Journal*. https://www.wsj.com/tech/ai/ai-tools-grading-teachers-students-396c2bfc

Reiser, B. J., Novak, M., McGill, T. A. W., & Penuel, W. R. (2021). Storyline units: An instructional model to support coherence from the students' perspective. *Journal of Science Teacher Education, 32*(7), 805–829. https://doi.org/10.1080/1046560X.2021.1884784

Rosebery, A. S., Warren, B., & Tucker-Raymond, E. (2016). Developing interpretive power in science teaching. *Journal of Research in Science Teaching, 53*(10), 1571–1600. https://doi.org/10.1002/tea.21267

Schulz, C., Meyer, C. M., & Gurevych, I. (2019). Challenges in the automatic analysis of students' diagnostic reasoning. *Proceedings of the AAAI Conference on Artificial Intelligence, 33*(01)., Article 01. https://doi.org/10.1609/aaai.v33i01.33016974

Tansomboon, C., Gerard, L. F., Vitale, J. M., & Linn, M. C. (2017). Designing automated guidance to promote productive revision of science explanations. *International Journal of Artificial Intelligence in Education, 27*(4), 729–757. https://doi.org/10.1007/s40593-017-0145-0

Tyson, A., Kennedy, B., & Funk, C. (2021). Gen Z, millennials stand out for climate change activism, social media engagement with issue. *Pew Research Center*. https://www.pewresearch.org/

science/2021/05/26/gen-z-millennials-stand-out-for-climate-change-activism-social-media-engagement-with-issue/

Vitale, J. M., McBride, E. A., & Linn, M. C. (2016). Distinguishing complex ideas about climate change: knowledge integration vs. specific guidance. *International Journal of Science Education, 38(9), 1548–1569.* https://doi.org/10.1080/09500693.2016.1198969

Wang, M.-T., & Eccles, J. S. (2012). Adolescent behavioral, emotional, and cognitive engagement trajectories in school and their differential relations to educational success. *Journal of Research on Adolescence, 22*(1), 31–39. https://doi.org/10.1111/j.1532-7795.2011.00753.x

White, B., & Frederiksen, J. (2000). Metacognitive facilitation: An approach to making scientific inquiry accessible to all. In J. Minstrell & E. van Zee (Eds.), *Inquiring into inquiry learning and teaching in science* (pp. 331–370). American Association for the Advancement of Science.

Yim, I. H. Y., & Su, J. (2025). Artificial intelligence (AI) learning tools in K-12 education: A scoping review. *Journal of Computers in Education, 12*(1), 93–131. https://doi.org/10.1007/s40692-023-00304-9

Part III
Educational Procedures

Chapter 6
Science Teaching and Learning with Artificial Intelligence

Knut Neumann, Xiaoming Zhai, and Marcus Kubsch

6.1 Introduction

Artificial Intelligence (AI) is changing the future of work. Medical doctors utilize AI to analyze and interpret medical images, system engineers to diagnose and troubleshoot errors, and lawyers to research reference cases and draft court documents. Teachers, as well as learners, are no exception to these changes. As digital technologies such as tablet computers make their way into classrooms, AI has become readily available for use by teachers and students. AI can, for example, help score students' essays, support teachers in creating instructional materials, and even help in the planning of instruction. Students can engage with AI as a peer tutor, asking it questions during instruction or using it to complete homework assignments. Beyond improving traditional methods of science education, AI has the potential to transform teaching and learning. AI based tutors allow for making room for students to take a stronger agency in their learning, allowing for a more self-directed teaching and learning of science where the teacher is drawing on AI to synthesize information about students learning such as whether students are engaging in collaborative learning as planned to engage with student groups or even students individually to

K. Neumann (✉)
Department of Physics Education, IPN-Leibniz-Institute for Science and Mathematics Education, Kiel, Germany
e-mail: neumann@leibniz-ipn.de

X. Zhai
AI4STEM Education Center, University of Georgia, Athens, GA, USA
e-mail: xiaoming.zhai@uga.edu

M. Kubsch
Department of Physics, Freie Universität Berlin, Berlin, Germany
e-mail: m.kubsch@fu-berlin.de

X. Zhai, K. J. Crippen (eds.), *Advancing AI in Science Education*, Advances in Technology-Rich Science Education 3,
https://doi.org/10.1007/978-3-032-16871-9_6

support them where needed. The actual uses of AI in science classrooms are innumerable, and the power of AI for transforming science teaching and learning to become more effective as well as efficient appears (nearly) unlimited.

However, with great power must also come great responsibility. Like Peter Parker, better known as Spiderman, ignores the burglar who then mugs his uncle Ben, ignoring the threats that come with the use of AI in science classrooms will do significant harm to teachers and learners. What if inaccurate scoring algorithms are yielding false scores? What if teachers fail to recognize the false scores because they trust in the AI, because, well it is an AI? Aware of what's at stake, learners tend to quickly adapt and provide the AI with the desired responses to earn high scores. The consequence is that instead of increasing students' agency over their learning, the use of AI restricts their learning to the memorization of rote knowledge or the training of procedural skills. However, this type of learning leads to outcomes that were thought to have long been left behind. Even when students engage in learning with the AI as a peer tutor, it is easy for them to ask the AI to provide the solution, essentially bypassing the learning process. Hence, we cannot simply rely on the use of AI in science classrooms being meaningful. Instead, the use of AI must be actively managed, including educating teachers and students on the responsible and ethical use of AI.

This chapter starts by illustrating several classroom scenarios showcasing the changes AI has brought to science teaching and learning. We then move on to discuss what it takes to responsibly and ethically use AI in these scenarios and derive specific conclusions about steps that need to be taken to engage teachers and students alike in a responsible and ethical use of AI in science classrooms. In doing so, we assume that the use of AI in classrooms is inevitable and that the question is how we can orchestrate teaching and learning that makes the best use of it while avoiding the traps and pitfalls that come with such use. One important thing to note in this context is that the responsibility for how AI is used really lies with the users—that is, the teachers and the students (Zhai & Nehm, 2023). Teachers must take responsibility for the grading of students' essays with AI as much as they must when grading essays manually, and they must realize that the AI is as likely to err as they are. In the same vein, students must take responsibility for how they use AI. That is, they must understand that while the AI can help them learn, the learning itself (and thus all the effort that comes with it) cannot be delegated to an AI. The chapter concludes by outlining the skills or competencies teachers and students need to do so; both of which will be explored more deeply in later chapters (see Chap. 11).

- Key changes to science classrooms brought by AI that we discuss in this chapter involve
- The range of uses of AI for the individualization of teaching and learning in science classrooms for an increasingly diverse body of students, and
- the reshaped dynamics of science teaching and learning that emerge from the different uses of AI, from being a tool to serving as an extended agent or partner.

Based on an in-depth discussion of these changes, we derive the following key points about how these changes can be facilitated responsibly and ethically:

- New and refined instructional strategies need to be developed that leverage AI to address students' needs in science teaching and learning.
- Frameworks for a meaningful collaboration of humans and AI in the classroom need to be developed to ensure AI contributes to improving science teaching and learning.
- Specific policies and guidelines for an ethical and responsible use of AI in science classrooms are needed to ensure AI helps increase student agency over their learning.
- Teachers need professional learning opportunities to develop the competencies required to facilitate a responsible and ethical use of AI in science classrooms.

6.2 Changes Brought to Science Teaching and Learning by AI

One of the biggest challenges of current science teaching and learning is an increasingly diverse learner base. Students with diverse backgrounds come to the classroom with different interests and varying levels of competence. Some students are interested in learning about how science concepts can help them make sense of everyday phenomena like rainbows, while others may be interested in solving technical problems like poor Wi-Fi reception at home. Yet others may simply want to learn about science because they are genuinely interested in learning more about this domain. Similarly, some students may have developed high levels of competence due to their previous engagement in science, while others may struggle with even the simplest concepts like mass or volume. This increasing diversity presents a challenge to teachers that is increasingly difficult to overcome. An overwhelming amount of research shows that students' progression in developing competence in science as a function of instruction is messy (Alonzo, 2012; Duncan & Rivet, 2018; Gotwals & Songer, 2010). This is because instructions designed according to one hypothetical progression in making sense of a domain will not be able to meet the diverse needs of all students or even most students (Lehrer & Schauble, 2015). In fact, research also suggests that instruction meets the needs of a small fraction of students at best (e.g., Neumann et al., 2013). But how can teachers possibly address the diverse needs of students with classrooms of 20 or more students, teaching multiple classes per day, over 5 days a week?

The answer is AI. AI has been used for quite some time to analyze learners' behavior to make inferences about their past and predictions about their future learning. AI, more specifically predictive AI, is used, for example, to analyze learners' behavior in intelligent tutoring systems, exploratory learning environments, and, more recently, also in regular classrooms—using data produced through students' learning with digital tools. Intelligent tutoring **systems are, for example, used to support students in learning about scientific inquiry (e.g., Gobert,** 2023). As students work through examples, tutoring systems can analyze their challenges and provide them adaptively with instruction designed to help them overcome their struggles.

More recently, however, with the advent of generative AI, the possibilities that AI holds for the personalization of learning have grown exponentially. The main feature of generative AI is its capability to generate content. Generative AI can produce text, audio, images, and video—essentially any form of media that humans create. Like predictive AI, generative systems are built on machine learning algorithms, typically trained on vast datasets. Recent advances in neural network-based models have made it possible for computers to process and respond to natural language, opening new opportunities for interaction. Prominent examples include large language models such as ChatGPT, Bing Chat, Bard, and LLaMA, as well as text-to--image systems like Stable Diffusion, Midjourney, and DALL·E. In educational contexts, these systems hold significant promise. Their flexibility and responsiveness make them powerful tools for supporting learning in personalized and scalable ways (Wollny et al., 2021). Generative AI can, for example, help teachers create tasks or, more importantly, create different versions of the same tasks. That is, teachers can easily create a range of tasks supporting students to learn the same content, but, for example, in different contexts or ways. Similarly, AI can help teachers produce scaffolded learning aids to support students at various stages of the learning process.

The use of generative AI is, however, not limited to teachers. In fact, generative AI can help support students without the teacher in the middle. Students can, for example, ask the AI for help as they work through a task; they can also ask the AI for help with specific questions, different or additional visualizations of the problem, or even for a sample response. AI can help students understand a problem, researching further information on how things work, effectively allowing for more complex or even ill-defined problems. That is, issues are currently avoided in typical science instruction because of the amount of extra information (i.e., knowledge) needed to make sense of and solve them. Students can work with generative AI in a conversational style, asking the AI for questions or responding to questions the AI poses to them; almost like in a peer tutoring setting, where the AI can serve as the tutor as students learn, but also in the role of the tutee, asking questions, as students prepare for a summative, end of the year assessment. In this sense, generative AI enables a more self-directed, agentic style of learning—compared to regular classroom instruction. One of the bigger questions in this setting is that of the role of the teacher: as students are working more self-directedly with a conversational AI agent, what is the role of the teacher?

To sum up, AI is transforming not only how we learn, but also how we teach. As AI is increasingly utilized in science teaching and learning, one of the biggest questions to solve is about the role of the teacher. The collaboration between teachers and AI is commonly referred to as teacher-AI-co-orchestration. Teachers managing different aspects in the classroom environment to facilitate meaningful student learning are frequently referred to as classroom orchestration. Add AI to the mix, and it becomes teacher-AI-co-orchestration. That is, teacher-AI-co-orchestration is the management of both the teacher and the AI to more adequately support students in their learning. As working with an AI tutor provides every student with their own personal teacher, the role of the teacher is no longer to answer students' questions but rather to organize the classroom in a way that each student can learn most effectively.

In the remainder of this chapter, we discuss the potential of both predictive and generative AI for the individualization of students' learning at greater depth. By the example of different use scenarios, we highlight how AI can contribute to more individualized education. Eventually, we turn to how the use of AI to individualize students' learning leads to a reshaping of the dynamics of science teaching and learning. A transformation that does not remove teachers from the classroom or render them needless, but that requires a change in the teacher's role, with all the implications that role can have on teachers' self-esteem, need for professional development, and flexibility.

6.2.1 How AI Enables More Individualized Science Teaching and Learning

The idea of individualizing (science) teaching and learning using educational technologies is not new. In fact, the idea of providing students with the possibility to learn at their own pace and with the opportunity to access support in the form of additional instructional material or learning aids has already been coming up with personal computers (Neumann & Waight, 2020). With personal computers came computer-based training systems. Computer-based training (CBT) systems offered fixed sequences of tasks that students could work through at their own pace. The task sequences were based on respective learning models (e.g. Gagne's conditions of learning theory; Gagné, 1985). However, while CBT systems allowed students to work at their own pace, they offered little adaptivity.

6.2.1.1 Change 1: AI Promotes More Individualized Science Teaching and Learning

The next iteration of systems designed to give students greater agency over their own learning, while providing support based on their needs, was Intelligent Tutoring Systems (ITSs). Intelligent Tutoring Systems (ITS) are essentially a refined version of computer-based training systems (CBT). Compared to CBTs, in which tasks are presented in a predetermined order, ITSs are adaptive to learners' needs; that is, based on assessments of students' learning on one task, ITSs choose next steps or tasks to match learners' individual needs (VanLehn, 2006). They do so using two loops, an inner and an outer loop. The inner loop iterates over the steps students take in solving a multi-step problem (i.e., a task), which monitors students' learning while working on a task, potentially providing students with feedback. In parallel, the system updates the outer loop about students' progress in their learning. Based on this information, the outer loop iterates over tasks and chooses a new task appropriate to the students' learning.

ITSs monitor students' learning through a process called student modeling. Different approaches have been used for student modeling. Constraint-based modeling, for example, involves identifying the key features of the model solution for each group of tasks. Students' solutions to a task are then evaluated against the features of the model solution; students are given feedback about the extent to which their solution matches the required features, or which features it does not match. Following tasks are then chosen from the same group of tasks based on the features students' solutions failed to match. Once students reliably master the tasks from one group of tasks, the system moves to a new group of tasks. Once students consistently master one group of tasks, the system advances to the next group. In rule--based modeling, on the contrary, a set of production rules is defined. These rules include diagnostic ones that identify actions that lead to the correct solution and actions that lead to an incorrect solution. The system tracks students' actions, accepting actions that lead to the proper solution and providing feedback based on pedagogical or instructional rules, respectively. As students work through a group of tasks, the tutor keeps track of students' use of correct and incorrect actions. Once they begin to consistently use correct actions and avoid incorrect ones, the system moves to a new set of tasks. However, pre-defining the range of possible solutions or solution processes can be a problem in the context of more ill-defined tasks, such as complex problem-solving tasks or authentic scientific inquiry. Therefore, researchers have turned to using AI for more flexible student modeling.

AI-based student modeling in tutoring systems mainly draws on predictive methods. Predictive AI is a form of AI that leverages machine learning to predict future outcomes from previous behavior. Researchers have shown, for example, that predictive AI can be utilized to assess students' task performance (Zhai et al., 2020), to infer their knowledge (or skills) from their task performance (Gombert et al., 2023), or to predict students' competence development based on task performance (Wyrwich et al., 2025). Through the use of predictive AI, intelligent tutoring systems have become more flexible, allowing for more ill-defined tasks and a better, more flexible assessment of student learning (e.g., Rau et al., 2021; for an overview see Ouyan et al., 2023). In summary, AI has changed science education by enabling more individualized teaching and learning:

One example of an ITS that makes heavy use of AI is InqITS (Inquiry Intelligent Tutoring System, https://www.inqits.com, see also Gobert, 2023). InqITS was originally developed as a series of microworlds to support students in developing scientific inquiry skills, such as formulating hypotheses, designing and carrying out experiments, as well as interpreting data and drawing conclusions. InqITS now provides a series of labs designed to support students in developing an understanding of key science concepts such as energy (physical science), natural selection (life science), or seasons (earth science)—through *inquiry*. That is, InqITS engages students in the formulation of hypotheses, the design of experiments to test these hypotheses, the collection and analysis of data, and the communication of their findings (Fig. 6.1a). However, it is well established by research that inquiry learning is

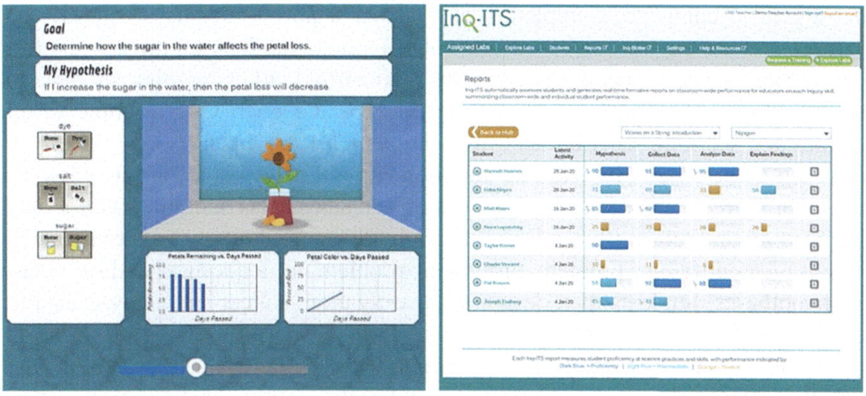

Fig. 6.1 Task from the inquiry-based intelligent tutoring system InqITS (**a**) and report about student competence in key inquiry activities (**b**) (https://www.inqits.com, see also Gobert, 2023)

effective when students are guided through the process by the teacher (for a discussion, see De Jong et al., 2023). InqITS uses AI to provide students with this guidance. More specifically, inqITS uses predictive AI (i.e., machine learning) to assess students' inquiry performance, monitor students' progress, and provide students with specific feedback, scaffolding interventions, and teachers with respective information. The AI has, for example, been trained on a range of student explanations based on the claim-evidence-reasoning (CER) framework and can subsequently assess the adequacy of students' claims, the extent to which students draw on evidence, and the logic of their reasoning (Li et al., 2017). This information is fed back to teachers through a specifically designed dashboard (Fig. 6.1b).

Although ITSs have been developed to support a more individualized teaching and learning experience for a range of areas across science domains, InqITS stands out for several reasons. For one, InqITS is an example of an ITS that supports, due to the use of (predictive) AI, more ill-defined tasks and hence a better, more flexible assessment of student learning. Compared to other ITSs, InqITS, with its range of labs, probably provides the broadest range of content. Most importantly, however, InqITS does not just focus on providing students with learning opportunities for select knowledge or skills but implements a domain-specific model of science teaching and learning: inquiry-based learning. Inquiry-based learning seeks to engage students in practices of science to acquire new knowledge and refine their skills to make sense of the world and solve problems. That is, InqITS provides a powerful tool to achieve individualization in authentic science teaching and learning.

6.2.1.2 Change 2: AI Supports More Authentic and Agentic Science Teaching and Learning

One way to provide students with an even more authentic learning experience is through exploratory learning environments (ELE). Such learning environments often present students with a game-like approach to learning. ELEs are open-ended digital environments that provide students with a simulated real-world environment and a goal to complete. Completion of the goal requires students to explore the environment and engage in a broad range of different tasks. Task solutions in exploratory learning environments can typically be manifold, allowing other students to pursue different approaches to meeting the expected goal. One example of an exploratory learning environment that implements an open-world game-type approach is the River City learning environment. River City is a learning environment that shares many of the features of a typical science curriculum. It is, for one, designed around (content) standards; it also implements an inquiry approach, with a focus on identifying a problem, turning the problem into a question, and formulating hypotheses. Students then design an investigation, collect and analyze data to test their hypothesis. They do so in a virtual world, however. In this world, students take the identity of an avatar, a virtual persona. Through this avatar, they can interact with the virtual world (e.g., communicate with other avatars, including computer-based ones). In River City, students' task is to figure out why the residents of the town have fallen ill. To do so, they can explore the city and pursue different strands of inquiry to find out what is going on (Fig. 6.2).

AI techniques have been used to track students' progress in ELEs, much like they have been used in the context of student modeling in ITS. Assessing students' progression in ELEs is significantly more complex. That is where the AI comes in.

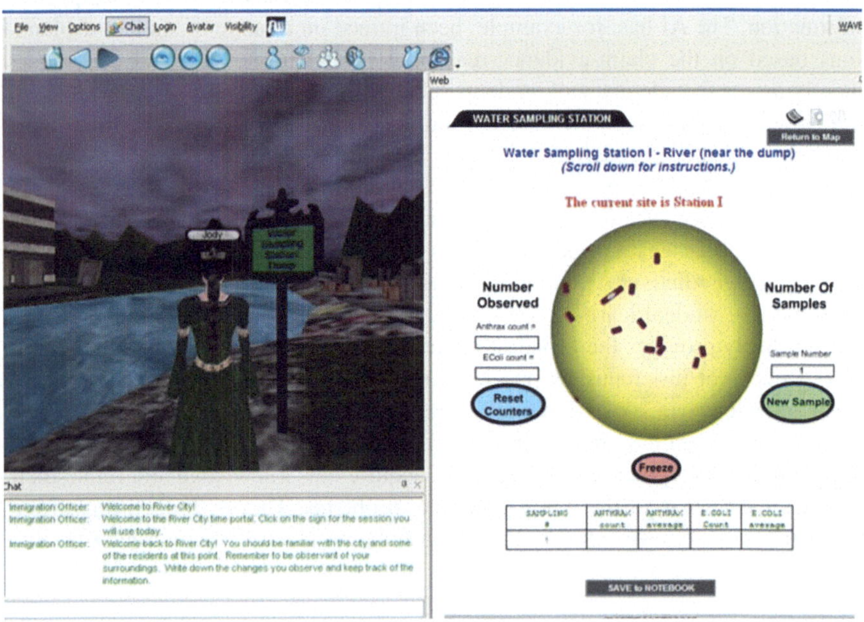

Fig. 6.2 Screenshot from the River City exploratory learning environment (taken from Clarke & Dede, 2009). Reprinted from Clarke, J., & Dede, C. (2009). Design for Scalability: A Case Study of the River City Curriculum. Journal of Science Education and Technology, 18, 353–365. https://doi.org/10.1007/s10956-008-9147-2 © Springer Nature 2009. Reprinted by permission of Springer Nature

Researchers have, for example, used predictive AI techniques to assess students' knowledge (Lester et al., 2013), skills (more specifically, twenty-first century skills, Shute, 2011), and even self-regulation skills (Sabourin et al., 2013). Amershi and Conati (2011) successfully combined different techniques to identify behaviors in the environment that are effective and ineffective with respect to the development of competence. That is, like in ITSs, AI allows tracking student learning or aspects of student learning in ELEs. The AI can monitor students' behavior, predict where students are in terms of their knowledge, and provide them with scaffolds through, for example, computer-based avatars or through other means in their interaction with the environment. In doing so, ELEs give a more natural, interactive, and non-pre-determined learning experience. In summary, although ELEs resemble many features of ITSs (e.g., InqITS), they enable a more authentic and agentic learning.

6.2.1.3 Change 3: AI Enables Real-Time Feedback and Instructional Decisions in Regular Science Teaching and Learning

Kubsch et al. (2022, 2024) take a slightly different approach to AI-supported science teaching and learning. The authors work off a regular classroom-based curriculum, except that this curriculum comes with a digital student workbook (Fig. 6.3). This workbook has students record the artifacts they typically produce in response to the tasks assigned to them. That is, students simply learn as they always do, except that instead of recording their responses on a worksheet, they do so in the digital workbook. This allows for analyzing the data using AI. The authors have demonstrated that predictive AI can be used to identify the learning performances,

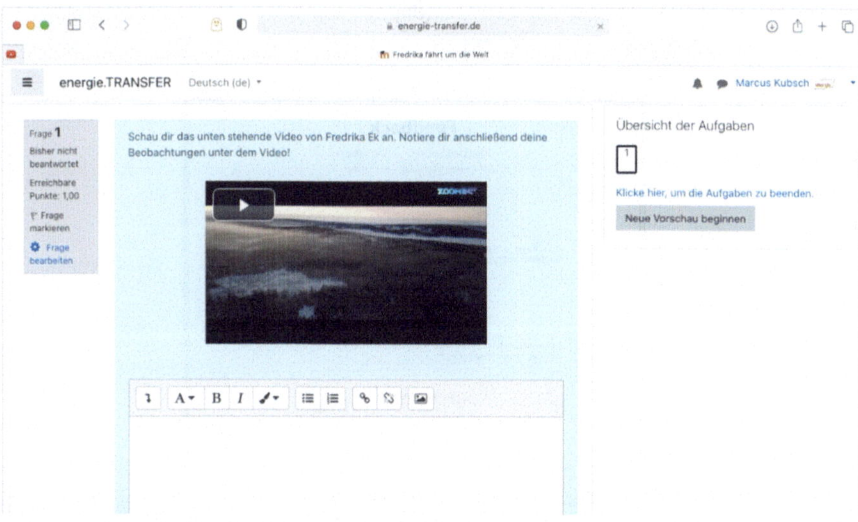

Fig. 6.3 Digital workbook for a regular curriculum unit on the concept of energy. The task asks students to watch a video and record their observations

as well as the knowledge elements students have shown in their responses to open-response questions (Gombert et al., 2023). In fact, the authors reach a remarkably high accuracy in determining students' knowledge elements. This can, for example, serve as a basis for identifying students' knowledge networks and their key characteristics, such as coherence (Kubsch et al., 2020). This information can also be utilized to track students' learning over time (Wyrwich et al., accepted pending minor revisions). Drawing on the ability to score students' learning performances and identifying the knowledge students have been showing, the authors created a teacher dashboard that would inform the teacher about individual students' learning performances per task, as well as which of the expected knowledge elements (based on the ideal solution) students demonstrated and which not. The dashboard allowed the teachers to send students feedback (Fig. 6.4; Karademir, Di Mitri, et al., 2024). The

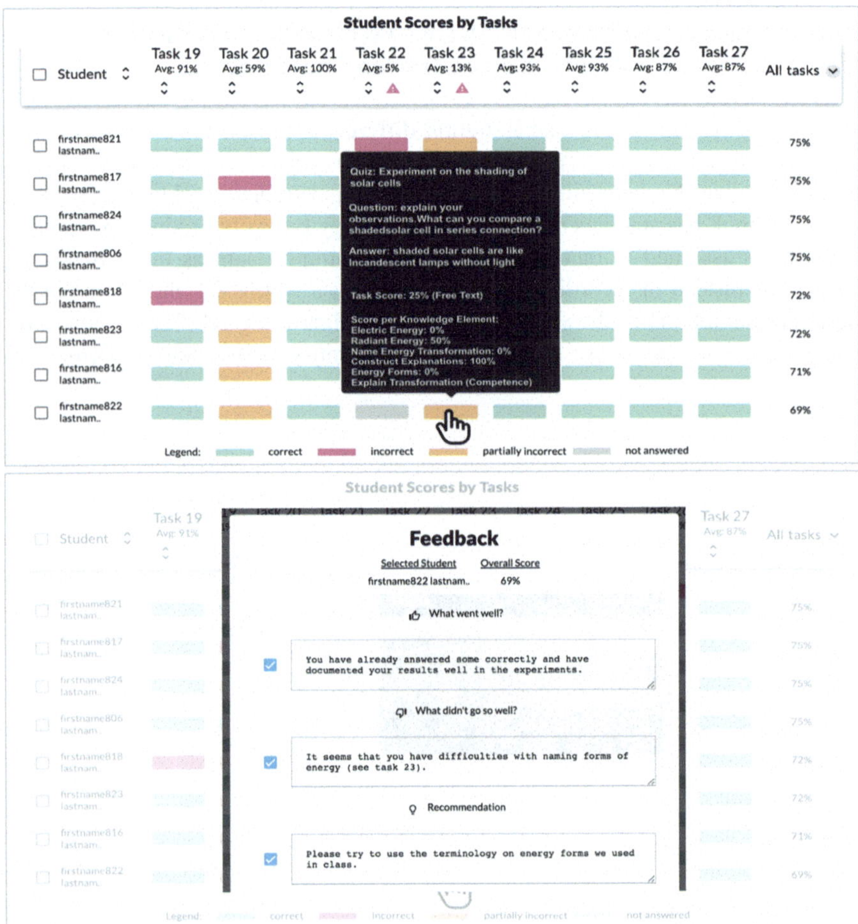

Fig. 6.4 Dashboard for teachers that shows students' performance on tasks, including the knowledge elements students have or have not demonstrated (**a**) and the opportunity for teachers to provide feedback (**b**)

findings from an accompanying study showed that students who learned with the curriculum and digital workbook, where teachers had the opportunity to give feedback, outperformed their counterparts who learned only with the curriculum and digital workbook (Karademir, Borgards, et al., 2024). This work shows that the potential of AI is not limited to the use of specific AI-based tools such as ITSs or ELEs as part of the regular curriculum. Still, it shows that AI can indeed be used to track students learning throughout regular science curriculum. However, currently the infrastructure for this kind of analysis is missing in most schools or classrooms respectively.

One challenge in the automated analysis of learning processes in the context of regular science classrooms is that monitoring the learning progress of each of their students, identifying potential issues they are struggling with, and providing them with respective feedback puts a substantial load on teachers. Research indicates, for example, that while teachers utilize information supplied by dashboards, they can also feel overwhelmed by the volume of data (Pozdniakov et al., 2023). But AI can help here as well.

Generative AI can help synthesize insights about students' learning progress, presenting them in a clear and comprehensible manner for teachers. For example, generative AI could provide information on prompts like: "What learning difficulties do students face who fail to meet the following learning objective: …?". Or, more specifically: "What misconceptions could students with the following responses hold?". Generative AI-based systems can also, if provided with the information on how students were expected to perform, how they have performed, and, optionally, how the predictive AI assesses students' performance, automatically generate feedback. For instance, Wedenig et al. (2023) found that ChatGPT can produce high-quality feedback for classic multiple-choice tests designed to assess medical students' knowledge. This feedback was not just about the (in)correctness of the response, but could explain why a response was correct or incorrect. Experts rated the quality of the explanations highly on the dimensions of content accuracy, linguistic clarity, and additional information provided. Similarly, Dai et al. (2023) noted that ChatGPT can generate more readable feedback compared to human tutors in data science projects, as human tutors often tend to provide short feedback only due to time constraints, while ChatGPT provided detailed, fluent feedback. At last, ChatGPT can create meaningful feedback on students' argumentative writing in science contexts (Jansen et al., 2024). Altogether, generative AI's potential in feedback generation lies in its ability to produce scalable, specific, and individualized feedback for learners, including open-ended formats like free-text responses.

Beyond synthesizing information, generative AI can also suggest materials or next steps in the learning process, through prompts like: "How can we address the following learning difficulty effectively: …?" or "Design a task for ….". Küchemann et al. (2023) demonstrated that ChatGPT can create tasks comparable to those in textbooks. ChatGPT can also produce learning materials, provide scaffolding, and evaluate students' solutions (Zhai, 2023b). Research by Zheng et al. (2023) suggests that AI can also generate entire learning plans. The use of generative AI introduces a new and underexplored field: AI-supported instruction. In this scenario, AI not only analyzes students' learning but also assists in planning and implementing lessons. AI may autonomously generate formative assessments, provide elaborative feedback, and offer conceptual scaffolds tailored to student needs—activities previously undertaken solely by educators (Lee & Zhai, 2024). Combining these capabilities, generative AI can create individualized learning plans, materials, scaffolding, and assessments, significantly relieving and supporting teachers. That is, generative AI can not only help teachers to monitor student learning or to provide students with feedback, but also recommend instructional strategies, and even generate instruction itself:

In summary, AI holds tremendous potential for individualizing students' learning. Predictive AI has enabled tools such as ITSs to assess students' learning, track their progression, and identify struggles hindering their progression. Researchers have also applied similar techniques to more open learning environments, including a regular curriculum. AI can transform science teaching and learning by summarizing and synthesizing information on students' learning, as well as recommending specific instructional strategies or assisting with their implementation. Much of this functionality is either already powered by LLMs or will experience significant improvements as LLMs are incorporated to perform this functionality. In doing so, AI has the potential to fundamentally improve the learning experience for many students.

Conclusion
AI changes science teaching and learning procedures through introducing advanced individualization tools—including intelligent tutors, open learning environments, and even regular classroom instruction. Powered by LLMs, these tools can further enhance individualized learning experiences by providing immediate feedback for students, as well as synthesizing critical information on student learning for the teacher, recommending adequate instructional strategies, and assisting with their implementation. This addresses a crucial challenge in science education: meeting the diverse needs of students from various backgrounds, something that traditional instructional methods struggle to achieve effectively.

6.2.2 How Human-AI Interaction Reshapes the Dynamics in the Science Classroom

AI has introduced a paradigm shift in the relational structures of science classrooms, which will fundamentally alter the roles and responsibilities of teachers, students, and technological agents. No longer confined to acting as ancillary supports, AI systems are becoming semi-autonomous actors in learning environments. They generate feedback, support instructional delivery, and shape interactions, thereby prompting reconsideration of human and non-human agency in classroom orchestration. This transformation is not only technological but socio-pedagogical, bringing changes to authority structures, epistemic ownership, and instructional routines (Zhai & Krajcik, 2024).

The teacher's role is changing as they are working with AI-powered systems (Shi & Choi, 2024). In a science classroom, in which instructional activities are shared with, co-created with, or created by AI, the role of the teacher is evolving from the primary transmitter of knowledge to a facilitator of hybrid learning environments, or extending from the facilitator of teaching and learning to an organizer and supervisor of multiple agents. Teachers must process information provided through learning analytics dashboards and adjust instructional strategies in real time (Haque Munim & Kim, 2023). In case of generative AI autonomously assessing students' learning, providing individualized feedback, and offering conceptual scaffolds tailored to student needs (Lee & Zhai, 2024), teachers must critically and continuously evaluate the AI for pedagogical appropriateness of its actions, evolving from an observer, adopter, collaborator toward an inventor (Zhai, 2025). As a result, teachers must cultivate new ways of organizing science teaching and learning.

For students, AI enables the creation of individualized learning trajectories and promotes greater autonomy in navigating content, regardless of where and when. However, this also introduces new challenges. As AI begins to suggest next steps in inquiry or recommend scientific resources, students must assess these prompts rather than follow them uncritically; they must take control of their own learning much more than in traditional classrooms. Students must learn to interact productively with algorithmic interfaces, interpret the rationale behind AI-generated feedback, and balance multiple sources of instructional guidance—including AI agents and human teachers. This complex teacher-student-AI triad requires rethinking student agency, emphasizing not only metacognitive strategies and self-regulation but also how students must develop digital and data literacy.

6.2.2.1 Change 4: AI Transforms the Dynamics in the Science Classroom

The integration of AI fundamentally disrupts the historically dyadic model of science instruction by embedding machine-driven logic into daily classroom interactions. It reshapes discourse patterns, authority flows, and participation norms, leading to increasingly distributed forms of agency among teachers, students, and AI systems. While these changes can foster more inclusive and differentiated learning, they require intentional design to ensure AI systems augment rather than constrain pedagogical goals.

The CLAIS (Collaborative Learning with AI Speakers) system developed by Lee et al. (2025) provides compelling evidence showing how AI reshapes classroom dynamics and how this requires intentional design. The CLAIS system was designed to integrate generative AI speakers as active participants in collaborative learning environments, positioning them not merely as tools but as epistemic peers in a group learning process. In the study, CLAIS was implemented in a science education course with pre-service elementary teachers, where each group of students included an AI speaker that engaged in the Jigsaw collaborative learning model (Fig. 6.5). Students interacted with AI speakers to retrieve specific content knowledge (e.g., points from Piaget's or Vygotsky's learning theories), asked questions, and used the AI-generated information to teach peers and solve problems collectively. The AI speaker provided structured responses derived from course materials, demonstrating the potential of AI to support distributed cognition and scaffold student understanding.

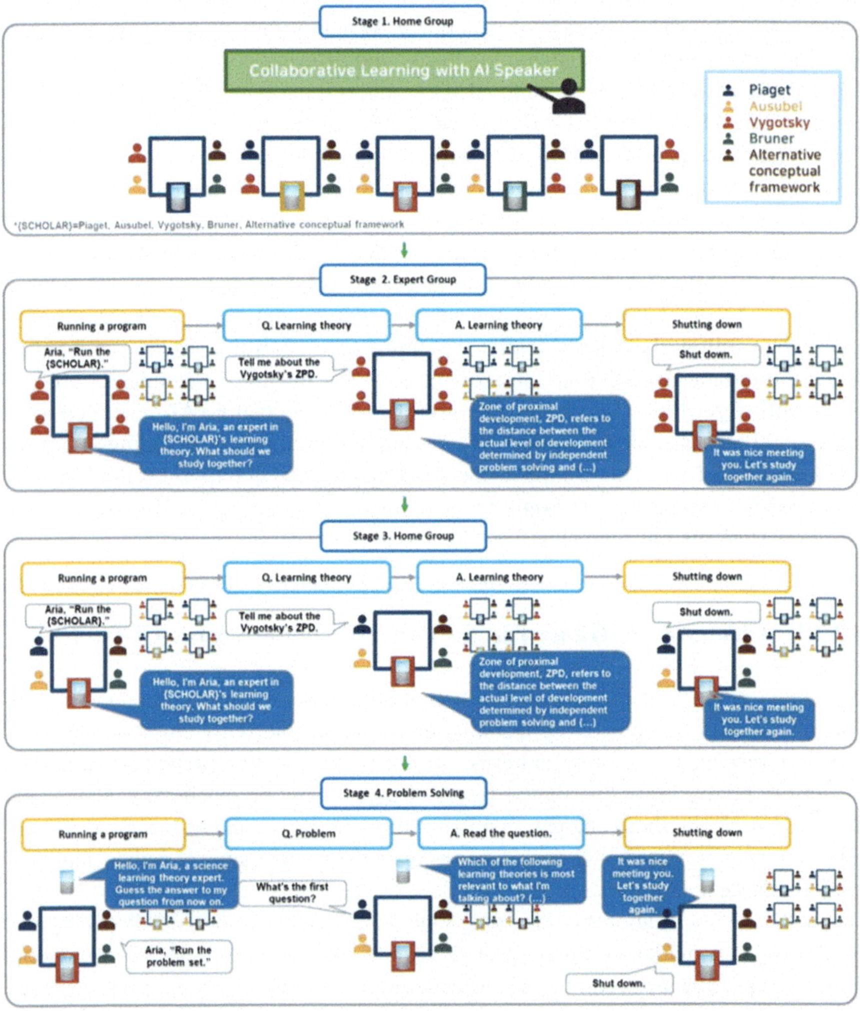

Fig. 6.5 Flow chart of the CLAIS-Jigsaw system (Lee et al., 2025). Reprinted from Lee, Mun, Shin, & Zhai, (2025). Collaborative Learning with Artificial Intelligence Speakers: Pre-service Elementary Science Teachers' Responses to the Prototype. Science & Education, 34, 847–875. https://doi.org/10.1007/s11191-024-00500-4 © Springer Nature 2025. Reprinted by permission of Springer Nature

Crucially, the introduction of AI into the learning group reshaped classroom dynamics and interactions. Students reported that the instructor's role shifted from content deliverer to facilitator of the AI-integrated learning system. Peer-to-peer interactions also changed, as students had to learn how to communicate effectively with and through the AI speaker. Survey data revealed increased perceptions of learning efficacy, altered roles of teachers and students, and enhanced engagement. For example, students noted that "the instructor does not deliver all the knowledge and students deliver with each other what they have learned through AI," indicating a redistribution of instructional authority and a more decentralized learning process. Thus, this case substantiates that AI, when embedded meaningfully, can transform classroom dynamics by altering epistemic roles, facilitating new forms of interaction, and prompting educators and learners to reconceptualize their positions in the learning process:

6.2.2.2 Change 5: AI Alters Epistemic Autonomy in Science Learning

AI's increasing role in science classrooms carries profound implications for epistemic autonomy—the capacity of learners to direct their own knowledge construction and critically engage with information sources. Epistemic autonomy is central to the goals of science education, as it empowers students to ask meaningful questions, develop explanations grounded in evidence, and engage in reflective judgment. The shift toward AI-supported learning environments introduces algorithmic mediation into essential epistemic practices, including hypothesis generation, evidence evaluation, argument construction, and explanatory reasoning. These applications extend beyond generating materials to serving as interactive partners for students, answering questions, or acting as tutors. In specific scenarios, generative AI could produce initial texts (e.g., scientific explanations) or arguments for students to review and refine (see Seßler et al., 2023). According to Zhai (2023a), this approach shifts the locus of control to students, making the learning process more student-centered and expanding opportunities for autonomous learning.

The study by Dai et al. (2025) provides robust empirical evidence on how AI alters epistemic autonomy in science teaching and learning by examining how generative AI-powered feedback systems influence students' learning behaviors, particularly their autonomy, in high school physics classrooms. In two randomized controlled trials, the researchers implemented AI-powered systems that provided either (1) compulsory personalized recommendations with heuristic solution hints, or (2) autonomous on-demand help (Fig. 6.6). These AI interventions were explicitly designed to scaffold students' learning by offering tailored feedback based on students' homework errors and self-reported needs. The systems included features such as conceptual explanations, solution strategies, and additional practice problems, generated through a retrieval-augmented AI pipeline. Dai et al. (2025) also revealed that AI systems significantly influenced students' self-regulated learning—a key component of epistemic autonomy. Notably, while low-achieving students experienced improved academic performance under compulsory AI-guided

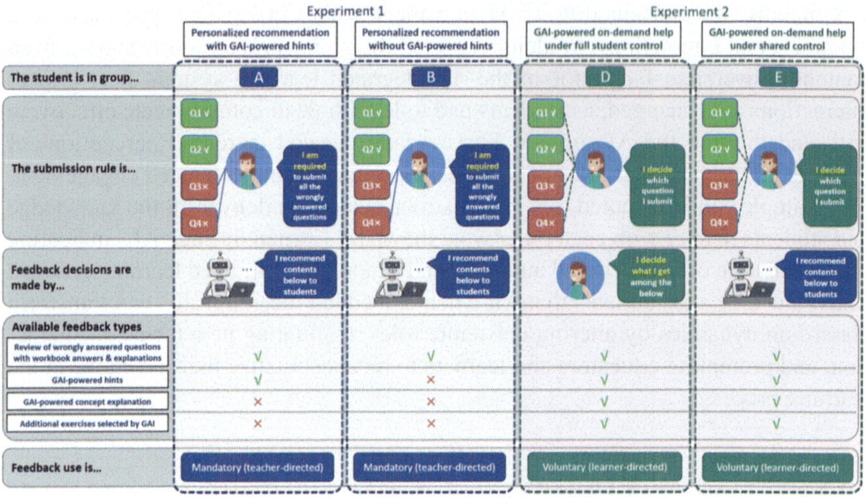

Fig. 6.6 Interventions in the two experimental designs (X. Dai et al., 2025)

recommendations, their autonomy remained unchanged. However, high-achieving students showed a significant decline in self-regulated learning when receiving the same AI support ($d = -0.477$, $p < 0.05$). Similarly, under autonomous on-demand conditions, high-achieving students benefited academically without a loss of autonomy. In contrast, lower-achieving students exhibited a notable decline in autonomy, particularly in the technical-psychological dimension ($d = -0.549$, $p < 0.05$).

The study indicates that AI can both support and undermine students' epistemic autonomy, depending on how it is used and who is using it. The AI systems shifted the locus of control in knowledge acquisition from the learner to the algorithm, particularly for students who lacked strong self-regulatory skills. This shift represents a profound change in epistemic dynamics: rather than constructing knowledge through independent inquiry, students may increasingly rely on AI to mediate their engagement with content. Thus, this case powerfully demonstrates that AI has the potential to reconfigure the conditions under which students exercise epistemic agency. While AI can scaffold learning, it may also inadvertently erode learners' capacity to take ownership of their knowledge-building processes—highlighting the need for careful pedagogical design when integrating AI into science classrooms.

More significantly, AI reconfigures traditional epistemic hierarchies by presenting itself as a non-human yet authoritative knowledge agent. Students may assume that AI outputs are inherently trustworthy, particularly when systems express confidence or lack transparency in their reasoning processes. This dynamic can weaken students' development of epistemic vigilance—the skill of evaluating the credibility, origin, and limitations of information across contexts (Markauskaite & Goodyear, 2017). Moreover, as students become habituated to AI as the primary guide for inquiry and problem-solving, they may disengage from the productive uncertainties and iterative reasoning processes that are hallmarks of authentic scientific exploration (Lee et al., 2025). This shift could inadvertently stifle students' curiosity, reduce their perseverance in solving complex problems, and limit their understanding of the tentative and evolving nature of scientific knowledge.

Nevertheless, when purposefully designed, AI can catalyze enhancing epistemic autonomy rather than diminishing it. Systems that embody principles of explainable AI (XAI; Wu et al., 2025), such as those that make their reasoning pathways visible, highlight uncertainty, or cite underlying data sources, can promote learners' metacognitive reflection and stimulate epistemic curiosity. These design features provide opportunities for students to question, challenge, and refine AI-generated content, thus transforming the learning process from one of consumption to one of critical engagement. Additionally, AI platforms that encourage learners to generate, revise, and critique alternative hypotheses or engage in debate around different interpretations of evidence contribute to a more dialogic and participatory learning culture.

In summary, generative AI holds transformative potential for science education. Beyond AI-driven learning analytics, it can support learning, automate material creation, and act as a personalized tutor. However, further research is essential to harness this potential effectively, focusing on the effectiveness of AI-generated feedback compared to traditional methods. Future studies must also explore the interplay between AI and teacher roles, effective teacher-AI collaboration, and the implications for teacher training and professional development. Moreover, the ethical, practical, and methodological challenges of integrating generative AI into education require systematic investigation to maximize its benefits and mitigate potential risks.

Ultimately, the transformation of epistemic autonomy in science education by AI technologies presents both opportunities and challenges. On one hand, AI can democratize access to high-quality learning resources and stimulate personalized inquiry; on the other hand, it can inadvertently undermine the development of students' independent thinking and critical reasoning if treated as an unquestioned authority. Educators, developers, and policymakers must therefore collaborate to ensure that AI is positioned as a co-participant in learning—one that complements rather than replaces human judgment—and that learners are equipped to exercise epistemic agency in increasingly complex digital learning landscapes. While AI systems can support these aims by scaffolding learning through explanations, curated resources, and modeled reasoning strategies, they also introduce the risk of displacing opportunities for independent sense-making, especially when implemented without pedagogical safeguards (Zawacki-Richter et al., 2019).

Tools such as automated concept map generators or AI-driven virtual labs can increase students' exposure to content and enable rapid iteration in experimental design and data interpretation. These tools can offer significant efficiencies, especially in resource-limited settings. However, these same conveniences may inadvertently encourage superficial engagement with scientific practices or foster an over-reliance on AI-generated interpretations if learners are not supported in developing the skills to critically examine algorithmic outputs and their underlying logic (Wu & Tsai, 2022). The danger lies in turning active learners into passive consumers of AI-curated knowledge.

Conclusion
AI reshapes classroom dynamics by acting as an interactive agent engaging both students and teachers. For students, AI can function as a supportive learning companion, providing guidance and tailored instruction. For teachers, AI serves as a valuable assistant, automating routine tasks and enabling educators to devote more time and attention to complex instructional activities and deeper student interactions.

6.3 Ensuring a Responsible and Ethical Use of AI in Science Teaching and Learning

In the preceding sections, we have outlined the kind of changes that AI brings to science classrooms, including a stronger individualization through AI-based intelligent tutors, authentic learning experiences, and automated feedback in regular classrooms. These changes hold the potential to address the increasing diversity amongst students. We have also acknowledged that the use of AI changes the dynamics in the classroom, effectively creating a classroom situation in which it is no longer the task of the teacher to orchestrate the learning for a larger number of students. Instead, as the AI will serve as an epistemic agent in its own right, both teachers and students need to work together to orchestrate teaching and learning with AI. That is, the classroom dynamics have changed from one teacher orchestrating a group of students to three agents co-orchestrating their interactions in the classroom. In this section, we discuss the necessities for the changes brought by AI to take place responsibly and ethically, drawing on the framework outlined in Chap. 3 by Krippen, Zhai, and Lee.

The first change we have discussed is that the use of AI enables a stronger individualization of science teaching and learning. Predictive AI currently only helps track students' learning and predict the (longer-term) productivity of students' learning trajectories. Generative AI can also support students in their learning, but has other limitations (Kasneci et al., 2023). More specifically, unless adequately prompted, generative AI may fail to support students in their learning (Li et al., 2025), due to a lack of pedagogical knowledge (Díaz & Nussbaum, 2024). AI can hence only play out its potential if integrated with the curriculum. That is, it is necessary to understand how AI can be integrated in the curriculum such that it not only supports an individualized learning of specific knowledge or skills, or provides students with responses to their questions but also systematically supports students in developing the competence they are envisioned to develop (e.g., NRC, 2012). ITSs, for example, can complement instruction by providing students with opportunities to further engage with specific content covered in the regular curriculum. An ITS helping students to learn Newtonian mechanics, for example, can aid students who have struggled to grasp some of the concepts through the regular curriculum if students are offered regular access to it (e.g., instead of homework). However, more research is needed on how ITSs can be integrated with regular science teaching and learning in such meaningful ways (e.g., Feng et al., 2021). This applies even more to the use of generative AI. Although the body of research on the potential of LLMs for science teaching and learning is rapidly increasing, much of this research is focused on the capabilities of the AI. For example, if LLMs can solve physics problems, if they can support students in learning problem-solving, or if they can support the creation of problems (Liang et al., 2023). There is little to no research so far on *how* LLMs must be used to support students, for example, in the individualization of problem-solving in science classrooms. While efforts have been made to integrate AI with regular science teaching and learning units (Kubsch et al., 2022, 2024, 2025), research on this integration is still in its infancy, representing more of a proof of concept than providing actual means for individualizing learning in the classroom. In summary, while AI can help individualize learning, there is a lack of instructional strategies on how to effectively integrate them with regular science teaching and learning.

Recommendation: Develop Instructional Strategies for Meaningful Integration of AI
Develop instructional strategies and/or revise existing instructional strategies for teachers to leverage AI-driven individualization to provide more flexible instruction and to better address diverse student needs.

The second change we discussed is that of a reshaped classroom dynamic with fundamental changes in authority structures, epistemic ownership, and instructional routines. As AI is increasingly used to perform tasks that were traditionally the responsibility of the teacher, the teachers' role changes from being a facilitator to that of an organizer or supervisor. Teachers must cultivate new forms of organizing science teaching and learning. Also, as students engage with an AI for a more individualized, autonomous learning experience, their role changes as well. Students must take greater responsibility for organizing their learning. So altogether substantial parts of the authority over the teaching and learning process that originally lay with teachers are, as AI is used in the classroom, shifted to the AI and the students as well.

Research suggests that teachers, AI, and students alike have complementary strengths and weaknesses. Teachers have a more comprehensive understanding of their students, including not only their cognitive abilities or their socio-emotional needs but also their personal circumstances. Teachers also possess rich heuristics to draw on, making them better suited to providing certain types of support—such as socio-emotional guidance (Holstein et al., 2019). AI can, as well as students, recognize aspects of student learning that are not immediately noticeable to teachers (e.g., student engagement, Shreeve et al., 2025). However, teachers, AI, and students have limitations when it comes to assessing and supporting, or organizing, aspects of student learning. Teachers, for instance, may be influenced in their assessment by irrelevant factors—such as students' background (Bonefeld & Dickhäuser, 2018). AI, on the other hand, may not have access to the information needed to accurately assess student learning or, more broadly, perform less well for certain groups or in specific contexts (Ocumpaugh et al., 2014). Students, more specifically low-achieving students, are well known to struggle to accurately assess their own achievements (Kruger & Dunning, 1999). In consequence, we suggest that the responsible and ethical use of AI in science teaching and learning is one that understands teachers, AI, and students as collaborators, capitalizing on each other's strengths while offsetting one another's weaknesses:

Recommendation: Establishing Teacher-Student-AI Collaboration Frameworks

Develop Teacher-Student-AI collaboration frameworks that clearly define the roles and responsibilities of teachers, learners, and AI to optimize collaboration, ensuring AI supplements rather than replaces teacher judgment and student learning, and that the AI can learn from teachers and students.

Despite the potential of teachers and AI tools having complemented strengths and weaknesses, there still are challenges arising from changes brought by AI that teachers or students cannot mitigate. These challenges do not emerge from a lack of frameworks for a co-orchestration of teachers, students and AI tools, or a lack of competencies of individual agents (i.e. teacher, students or AI), but are inherent to the use of AI.

One challenge of AI is, for example, training bias; that is, a bias in the data that has been used to train the AI (e.g., Grimm et al., under review; Guo & Caliskan, 2021). Training data bias stems from the underrepresentation of certain groups (e.g., students of color) in the training data. Since AI is trained better for groups well represented by the training data, it will achieve a greater accuracy for those groups, while the accuracy for underrepresented groups may be lower (Guo et al., 2025). Feedback to those students, or any instructional decision that affects these students, may not meet their needs. Inaccurate decisions made by an AI based on sensitive information such as ethnicity may also be referred to as unfair. One initial hope was that AI would help mitigate human bias in decision-making and hence make it fairer (Dressel & Farid, 2018). However, as it turns out, training data bias is often difficult to even recognize by humans and even more difficult to mitigate (Li et al., 2023).

Hence, the question is how to ensure fairness in AI-based educational interventions (see the discussion of Fairness in Chap. 3 by Crippen, Zhai, and Lee).

A second challenge is the non-transparent nature of many AI models. Especially neural network-based AI tools, such as large-language models, lack transparency to the extent that it is difficult to understand how the AI came to its answer or decision, respectively. In the context of education, intransparency is a challenge as the lack of transparency of how the decision was made also means there is a lack of justification for why the decision was made, or the answer was produced. Researchers have argued that this lack of transparency can lead to a lack of trust in results produced by the AI (Mienye & Swart, 2025). This could apply, for example, to students' trust in the assessment or feedback provided by the AI. Interestingly, the opposite effect may be reasonably assumed as well. A lack of transparency and justification may lead students to not reflect on the answer or decision, respectively, and instead accept it as a fact, especially as the AI does what it is expected to do. It gives an answer, or feedback, or proposes a decision. Where a Google search may not yield a result, the (generative) AI will always provide a result, misleading both students and teachers to think it (the AI) knows why (see O'Neil, 2017 for a broader discussion). In summary, although research on explainable AI has made substantial progress (Bilal et al., 2025), the question still is how AI can be transparently used in science education (see the discussion of Transparency in Chap. 3 by Crippen, Zhai, and Lee).

A third challenge is related to inclusivity. As AI tools increasingly find their way into classrooms, the question is who will benefit from the AI. Especially when it comes to working with the AI, the question that emerges is, will the AI equally support students in their learning? Research has shown that students learning individually with intelligent tutoring systems will yield (more or less) similar learning gains but will not close the gap. If, however, teachers are provided with analytics on their students' learning (i.e., analytics), students most in need will receive more attention from the teacher and show higher learning gains (Holstein et al., 2018). Interestingly, this did not happen at the expense of high-achieving students, who showed the same learning gains. This suggests that AI has the potential to create a more inclusive science education. However, the International Computer and Information Literacy Study (ICILS) also showed that students from advantaged backgrounds benefit more from the use of digital technologies, as they typically have access to such technologies and receive support in the use of such technologies at home (Fraillon, 2025). This suggests that although AI can increase inclusivity, it needs moderation by the teacher to unfold its full potential (see the discussion of Human oversight and Autonomy in Chap. 3 by Crippen, Zhai, and Lee).

In summary, the use of AI in science classrooms is not straightforward. AI can help individualize learning. It can, however, also simply engage students individually instead of supporting them according to their needs. Even worse, while AI can provide students with just-in-time feedback, it can also mislead students to continuously ask for feedback instead of engaging in learning in the first place. At last, AI can help address a more diverse group of students, again according to their needs, but it can also shield the teacher from the students' learning process. To avoid these issues while making the best use of AI in the classroom, there must be guidelines and policies for teachers (and students) to ensure an ethical and responsible use of AI in science teaching and learning:

Recommendation: Develop Policies and Guidelines for the Use of AI
Develop clear and actionable policies and guidelines to ensure responsible and ethical use of AI in science teaching and learning, emphasizing fairness, transparency, accountability, privacy, human oversight, benevolence, robustness, and respect in AI-based educational interventions (see the Framework for an ethical and responsible use of AI in Chap. 3 by Crippen, Zhai, and Lee).

The challenges outlined above highlight the need for teachers and students to develop specific competencies that will ideally enable them to make the best use of AI in their science teaching and learning. Consider, for example, the challenge that (generative) AI will always give one an answer. A teacher's online search about an experiment to introduce eighth graders to the phenomenon of refraction or a search by a student to find an explanation for the phenomenon, will yield different results. Commonly, these results will not provide a direct answer to the question. Experiments may not match the experimentation materials available at school. Explanations may not exactly match the specific phenomenon in question. Teachers as well as learners will need to make sense of the search results, in that they will need to check the information against the backdrop of their own knowledge. *They will learn.* Generative AI will, unless prompted correctly, simply provide an (one!) answer. An answer that students (as well as teachers) may tend to accept as is and copy, for the sake of efficiency, effectively bypassing the learning process. *They will not learn.* In addition, the answer, as well as any single result from a web search or even information from a university textbook, can be wrong or erroneous, respectively. If there are multiple answers, there is a chance that, to find the best matching answer, students and teachers will perform a cross-check. In case of a single answer from a generative AI, they may simply take it as true. These examples underscore the necessity of teachers and students to develop competence in using AI in their teaching and learning, respectively. This includes, for example, an understanding of how decisions were made by (predictive) AI (e.g., Jang et al., 2022), how to prompt (generative) AI to optimally support students' learning (e.g., Pankiewicz & Baker, 2023), and recognize the challenges that come with the use of AI in the classroom.

Teachers as well as students must be prepared not only to use AI effectively but to understand affordances, recognize constraints, and evaluate the epistemic legitimacy of the AI or the information generated by the AI. This includes instruction on how algorithms are trained, what data they rely on, how biases can be encoded, and how interpretability affects trustworthiness. By fostering this form of AI literacy, educators can help students develop into thoughtful, reflective, and autonomous thinkers who are capable of navigating an increasingly algorithmic and data--saturated world. Furthermore, the widespread deployment of AI in classrooms necessitates, in addition to policies and guidelines, institutional support structures and sustained professional development for educators (Nyaaba & Zhai, 2024). Providing these supports is essential to harnessing AI's potential while maintaining an ethical and responsible use:

Recommendation: Provide Teacher Professional Development
Develop and implement comprehensive training programs to build teachers' competencies in using AI, ensuring they can integrate AI effectively, ethically, and responsibly into science teaching and learning.

6.4 Summary and Conclusions

There is no question that AI is changing the workplace in science and beyond. This development does not stop in the science classroom. The idea of individualizing learning through digital technologies is not new, though. It has, however, experienced a renaissance through the developments in the field of AI. Increased computing power, together with technological innovations, has enabled powerful predictive and, probably more importantly, generative AI models. Intelligent tutoring systems as a means of individualization, historically limited to areas (very) well defined, with the help of AI, have been developed for ill-defined domains such as scientific inquiry (e.g., Gobert, 2023), or more interactive—open—learning environments; up to the point of recent developments to use predictive AI to analyze students' learning and predict their progression in developing competence in science (e.g., Kubsch et al., 2025; Wyrwich et al., 2025). These tools alone do not support students' learning, especially not in the long run, without being embedded in quality science teaching and learning materials (i.e., curriculum). ITSs, even if they address less well-defined domains than they traditionally did, need to be orchestrated together with other forms of learning. The same applies to more open learning environments such as Open World games. Even a regular curriculum that embeds AI does not guarantee individualization. Sure, the system can provide students and teachers with feedback, but at a higher level of abstraction, teachers still need to define the course of instruction. Tracking students' learning using AI, for example, teachers must still process the information, synthesize it to identify students at risk, and then plan and implement additional learning opportunities for these students. The potential of generative AI is also largely untapped. Much of the current research focuses on exploring the potential itself instead of exploring how the potential can be tapped for science teaching and learning, and to what effect (cf., Meyer et al., 2024). This applies in particular to the challenges that come with the use of AI. To ensure a meaningful use of AI, frameworks are needed that orchestrate the collaboration of teachers, students, and the AI in the classroom. Within these Frameworks, teachers need to be provided with instructional strategies of how to use AI for improved science teaching and learning.

Like with any other aspects of science education, the general principle that there is no holy grail applies also to AI, meaning that AI is not the solution for everything, everyone, or all the time. Research shows that there are inherent challenges of AI that students and teachers alike need to be aware of. One such challenge is the accuracy of AI. As much as researchers have tried to improve the accuracy of AI, there will always be a level of inaccuracy. As much as with any teacher. The question is, though, what level of accuracy is acceptable for use in education? Does it suffice if AI meets the level of agreement that is identical to that between humans? These are normative questions for which teachers and students need to be provided with guidance through general policies (such as the EU AI Act), but also specific guidelines for the use in science teaching and learning. More generally, teachers and students need to understand how AI is not a perfect source of information. The AI is likely best viewed as an epistemic agent in its own right. It is part of the epistemic network of students and teachers (see Chap. 2 by Nehm & Kubsch this volume). To his end, AI, like any epistemic agent, has its strengths and weaknesses. There is no absolute truth. There is a probability that the response or decision made by the AI is correct. The feedback the AI provides may be based on an erroneous assessment of student learning. While this is not ideal, it does happen with the teacher, too. If students are to learn how to live in a world that will, without any question, be driven by AI in the not-too-distant future, students need to develop literacy in using AI or working with or collaborating on AI projects. As AI is used in science classrooms, science teaching and learning must contribute to developing such AI literacy. This necessitates the teacher being literate in the use of AI, its potentials and limitations, as well as policies or guidelines to follow in working with an AI to ensure the best possible outcomes. For this purpose, teachers, who for the most part, were not prepared for the use of AI in their professional education, need professional learning or development opportunities. These will need to be developed and rolled out quickly to avoid falling behind technological developments.

As a broader conclusion, research has documented that AI holds great potential for improving science teaching and learning. Yet the potential is largely untapped in regular science classrooms, which is likely due to a lack of guidance for teachers and students, such as teaching and learning materials showing how to use AI. So, more research is needed in science education, specifically seeking to develop and explore broader concepts of using AI in science classrooms.

Disclosure The authors wrote the paper. Parts of it were written by the authors with the assistance of generative AI tools for language editing or paragraphing. The authors reviewed, revised, and verified all AI-supported content for accuracy, appropriateness, and originality. No confidential or sensitive data was entered into AI systems. The final responsibility for the content rests solely with the authors.

Acknowledgement This material is based upon work supported by the National Science Foundation under Grant No. 2332964 (PI Zhai). Any opinions, findings, and conclusions or recommendations expressed in this material are those of the author(s) and do not necessarily reflect the views of the National Science Foundation.

References

Alonzo, A. C. (Ed.). (2012). *Learning progressions in science: Current challenges and future directions*. Sense Publ.

Amershi, S., & Conati, C. (2011). Automatic recognition of learner types in exploratory learning environments. In C. Romero, M. Ventura, M. Pechenizkiy, & R. S. Baker (Eds.), *Handbook of educational data mining* (pp. 213–230). CRC Press.

Bilal, A., Ebert, D., & Lin, B. (2025). *LLMs for explainable AI: A comprehensive survey* (version 1). arXiv. https://doi.org/10.48550/ARXIV.2504.00125

Bonefeld, M., & Dickhäuser, O. (2018). (Biased) grading of students' performance: Students' names, performance level, and implicit attitudes. *Frontiers in Psychology, 9*, 481. https://doi.org/10.3389/fpsyg.2018.00481

Clarke, J., & Dede, C. (2009). Design for scalability: A case study of the River City curriculum. *Journal of Science Education and Technology, 18*(4), 353–365. https://doi.org/10.1007/s10956-009-9156-4

Dai, W., Lin, J., Jin, F., Li, T., Tsai, Y.-S., Gasevic, D., & Chen, G. (2023). *Can large language models provide feedback to students? A case study on ChatGPT* [preprint]. EdArXiv. https://doi.org/10.35542/osf.io/hcgzj

Dai, X., Wen, Z., Jiang, J., Liu, H., & Zhang, Y. (2025). *How students use AI feedback matters: Experimental evidence on physics achievement and autonomy* (no. arXiv:2505.08672). arXiv. https://doi.org/10.48550/arXiv.2505.08672

De Jong, T., Lazonder, A. W., Chinn, C. A., Fischer, F., Gobert, J., Hmelo-Silver, C. E., Koedinger, K. R., Krajcik, J. S., Kyza, E. A., Linn, M. C., Pedaste, M., Scheiter, K., & Zacharia, Z. C. (2023). Let's talk evidence—The case for combining inquiry-based and direct instruction. *Educational Research Review, 39*, 100536. https://doi.org/10.1016/j.edurev.2023.100536

Díaz, B., & Nussbaum, M. (2024). Artificial intelligence for teaching and learning in schools: The need for pedagogical intelligence. *Computers & Education, 217*, 105071. https://doi.org/10.1016/j.compedu.2024.105071

Dressel, J., & Farid, H. (2018). The accuracy, fairness, and limits of predicting recidivism. *Science Advances, 4*(1), eaao5580. https://doi.org/10.1126/sciadv.aao5580

Duncan, R. G., & Rivet, A. E. (2018). Learning progressions. In F. Fischer, C. E. Hmeleo--Silver, S. R. Goldman, & P. Reimann (Eds.), *International handbook of the learning sciences* (pp. 422–432). Routledge.

Feng, S., Magana, A. J., & Kao, D. (2021). A systematic review of literature on the effectiveness of intelligent tutoring systems in STEM. *2021 IEEE frontiers in education conference (FIE)* (pp. 1–9). https://doi.org/10.1109/FIE49875.2021.9637240

Fraillon, J. (2025). *An international perspective on digital literacy*. Springer.

Gagné, R. M. (1985). *The conditions of learning and theory of instruction* (4th ed.). Holt, Rinehart and Winston.

Gobert, J. D. (2023). Inq-ITS: Creating rigorous assessment and real-time support of science learning. *XRDS: Crossroads, The ACM Magazine for Students, 29*(3), 36–40. https://doi.org/10.1145/3589647

Gombert, S., Di Mitri, D., Karademir, O., Kubsch, M., Kolbe, H., Tautz, S., Grimm, A., Bohm, I., Neumann, K., & Drachsler, H. (2023). Coding energy knowledge in constructed responses with explainable NLP models. *Journal of Computer Assisted Learning, 39*(3), 767–786. https://doi.org/10.1111/jcal.12767

Gotwals, A. W., & Songer, N. B. (2010). Reasoning up and down a food chain: Using an assessment framework to investigate students' middle knowledge. *Science Education, 94*(2), 259–281. https://doi.org/10.1002/sce.20368

Guo, S., Wang, Y., Yu, J., Wu, X., Ayik, B., Watts, F. M., Latif, E., Liu, N., Liu, L., & Zhai, X. (2025). Artificial intelligence bias on English language learners in automatic scoring. In A. I. Cristea, E. Walker, Y. Lu, O. C. Santos, & S. Isotani (Eds.), *Artificial intelligence in education* (Vol. 15881, pp. 268–275). Springer. https://doi.org/10.1007/978-3-031-98462-4_34

Guo, W., & Caliskan, A. (2021). Detecting emergent intersectional biases: Contextualized word embeddings contain a distribution of human-like biases. *Proceedings of the 2021 AAAI/ACM conference on AI, ethics, and society* (pp. 122–133). https://doi.org/10.1145/3461702.3462536

Haque Munim, Z., & Kim, T.-E. (2023). A review of learning analytics dashboard and a novel application in maritime simulator training. *14th International conference on applied human factors and ergonomics* (AHFE 2023). https://doi.org/10.54941/ahfe1003158

Holstein, K., McLaren, B. M., & Aleven, V. (2018). Student learning benefits of a mixed-reality teacher awareness tool in AI-enhanced classrooms. In C. Penstein Rosé, R. Martínez--Maldonado, H. U. Hoppe, R. Luckin, M. Mavrikis, K. Porayska-Pomsta, B. McLaren, & B. Du Boulay (Eds.), *Artificial intelligence in education* (Vol. 10947, pp. 154–168). Springer International Publishing. https://doi.org/10.1007/978-3-319-93843-1_12

Holstein, K., McLaren, B. M., & Aleven, V. (2019). Designing for complementarity: Teacher and student needs for orchestration support in AI-enhanced classrooms. In S. Isotani, E. Millán, A. Ogan, P. Hastings, B. McLaren, & R. Luckin (Eds.), *Artificial intelligence in education* (Vol. 11625, pp. 157–171). Springer International Publishing. https://doi.org/10.1007/978-3-030-23204-7_14

Jang, Y., Choi, S., Jung, H., & Kim, H. (2022). Practical early prediction of students' performance using machine learning and eXplainable AI. *Education and Information Technologies, 27*(9), 12855–12889. https://doi.org/10.1007/s10639-022-11120-6

Jansen, T., Höft, L., Bahr, L., Fleckenstein, J., Möller, J., Köller, O., & Meyer, J. (2024). Comparing generative AI and expert feedback to students' writing: Insights from student teachers. *Psychologie in Erziehung und Unterricht, 71*(2). https://doi.org/10.2378/peu2024.art08d

Karademir, O., Borgards, L., Di Mitri, D., Strauß, S., Kubsch, M., Brobeil, M., Grimm, A., Gombert, S., Rummel, N., Neumann, K., & Drachsler, H. (2024). Following the impact chain of the LA cockpit: An intervention study investigating a teacher dashboard's effect on student learning. *Journal of Learning Analytics, 11*(2), 215–228. https://doi.org/10.18608/jla.2024.8399

Karademir, O., Di Mitri, D., Schneider, J., Jivet, I., Allmang, J., Gombert, S., Kubsch, M., Neumann, K., & Drachsler, H. (2024). I don't have time! But keep me in the loop: Co-designing requirements for a learning analytics cockpit with teachers. *Journal of Computer Assisted Learning, 40*(6), 2681–2699. https://doi.org/10.1111/jcal.12997

Kasneci, E., Sessler, K., Küchemann, S., Bannert, M., Dementieva, D., Fischer, F., Gasser, U., Groh, G., Günnemann, S., Hüllermeier, E., Krusche, S., Kutyniok, G., Michaeli, T., Nerdel, C., Pfeffer, J., Poquet, O., Sailer, M., Schmidt, A., Seidel, T., … Kasneci, G. (2023). ChatGPT for good? On opportunities and challenges of large language models for education. *Learning and Individual Differences, 103*, 102274. https://doi.org/10.1016/j.lindif.2023.102274

Kruger, J., & Dunning, D. (1999). Unskilled and unaware of it: How difficulties in recognizing one's own incompetence lead to inflated self-assessments. *Journal of Personality and Social Psychology, 77*(6), 1121–1134. https://doi.org/10.1037/0022-3514.77.6.1121

Kubsch, M., Czinczel, B., Lossjew, J., Wyrwich, T., Bednorz, D., Bernholt, S., Fiedler, D., Strauß, S., Cress, U., Drachsler, H., Neumann, K., & Rummel, N. (2022). Toward learning progression analytics—Developing learning environments for the automated analysis of learning using evidence centered design. *Frontiers in Education, 7*, 981910. https://doi.org/10.3389/feduc.2022.981910

Kubsch, M., Grimm, A., Neumann, K., Drachsler, H., & Rummel, N. (2024). Using evidence--centered design to develop an automated system for tracking students' physics learning in a digital learning environment. In X. Zhai & J. Krajcik (Eds.), *Uses of artificial intelligence in STEM education* (1st ed., pp. 230–249). Oxford University Press. https://doi.org/10.1093/oso/9780198882077.003.0011

Kubsch, M., Strauß, S., Grimm, A., Gombert, S., Drachsler, H., Neumann, K., & Rummel, N. (2025). Self-regulated learning in the digitally enhanced science classroom: Toward an early warning system. *Educational Psychology Review, 37*(2), 34. https://doi.org/10.1007/s10648-025-10011-9

Kubsch, M., Touitou, I., Nordine, J. C., Fortus, D., Neumann, K., & Krajcik, J. (2020). Transferring knowledge in a knowledge-in-use task—Investigating the role of knowledge organization. *Education Sciences, 10*(1), 20. https://doi.org/10.3390/educsci10010020

Küchemann, S., Steinert, S., Revenga, N., Schweinberger, M., Dinc, Y., Avila, K. E., & Kuhn, J. (2023). Can ChatGPT support prospective teachers in physics task development? *Physical Review Physics Education Research, 19*(2), 020128. https://doi.org/10.1103/PhysRevPhysEducRes.19.020128

Lee, G.-G., Mun, S., Shin, M.-K., & Zhai, X. (2025). Collaborative learning with artificial intelligence speakers: Pre-service elementary science teachers' responses to the prototype. *Science & Education, 34*(2), 847–875. https://doi.org/10.1007/s11191-024-00526-y

Lee, G.-G., & Zhai, X. (2024). Using ChatGPT for science learning: A study on pre-service teachers' lesson planning. *IEEE Transactions on Learning Technologies, 17*, 1643–1660. https://doi.org/10.1109/TLT.2024.3401457

Lehrer, R., & Schauble, L. (2015). Learning progressions: The whole world is NOT a stage. *Science Education, 99*(3), 432–437. https://doi.org/10.1002/sce.21168

Lester, J. C., Ha, E. Y., Lee, S. Y., Mott, B. W., Rowe, J. P., & Sabourin, J. L. (2013). Serious games get smart: Intelligent game-based learning environments. *AI Magazine, 34*(4), 31–45. https://doi.org/10.1609/aimag.v34i4.2488

Li, H., Gobert, J., & Dickler, R. (2017). *Automated assessment for scientific explanations in on--line science inquiry*. International Educational Data Mining Society.

Li, L., Sha, L., Li, Y., Raković, M., Rong, J., Joksimovic, S., Selwyn, N., Gašević, D., & Chen, G. (2023). Moral machines or tyranny of the majority? A systematic review on predictive bias in education. *LAK23: 13th International learning analytics and knowledge conference* (pp. 499–508). https://doi.org/10.1145/3576050.3576119

Li, P.-H., Lee, H.-Y., Lin, C.-J., Wang, W.-S., & Huang, Y.-M. (2025). InquiryGPT: Augmenting ChatGPT for enhancing inquiry-based learning in STEM education. *Journal of Educational Computing Research, 62*(8), 1937–1966. https://doi.org/10.1177/07356331241289824

Liang, Y., Zou, D., Xie, H., & Wang, F. L. (2023). Exploring the potential of using ChatGPT in physics education. *Smart Learning Environments, 10*(1), 52. https://doi.org/10.1186/s40561-023-00273-7

Markauskaite, L., & Goodyear, P. (2017). *Epistemic fluency and professional education: Innovation, knowledgeable action and actionable knowledge*. Springer.

Meyer, J., Jansen, T., Schiller, R., Liebenow, L. W., Steinbach, M., Horbach, A., & Fleckenstein, J. (2024). Using LLMs to bring evidence-based feedback into the classroom: AI-generated feedback increases secondary students' text revision, motivation, and positive emotions. *Computers and Education: Artificial Intelligence, 6*, 100199. https://doi.org/10.1016/j.caeai.2023.100199

Mienye, I. D., & Swart, T. G. (2025). ChatGPT in education: A review of ethical challenges and approaches to enhancing transparency and privacy. *Procedia Computer Science, 254*, 181–190. https://doi.org/10.1016/j.procs.2025.02.077

National Research Council [NRC]. (2012). *A framework for K-12 science education: Practices, crosscutting concepts, and core ideas*. The National Academies Press.

Neumann, K., Viering, T., Boone, W. J., & Fischer, H. E. (2013). Towards a learning progression of energy. *Journal of Research in Science Teaching, 50*(2), 162–188. https://doi.org/10.1002/tea.21061

Neumann, K., & Waight, N. (2020). The digitalization of science education: Déjà vu all over again? *Journal of Research in Science Teaching, 57*(9), 1519–1528. https://doi.org/10.1002/tea.21668

Nyaaba, M., & Zhaı, X. (2024). Generative AI professional development needs for teacher educators. *Journal of AI, 8*(1), 1–13. https://doi.org/10.61969/jai.1385915

O'Neil, C. (2017). *Weapons of math destruction: How big data increases inequality and threatens democracy*. Penguin Books.

Ocumpaugh, J., Baker, R., Gowda, S., Heffernan, N., & Heffernan, C. (2014). Population validity for educational data mining models: A case study in affect detection. *British Journal of Educational Technology, 45*(3), 487–501. https://doi.org/10.1111/bjet.12156

Ouyan, F., Jiao, P., McLaren, B. M., & Alavi, A. H. (Eds.). (2023). *Artificial intelligence in STEM education: The paradigmatic shifts in research, education, and technology* (1st ed.). CRC Press. https://doi.org/10.1201/9781003181187

Pankiewicz, M., & Baker, R. S. (2023). *Large language models (GPT) for automating feedback on programming assignments* (version 1). arXiv. https://doi.org/10.48550/ARXIV.2307.00150

Pozdniakov, S., Martinez-Maldonado, R., Tsai, Y.-S., Srivastava, N., Liu, Y., & Gasevic, D. (2023). Single or multi-page learning analytics dashboards? Relationships between teachers' cognitive load and visualisation literacy. In O. Viberg, I. Jivet, P. J. Muñoz-Merino, M. Perifanou, & T. Papathoma (Eds.), *Responsive and sustainable educational futures* (Vol. 14200, pp. 339–355). Springer. https://doi.org/10.1007/978-3-031-42682-7_23

Rau, M. A., Zahn, M., Misback, E., Herder, T., & Burstyn, J. (2021). Adaptive support for representational competencies during technology-based problem solving in chemistry. *Journal of the Learning Sciences, 30*(2), 163–203. https://doi.org/10.1080/10508406.2021.1888733

Sabourin, J. L., Shores, L. R., Mott, B. W., & Lester, J. C. (2013). Understanding and predicting student self-regulated learning strategies in game-based learning environments. *International Journal of Artificial Intelligence in Education, 23*(1–4), 94–114. https://doi.org/10.1007/s40593-013-0004-6

Seßler, K., Xiang, T., Bogenrieder, L., & Kasneci, E. (2023). PEER: Empowering writing with large language models. In O. Viberg, I. Jivet, P. J. Muñoz-Merino, M. Perifanou, & T. Papathoma (Eds.), *Responsive and sustainable educational futures* (Vol. 14200, pp. 755–761). Springer. https://doi.org/10.1007/978-3-031-42682-7_73

Shi, L., & Choi, I. (2024). A systematic review on artificial intelligence in supporting teaching practice: Application types, pedagogical roles, and technological characteristics. In X. Zhai & J. Krajcik (Eds.), *Uses of artificial intelligence in STEM education* (1st ed., pp. 321–347). Oxford University Press. https://doi.org/10.1093/oso/9780198882077.003.0015

Shreeve, K., Jessen Eller, K., Cassidy, M., Price, B., Jackson, B., Perry, A., Celi, L., Lourentzou, I., & Hendrik, L. (2025). Development of a machine-learning-driven digital teaching assistant that utilises student engagement data to improve access to and success in K-12 STEM education. *Learning Letters, 4*, 31. https://doi.org/10.59453/ll.v4.31

Shute, V. J. (2011). Stealth assessment in computer-based games to support learning. *Computer Games and Instruction, 55*(2), 503–524.

VanLehn, K. (2006). The behavior of tutoring systems. *International Journal of Artificial Intelligence in Education, 16*(3), 227–265.

Wedenig, H. H., Sporn, F., Franz, A., Kaminski, J. J., Holzhausen, Y., & Peters, H. (2023). *AI-generated answer specific feedback for test-enhanced learning with multiple-choice questions*. F1000 Research Limited. https://doi.org/10.21955/MEP.1115392.1

Wollny, S., Schneider, J., Di Mitri, D., Weidlich, J., Rittberger, M., & Drachsler, H. (2021). Are we there yet? - A systematic literature review on Chatbots in education. *Frontiers in Artificial Intelligence, 4*, 654924. https://doi.org/10.3389/frai.2021.654924

Wu, J.-Y., & Tsai, C.-C. (2022). Harnessing the power of promising technologies to transform science education: Prospects and challenges to promote adaptive epistemic beliefs in science learning. *International Journal of Science Education, 44*(2), 346–353. https://doi.org/10.1080/09500693.2022.2028927

Wu, X., Zhao, H., Zhu, Y., Shi, Y., Yang, F., Hu, L., Liu, T., Zhai, X., Yao, W., Li, J., Du, M., & Liu, N. (2025). *Usable XAI: 10 strategies towards exploiting explainability in the LLM era* (no. arXiv:2403.08946). arXiv. https://doi.org/10.48550/arXiv.2403.08946

Wyrwich, T., Domenichini, D., Gombert, S., Kubsch, M., & Neumann, K. (2025). Characterizing students' energy learning trajectories. *Disciplinary and Interdisciplinary Science Education Research, 7(1), 23.* https://doi.org/10.1186/s43031-025-00141-z

Wyrwich, T., Kubsch, M., Drachsler, H., & Neumann, K. (2025). Tracking students' progression in developing understanding of energy using AI technologies. *Physical Review Physics Education Research, 21*(1), 010152. https://doi.org/10.1103/PhysRevPhysEducRes.21.010152

Zawacki-Richter, O., Marín, V. I., Bond, M., & Gouverneur, F. (2019). Systematic review of research on artificial intelligence applications in higher education—Where are the educators? *International Journal of Educational Technology in Higher Education, 16*(1), 39. https://doi.org/10.1186/s41239-019-0171-0

Zhai, X. (2023a). ChatGPT and AI: The game changer for education. *Shanghai Education*, 16–17. https://ssrn.com/abstract=4389098

Zhai, X. (2023b). ChatGPT for next generation science learning. *XRDS: Crossroads, The ACM Magazine for Students, 29*(3), 42–46. https://doi.org/10.1145/3589649

Zhai, X. (2025). Transforming teachers' roles and agencies in the era of generative AI: Perceptions, acceptance, knowledge, and practices. *Journal of Science Education and Technology, 34*, 1323–1333. https://doi.org/10.1007/s10956-024-10174-0

Zhai, X., & Krajcik, J. S. (2024). *Uses of artificial intelligence in STEM education*. Oxford University Press.

Zhai, X., & Nehm, R. H. (2023). AI and formative assessment: The train has left the station. *Journal of Research in Science Teaching, 60*(6), 1390–1398. https://doi.org/10.1002/tea.21885

Zhai, X., Yin, Y., Pellegrino, J. W., Haudek, K. C., & Shi, L. (2020). Applying machine learning in science assessment: A systematic review. *Studies in Science Education, 56*(1), 111–151. https://doi.org/10.1080/03057267.2020.1735757

Zheng, Z., Sun, Y., Song, X., Zhu, H., & Xiong, H. (2023). Generative learning plan recommendation for employees: A performance-aware reinforcement learning approach. *Proceedings of the 17th ACM conference on recommender systems* (pp. 443–454). https://doi.org/10.1145/3604915.3608795

Chapter 7
Capitalizing on Artificial Intelligence to Create Critical Entry Points That Support Science Teacher Learning and the Work of Science Teaching

Jamie N. Mikeska, Natalie S. King, and Yizhu Gao

7.1 Introduction

Artificial intelligence (AI) is significantly reshaping the learning opportunities that teachers have and how they prepare for and engage in the work they do as educators. It is transforming how educators develop professionally, learn new teaching skills, and create content for and support their students by helping them to be more responsive and effective practitioners who can support their students' engagement and learning. The last few years have seen an explosion in AI tools available to science teachers and teacher educators (Ouyang et al., 2022; Wang, Wang et al., 2024). These tools can provide students with personalized support and offer opportunities for students to engage in interactive learning and scientific experimentation (Zawacki-Richter et al., 2019). For example, AI-supported simulations can help students visualize complex scientific phenomena and be incorporated into virtual science labs to support student learning (Celik et al., 2022; Quellmalz et al., 2012). Likewise, AI tools can support teachers' lesson planning, assessment, and professional development (Walter, 2024). For example, AI chatbot assistants can be used to help teachers create scoring rubrics, brainstorm lesson activities to address specific learning goals, and develop formative assessment tasks. AI simulations can

J. N. Mikeska (✉)
ETS, Princeton, NJ, USA
e-mail: jmikeska@ets.org

N. S. King
Georgia State University, Atlanta, GA, USA
e-mail: natalieking@gsu.edu

Y. Gao
University of Georgia, Athens, GA, USA
e-mail: yizhu.gao@uga.edu

X. Zhai, K. J. Crippen (eds.), *Advancing AI in Science Education*, Advances in Technology-Rich Science Education 3,
https://doi.org/10.1007/978-3-032-16871-9_7

also be used as online practice spaces where teachers can develop their ability to engage in science teaching practices, while AI-powered feedback can help teachers identify their instructional strengths and areas for growth (Liu et al., 2014). Collectively, AI-supported tools offer several benefits in science education, including personalized feedback and learning tasks, improved engagement, and reduced teacher workload (Kasneci et al., 2023).

Despite these opportunities and benefits, using AI tools within K-12 science classrooms and within professional learning contexts also raises several challenges and concerns. Two important ones relate to access and fairness. Students and teachers in under-resourced schools may be less likely to have adequate access to hardware, infrastructure, and training to learn how to use AI tools productively (NASEM, 2023). In addition, AI tools also need to be consistently monitored for bias, and strategies need to be employed to minimize bias in AI algorithms to ensure fair outcomes across users. Another ongoing challenge focuses on ensuring that teachers have sufficient training and support so they are prepared to adapt and integrate AI tools into their science instruction to meet their students' learning needs. Finally, ethical concerns abound regarding understanding how AI tools work and make decisions, which results in increased skepticism about the value of AI tools and questions about when and how they should be used.

In this chapter, we highlight three main purposes by which AI technologies can be used to create critical entry points to support science teacher learning and the work of science teaching. By entry points, we mean the various strategies, approaches, and tools that are powered by AI and used to support K-12 students' or teachers' learning and engagement. First, integrating AI into K-12 science education contexts has the potential to increase science teachers' efficiency as they work to prepare for, engage in, and reflect on their instruction (Wiuggansa, 2024). Second, AI technologies show promise for promoting science teachers' professional learning and helping them develop their ability to engage in key science teaching competencies (Mikeska, Halder, et al., 2025; Nazaretsky et al., 2023). Finally, AI technologies can provide science teachers with timely and actionable insights into their students' knowledge and areas of struggle (de Araujo et al., 2025; Gobert et al., 2023; He et al., 2024). Such information enables science teachers to provide more customized and targeted support to their students in future instruction. For each purpose, we provide an overview of its key considerations, describe recent and ongoing initiatives or research efforts aligned with the purpose, and discuss the promise and challenges, including ethical concerns, that science educators must navigate when using or interacting with specific AI tools geared towards that purpose. We also use short vignettes to illustrate the potential of various AI tools within the context of science instruction or professional development. We argue that productively integrating AI into K-12 science education contexts to support student and teacher learning requires thoughtful implementation, comprehensive support, and careful consideration of ethical implications.

7.2 Changes Brought to Science Teacher Education by Artificial Intelligence

7.2.1 Using AI Tools to Increase Science Teachers' Efficiency in the Classroom

AI has moved from buzzword to daily reality in K-12 science education. For teachers who are stretched thin, its most immediate promise is pragmatic: AI can make core parts of the job, such as planning coherent lessons, facilitating inquiry-based lessons, and reflecting on student thinking, significantly more efficient without hollowing out the human craft of teaching. When implemented thoughtfully, AI has the potential to help teachers create standards-aligned materials in minutes instead of hours, deliver just-in-time supports to diverse learners during investigations, and surface actionable insights about students' progress that would otherwise be buried in notebooks, lab reports, or exit tickets. This section discusses the promises of integrating AI to increase science teachers' efficiency through planning, teaching, and reflection.

Efficiency is critical for highly effective science teaching because teaching is a cognitively and logistically demanding profession. Science teachers regularly align phenomena and anchoring experiences to standards, design investigations with safety and accessibility in mind, differentiate supports for multilingual learners and students with disabilities, and construct rubrics that capture scientific practices (e.g., modeling, argumentation, data analysis). Day to day, they must also synthesize evidence of learning fast enough to adapt to tomorrow's instruction. AI has the potential to streamline these high-leverage but time-intensive tasks. Recent syntheses chart this shift from labor-heavy workflows to AI-supported ones. In STEM classrooms, AI systems now (a) recommend lesson resources keyed to standards and learning goals, (b) auto-generate or adapt formative assessment items, (c) provide immediate, criterion-aligned feedback on student explanations and arguments, and (d) offer teachers real-time dashboards that make student misconceptions visible while an activity is still unfolding (Ouyang, et al., 2022; Wang et al., 2024). Crucially, these tools do not replace professional judgment but rather amplify it by reducing routine load and making patterns in student thinking easier to see and act upon.

Education remains a fundamentally human endeavor, and we should always prioritize humans first and technology second. Even as AI accelerates planning and feedback cycles, teachers must remain the ones who frame problems, interrogate how tools work, notice privacy risks, and keep space for creativity, serendipity, and culturally responsive practice (Popenici & Kerr, 2017). Simply put, AI can increase efficiency in doing the work, but teachers are the ones who give the work meaning. One powerful example of this is the AI4K12 Initiative, which provides national guidelines to support educators in introducing AI concepts across subjects. While the framework offers structure and content, its impact depends entirely on how teachers interpret, adapt, and implement it within their unique classroom contexts.

Teachers use AI4K12 not as a script but as a springboard to bring their pedagogical values, cultural awareness, and professional judgment to bear on how students learn about AI (Jong et al., 2022; Kong & Yang, 2024). In this way, even the adoption of AI frameworks becomes a deeply human act, guided by educators who ensure that technological literacy serves broader educational and ethical goals (Lin & Van Brummelen, 2021).

In considering efficiency as it relates to preparation, AI is useful in standards-aligned lesson design and considering formative assessments. Research compiled in *Artificial Intelligence in STEM Education* documents tools that map curriculum standards to activity banks and assessments, reducing planning time while preserving alignment and coherence (Next Generation Science Standards (NGSS), 2013; Ouyang et al., 2022). In practical terms, teachers input a performance expectation (e.g., "MS-PS1-4: Develop a model that predicts and describes changes in particle motion, temperature, and state of a pure substance when thermal energy is added or removed") and receive phenomenon prompts, data sets, and candidate scaffolds. Instead of starting from scratch, teachers begin from a vetted draft they can localize to students' lived realities and free up time to think about equity moves, language supports, or place-based examples (Kong & Yang, 2024).

AI also enhances formative assessment through natural language processing (NLP) systems that generate short-answer prompts and evaluate student responses using science-specific rubrics. Tools like ETS's *c-rater-ML* provide immediate, rubric-aligned feedback on scientific explanations, giving teachers a quick snapshot of class understanding and allowing them to prioritize misconceptions that require intervention (Liu et al., 2014; Wang et al., 2024). Virtual labs such as *SimScientists* go a step further by combining hands-on simulation with automated feedback on modeling, data interpretation, and argumentation. These tools improve feedback quality while reducing the cognitive and logistical load of managing labs and collecting assessment data (Quellmalz & Silberglitt, 2017).

Classroom orchestration also benefits from AI-powered dashboards and intelligent tutoring systems. These tools help teachers monitor progress, surface student difficulties in real time, and focus instructional support more strategically. For instance, the *Lumilo* smart glasses project demonstrated that teachers could intervene more precisely and frequently when supported by real-time analytics, improving classroom flow and instructional timing (Holstein et al., 2020). Intelligent tutors like *Andes Physics Tutor* and *Betty's Brain* provide students with immediate, step-level feedback on scientific tasks, enabling teachers to spend more time facilitating deeper conceptual conversations and less time troubleshooting procedures (VanLehn, 2020).

As science teachers juggle complex instructional demands, including labs, modeling, discourse, and equity-focused differentiation, AI systems help preserve the integrity of ambitious teaching while reducing manual load. Advanced NLP tools can now analyze student explanations at scale, clustering them by common misconceptions or conceptual gaps. This enables science teachers to design targeted mini-lessons or small group supports with less delay, compressing the usual cycle of reflection, planning, and reteaching from several hours to a single prep period

(Dillenbourg & Hillaire, 2023). When integrated into dashboards or learning management systems, these tools make it easier for teachers to take immediate, data-informed instructional actions. In this way, AI enhances both the efficiency and effectiveness of science teachers by allowing them to concentrate their expertise where it matters most—on fostering deep learning and responsive instruction.

There are some guardrails to professional learning, ethics, and responsibility as we increasingly embed AI tools into our field and K-12 schools. The introduction of AI into classrooms represents a break with purely conventional methods by enabling responsive, personalized support for students with disabilities and multilingual learners, but it also demands robust teacher learning and curriculum adaptation (Walter, 2024; Zawacki-Richter et al., 2019). Without intentional design, AI can reproduce patterns of exclusion already present in society. The National Academies (2023) call for scrutinizing AI systems for the ways entrenched practices and beliefs manifest as bias in datasets and models, especially in STEM, where access and representation gaps are longstanding. OECD's AI Principles similarly foreground fairness, inclusivity, transparency, and accountability; these values must move from posters to policy and classroom routines (OECD, 2024). Practically, that means: (1) building teacher AI literacy that includes how models work, where they fail, and how to audit outputs; (2) offering professional development (PD) that models AI as a learning partner rather than an answer machine (e.g., letting students use a chatbot to explain missed test items rather than simply reveal answers); and (3) anchoring decisions to short- and long-term equity goals that the school system is willing to resource (devices, connectivity, accessible materials, time for teacher collaboration). If we ignore AI because of fears of cheating, students will still use it, but use it badly. If we teach *with* AI, we can position it to deepen inquiry, make reasoning visible, and expand access to rich science experiences.

Emerging tools demonstrate how AI can support science teacher effectiveness by improving efficiency across the instructional cycle from preparation to classroom orchestration to formative assessment. However, realizing this potential depends on whether teachers are introduced to AI tools in ways that build their capacity and align with ethical, student-centered goals. Integrating AI into K-12 science education has the power to support science teachers' efficiency, but it must be coupled with professional learning that equips educators to understand how these tools function and how to apply them thoughtfully in diverse classrooms (Bougdira & Al Murshidi, 2025). At the same time, the use of AI raises serious ethical considerations. Many of these systems rely on student data to generate insights or feedback, which introduces risks related to privacy, surveillance, and data misuse. Educators must be aware of how student information is collected and ensure compliance with privacy laws and ethical guidelines (Rahman & Shuvo, 2025). Furthermore, algorithmic bias remains a critical concern. AI tools trained on non-representative data sets can unintentionally reinforce disparities by privileging certain language styles, cultural references, or learning trajectories. Teachers need not only to use AI tools but to critically examine their underlying assumptions, training data, and impacts on equity (John Jr., 2025).

In the context of science education, where inquiry, reasoning, and diversity of thought are central, it is especially important to ensure that AI accelerates what we value most—curiosity, evidence-based thinking, and connections to students' lived experiences. If implemented well, AI can streamline the procedural aspects of teaching while creating more space for the human work that cannot be automated: listening carefully, asking powerful questions, and cultivating learning environments rooted in justice, joy, and scientific wonder. Several ongoing initiatives are helping to build teacher capacity in this area by offering professional learning focused on both AI literacy and practical applications that enhance instructional efficiency. One such initiative will be described in the vignette below.

Vignette 7.1: Empowering Teachers with AI Skills and Partnerships
When Ms. Lwin, a 10th-grade chemistry teacher at Oak Ridge High School in Riverbend County Schools (all pseudonyms), first encountered the idea of integrating AI into her teaching, she was hesitant. Like many educators, she worried that AI might encourage shortcuts or undermine student thinking. But after attending an AI professional development workshop co-hosted by Georgia State University, Microsoft, and World Wide Technology, her mindset began to shift. "I feel more confident and encouraged to use AI after this workshop," she reflected. "It's not about replacing what I do. It's about saving time so I can focus more on my students." The workshop series, attended by 60 teachers across the metro area, focused on practical classroom applications of tools like Microsoft Copilot. Participants used AI to solve real instructional challenges, such as generating lab report variations for diverse learners, drafting rubrics aligned with district standards, and analyzing student data more efficiently. Ms. Lwin, who once spent entire afternoons creating differentiated assignments, shared how AI allowed her to produce high-quality drafts in minutes: "Now I can use my planning time to personalize and improve, not just to produce."

Sessions also addressed the ethical dimensions of AI use. Teachers explored how to guide students in recognizing bias, protecting privacy, and using AI responsibly. Facilitators emphasized that AI should extend, not replace, teacher judgment and creativity. Ms. Schrage, a former elementary teacher and now a program lead at World Wide Technology, explained that the curriculum was co-designed with teacher input: "We asked teachers where they lose time and what they wish they could do more of. Then we focused the AI training on those exact pain points." For Mr. Jamal, an eighth-grade physics teacher at Eastview Middle School, the collaboration was just as valuable as the technology. "I walked away with better prompts and better lesson ideas. I saw how AI could help me keep the rigor but meet students where they are." By the final session, participants had created personalized AI integration plans, with many planning to use AI for lesson planning, feedback, and formative assessment in the coming few months. This initiative reflects the power of cross-sector partnerships. University researchers brought evidence-based design, industry partners contributed tools and training, and teachers shaped the work with their lived classroom realities. Together, they created a professional learning model that shows how AI can be a practical, empowering tool; a tool not for replacing teachers, but for helping them do more of what matters.

This vignette illustrates how intentional professional learning and cross-sector collaboration can help teachers move from skepticism to strategic use of AI. As the teachers in this program discovered, AI can significantly reduce the time spent on repetitive tasks while enhancing the precision and responsiveness of instruction. For science educators, who are often tasked with balancing complex content, labs, and differentiated support, this kind of efficiency can make the difference between surviving and thriving in the classroom. However, as the workshops emphasized, efficiency must not come at the cost of equity or ethics. Teachers must be equipped not just with tools, but with the skills and critical understanding to use AI responsibly. When introduced thoughtfully, AI can amplify what teachers value most: engaging students, deepening inquiry, and ensuring all learners are seen and supported. The success of this initiative reinforces the core argument that integrating AI into K-12 science education begins not with software, but with teachers and with the systems that empower them to use technology purposefully and ethically.

7.2.2 Using AI Tools to Provide Science Teachers with Practice-Based Learning Opportunities and Personalized Feedback

As noted by Crippen et al. in Chap. 3 of this volume, meaningful science education supported by AI requires attention to both responsible and ethical principles to ensure that AI's use supports learning for all. Two key principles—beneficial and meaningful use and robustness and scientific integrity—are paramount to ensuring that AI can be used to improve learner outcomes. These principles are especially salient when using AI to provide science teachers with personalized learning opportunities and feedback, as AI systems should enhance science teacher learning by promoting teacher engagement and participation in authentic practices of science teaching and should be accurate and aligned with the work that science teachers engage in to support their students' learning. To this end, recent efforts in science education have explored how AI can be used to support K-12 teachers—both pre- and in-service teachers—in developing their knowledge and their ability to engage in key teaching competencies or instructional skills (Gunawan et al., 2021; Halder & Mikeska, 2025; Lee et al., 2025; Pellas, 2025; Spasopoulos et al., 2025). Research has shown that being able to engage in core teaching practices, such as knowing how to facilitate productive science discussions, elicit students' scientific thinking and conceptions, or select useful scientific models to address specific learning goals, is critical to enhancing student learning and engagement. Two key areas where recent efforts have focused include: (a) examining how NLP, machine learning (ML), and generative AI can be used to analyze science teachers' instruction and provide feedback on their instructional strengths and areas for growth and (b) exploring how generative AI can be used to power interactive, digital teaching simulations where science teachers can practice aspects of the work of teaching, but in reduced complexity settings. In both use cases, it is critical that AI is used in ways that provide science teachers with meaningful learning opportunities where they can engage in and reflect on their science teaching practices and consider how they can become effective practitioners who can support their students' learning and engagement.

7.2.2.1 Using AI to Provide Personalized Feedback to Science Teachers

Feedback is an essential aspect of professional learning for teachers and can serve several purposes within professional development (Krasniqi & Ismaj, 2025; Scheeler et al., 2004). Providing teachers with feedback about their instruction supports them in reflecting on their practice. Teachers can also use feedback to identify their instructional strengths and areas for growth. It also allows them to adapt and refine the instructional strategies and approaches that they use to improve student learning and engagement in the future. Using feedback within professional learning contexts for teachers is a critical mechanism that supports teacher learning (Rodgers et al., 2019).

However, research has suggested that not all feedback is created equal. Feedback needs to be specific and targeted to maximize its impact (Erdemir & Yeşilçınar, 2021; Hellrung & Hartig, 2013; Jin et al., 2025). To support their growth, the feedback that teachers receive should be focused on observable behaviors and linked to student learning goals. In addition, feedback should be provided within close proximity to when the specific behaviors have been observed, and offer teachers insights on how they can improve their practice.

Typically, the provision of specific, targeted, and actionable feedback on teachers' instructional practice has required copious human time and effort to review the instruction, identify the strengths and areas for growth, and document these ideas for sharing with the teacher. To date, most feedback has been provided by peers, school leaders, K-12 students themselves, and through self-reflection (Ovando, 2005; Rodriguez et al., 2020; Watkins, 2004). More recently, researchers have been exploring how to leverage AI to provide such feedback in a timely way to support teachers' instructional improvement (Demszky et al., 2025; Mikeska, Beigman Klebanov, et al., 2024; Nazaretsky et al., 2023; Suresh et al., 2021). AI can be used to analyze written transcripts and videos from teachers' instruction—either in real classrooms or simulated classrooms—and provide teachers with specific and timely feedback about how well they are doing at engaging in specific instructional practices.

One such effort to explore how AI can be leveraged to provide personalized feedback to support science teacher learning has been conducted by a cross-disciplinary team of researchers, assessment developers, and teacher education experts at ETS, a non-profit educational organization. This work is part of a multi-year research effort to explore how AI can be used to provide personalized feedback to science and mathematics teachers regarding their ability to engage in key teaching competencies (Asano et al., in review; Beigman Klebanov et al., 2025; Jorgenson et al., 2024; Ilagan et al., 2024; Mikeska, Tierney et al., 2024; Nazaretsky et al., 2023).

Fig. 7.1 Image of Mursion's upper elementary, human-powered simulation classroom (image provided by Mursion, Inc.)

In this line of research, the team has explored how NLP techniques and large language models (LLMs) can be used to conduct automated evaluation of various features of teachers' science or math discussions and translate the evaluation into personalized feedback reports that help teachers identify their strengths and areas for growth when facilitating discussions. One unique aspect of the discussions is that they are conducted in an online practice space where teachers can try out and receive feedback on their ability to engage in specific instructional practices—Mursion's upper elementary simulated classroom (see Fig. 7.1), consisting of five student avatars who can interact verbally in real time with the teacher and with each other. To date, this automated evaluation has focused on a few key instructional skills, including teachers' ability to elicit meaningful student contributions, use student ideas to make progress towards addressing the discussion's learning goal, facilitate a coherent discussion, and engage students in direct peer interaction during the discussion.

Figure 7.2 shows an image of one component of a personalized feedback report (Jorgenson et al., 2025); this component provides feedback to a teacher about how well they elicited meaningful student ideas when facilitating a science discussion with five student avatars in the simulated classroom. The team developed and refined AI models that could evaluate each student's utterance in the science discussion as an example or non-example of a meaningful student idea. The personalized feedback report provides the teacher with a holistic picture of the extent to which the five students provided meaningful ideas throughout the discussion. Other parts of the feedback report include: (a) an Examples tab that gives details showing actual example utterances of meaningful student ideas, (b) various graphs showing the extent of meaningful and non-meaningful contributions by students for their discussion and for another teacher who facilitated the same discussion that was rated as a high-quality one, and (c) a Next Steps tab that provides ideas about various strategies they can use to productively elicit meaningful student contributions in future discussions.

Across various studies, research findings have suggested that teachers understand and use the personalized feedback to identify their strengths and areas for growth and appreciate the detailed nature of the feedback that connects their performance to specific evidence and provides ideas for how they can improve (Jorgenson et al., 2024; Mikeska, Beigman Klebanov, et al., 2024; Mikeska, Tierney, et al., 2024). These AI-powered personalized reports and the AI models undergirding

Eliciting Meaningful Student Contributions

A **Meaningful Contribution (MC)** is a statement or question made by a student, focused on key ideas, that allows the discussion to move toward a shared understanding of the learning goal. MCs often involve students elaborating on their ideas or responding to the ideas of other students. It is **not** expected that every student contribution will be classified as meaningful.

Some examples of contributions that are **not** meaningful **(NonMC)** are (1) restatements of information from students' previous work that do not add any additional thinking and (2) answers to closed-ended questions (e.g., yes/no questions, basic fact questions with one correct answer).

Comparing Typical High-Quality Discussions and Your Discussion

	Typical High-Quality Discussions	**Your Discussion**
Number of student contributions	53 – 64	60
Number of student contributions that were MCs	25 – 35	18
Percent of student contributions that were MCs	46% – 55%	30%
Number of students who provided at least 2 MCs	5 of 5	4 of 5
Percents at which students provided MCs	Proportionate	Disproportionate

Fig. 7.2 Example of one component of an AI-powered personalized feedback report for science teachers (image reproduced from Jorgenson et al., 2025)

them can be developed to support science teachers as they learn to engage in different teaching competencies across various science content topics and integrated into a variety of science teacher education professional learning settings.

7.2.2.2 Using AI to Provide Science Teachers with Opportunities to Improve Their Instructional Skills

Ensuring equitable and inclusive education requires a highly qualified teaching workforce (Boyd et al., 2005; Cowan et al., 2016; Fuller & Pendola, 2019; Ingersoll & May, 2011, 2012; Wolf, 2015). High-quality PDD is needed to ensure that all teachers are effective practitioners who can engage in core teaching practices to support teacher learning. Preservice and in-service teachers need opportunities to learn how to support students as they engage in the science and engineering practices noted as critical for student learning in the Next Generation Science Standards (Davis & Palincsar, 2023; NASEM, 2021; National Research Council, 2012; NGSS Lead States, 2013). Research has suggested that approximations of practice, where teachers try out aspects of instruction in situations of reduced complexity, are productive for developing teachers' beliefs, knowledge, and skills (Benedict et al., 2016; Benedict-Chambers et al., 2020; Davis et al., 2017; Grossman et al., 2009; Masters, 2020).

Digital teaching simulations have become more widely used as approximations to support teacher learning (Bondie et al., 2021; Dieker et al., 2014; Mikeska, Cross Francis et al., 2025; Reich, 2022; Straub et al., 2015). Previous research has indicated that human-powered digital teaching simulations are productive tools that can be integrated into educator preparation programs and PD to improve educators' instructional skills and beliefs (Bondie et al., 2021; Reich, 2022; Straub et al., 2015). However, the current teaching simulations are expensive to implement and require a significant amount of time and resources since a human actor needs to actively engage in training to play the role of the student avatars during the practice teaching sessions. For example, TeachLivE and Mursion simulated classrooms (Fig. 7.1) rely on a human puppeteer to respond as the student avatars during real-time interactions with teachers. One promising approach that is currently being explored is the use of automatically simulated students, powered by generative AI (GenAI; Käser & Alexandron, 2023; Kasneci et al., 2023). Using GenAI to power automated students can leverage GenAI's ability to play different roles in realistic and flexible conversations with humans across varied subject matter contexts (Mollick & Mollick, 2023; Ray, 2023; Urmeneta & Romero, 2024), while addressing its limitations, such as GenAI's difficulty in maintaining context and coherence in an extended conversation (Ray, 2023).

Vignette 7.2: Using Interactive Chatbots to Develop Science Teachers' Instructional Skills

At Clearwater Elementary School, the elementary science specialist is facilitating a professional learning community (PLC) with 10 in-service teachers who currently teach science at the upper elementary level. The goal of the PLC is to help the teachers improve their ability to engage in the core teaching practice of eliciting and attending to student ideas in their science classrooms. The elementary science specialist has been using GenAI teaching simulations as online practice spaces, whereby the teachers can prepare for and try out this teaching practice and then reflect on their strengths and areas for growth.

Recently, the elementary science specialist facilitated a series of sessions to help the teachers in preparing for, engaging in, and reflecting on how they elicited student thinking when using one GenAI teaching simulation called the Mystery Substance science task. In this task, teachers learn that a class of fifth-grade students has been working on learning about properties of matter, and one student named Caleb (he is the generative AI (GenAI) fifth-grade student) recently completed a science investigation where he worked to identify a mystery substance by comparing its properties to the properties of other known substances. The teacher's goal in the GenAI teaching simulation is to: (1) ask questions to elicit what the student (Caleb) did to produce the claim provided and (2) probe to understand why the student (Caleb) performed the particular steps in his investigation and what conceptual understanding the student has and does not have regarding how to identify the mystery substance based on comparing its properties to the other substances they tested. Figure 7.3 shows images of the GenAI Mystery Substance teaching simulation task.

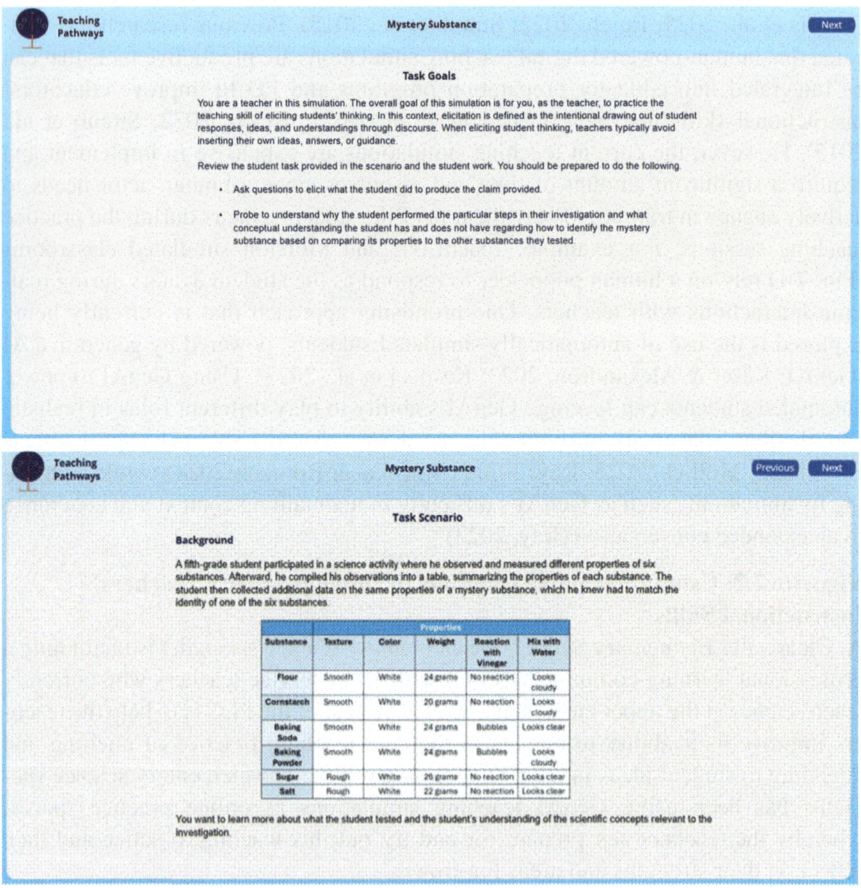

Fig. 7.3 Images of the mystery substance eliciting student thinking GenAI simulation interface (image reproduced from Mikeska, Halder, et al., 2025)

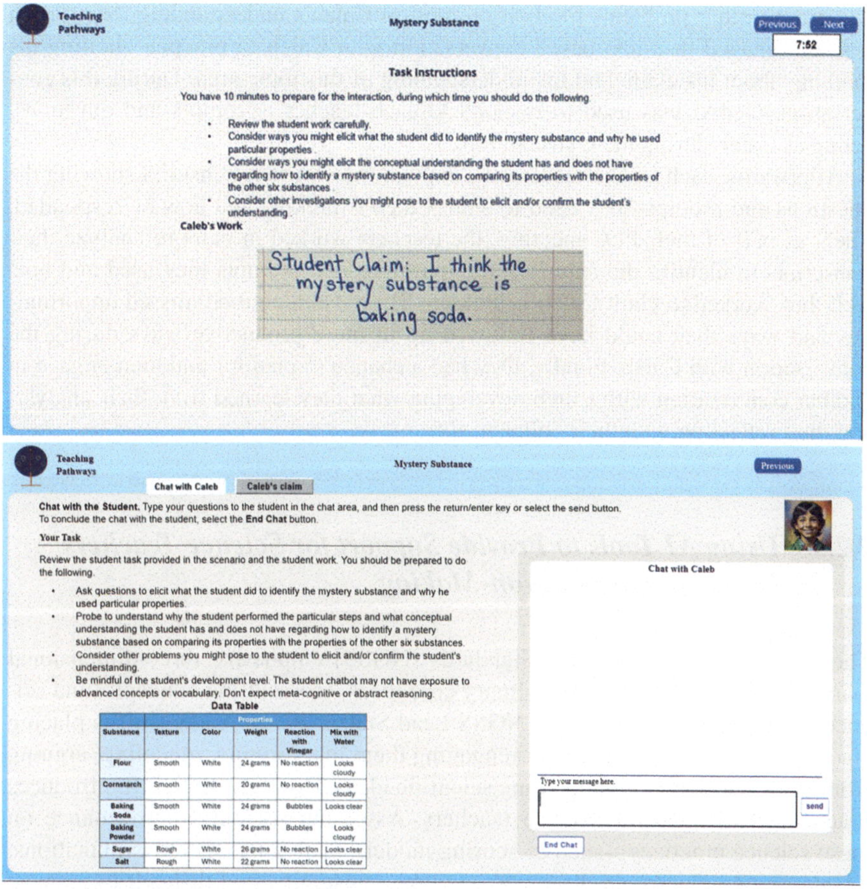

Fig. 7.3 (continued)

As part of their preparation, the elementary science specialist had the teachers work in small collaborative groups to conduct the same Mystery Substance science investigation; part of this work involved having the teachers generate claims and evidence-based reasoning about the mystery substance's identity and determine which properties were most useful to identify the mystery substance, just like Caleb had to do in his work. Then, the elementary science specialist had the teachers review Caleb's written work (see Fig. 7.3) and create a list of question prompts that they could ask Caleb to elicit his thinking about what he thinks the mystery substance is and why, and his conceptual understanding of properties of matter. In their brainstorming, the teachers were encouraged to consider ways they could elicit the following: Caleb's process for identifying the mystery substance, his reasoning for using particular properties, his conceptual understanding of identifying a mystery substance by comparing its properties with the properties of six known substances, and posing other problems to elicit or confirm Caleb's understanding. Next, each teacher engaged in a text-based conversation with Caleb to practice eliciting his thinking about his claim and his understanding of this topic area. During this conversation, GenAI was used to respond to each teacher's prompts and questions, acting as Caleb during the conversation.

Afterwards, each teacher accessed a copy of their written transcript showing the questions and prompts they used to elicit Caleb's thinking and how he responded. Then, as part of their PLC meeting, the teachers worked in pairs to analyze their transcripts to identify the nature of the questions and prompts they used and how well they worked to elicit Caleb's thinking. They also identified missed opportunities and ways they could have followed up in more productive ways during the conversation with Caleb. Finally, they had a chance to plan for and then engage in another conversation with Caleb, leveraging what they learned from their analysis and their reflection with their colleagues.

7.2.3 Using AI Tools to Provide Support for Science Teachers' Instructional Decision-Making

The Next Generation Science Standards (NGSS) emphasize three-dimensional learning, which integrates disciplinary core ideas, crosscutting concepts, and science and engineering practices (NGSS Lead States, 2013). A key shift is placing students in the role of scientists—engaging them in reasoning, modeling, arguing from evidence, and communicating scientific ideas. However, this shift introduces practical challenges for science teachers. Assessing students' performance on these science practices—such as scoring models, analyzing written explanations, or interpreting classroom discourse—is time-consuming and difficult to manage during busy instructional schedules. AI is emerging as a powerful support tool to address these challenges. For example, machine learning-based scoring systems can evaluate written responses and diagrams quickly and reliably, reducing the burden of manual grading. Additionally, AI-powered collaborative conversational agents can guide and analyze students' real-time scientific argumentation, capturing critical components in student thinking and offering immediate feedback (de Araujo et al., 2025). By integrating these tools, science teachers can gain timely, actionable insights into students' engagement with science practices—enabling more responsive instruction, personalized support, and stronger alignment with NGSS goals.

7.2.3.1 Automatically Scored Science Assessment Responses

In science classrooms, open-ended tasks are essential tools in prompting students' thinking. While multiple-choice questions are convenient to score and analyze, they often fall short in revealing students' reasoning (Black & Wiliam, 1998). In contrast, open-ended formats—like written explanations, sketches, and models—offer much richer insights into students' ideas. However, assessing these responses can be time-consuming. In science classrooms, as a result, valuable student thinking often goes unexamined, and opportunities for timely feedback may be lost.

To address these challenges, three important developments in AI are particularly promising for classroom use. First, we are seeing a shift from task-specific machine learning models—which require customization—to more flexible tools like LLMs that can adapt to a wide range of science tasks with minimal setup. For example, teachers can now use an LLM like GPT-4 to help evaluate students' short written responses and explain why an answer is strong or weak. These tools do not just mark students' answers as right or wrong—they can provide scaffolded reasoning, helping teachers see where students are on their learning journey. A series of studies by Zhai and his collaborators (Latif & Zhai, 2024; Liu et al., 2023; Wu et al., 2023) demonstrated that LLMs, when guided by effective prompting, can approach the scoring accuracy of traditional task-specific models (e.g., fine-tuned Bidirectional Encoder Representations from Transformers). At the same time, LLMs provide broader applicability across varied topics and response formats.

Second, AI tools are expanding beyond written responses. Science learning is multimodal—students use words, graphs, diagrams, tables, and even gestures to communicate their ideas (Tang & Moje, 2010). Until recently, most automated tools could only handle typed text, leaving other types of work unassessed. But new multimodal LLMs that can interpret images, speech, and video are changing that. For example, Lee and Zhai (2025) used GPT-4 with vision capabilities to score student-drawn scientific models. The AI system was able to evaluate sketches for key content, identify errors, and suggest areas for improvement—all in near real-time. Imagine being able to upload a photo of a student's cell diagram and receiving instant feedback that highlights missing organelles or incorrect labels. This kind of tool could be a game-changer for teachers, especially in tasks that involve scientific modeling or visual explanations.

In addition to evaluating final products like written explanations and models, recent advances in AI are beginning to illuminate the inquiry process itself—a critical dimension of science learning. Science is not only about what students conclude, but also about *how* they arrive at those conclusions. In digital science simulations, for example, students engage in rich sequences of inquiry actions: selecting variables, designing tests, analyzing outcomes, and refining ideas. Emerging AI tools can now analyze these process-level behaviors by mining log data and identifying patterns in students' investigative strategies (Gobert et al., 2012; Radmehr et al., 2025). Instead of only seeing a student's final answer, teachers can gain insight into whether the student used a systematic approach or engaged in trial-and-error experimentation. These real-time analytics enable teachers to identify productive or unproductive inquiry strategies, provide timely feedback, and support students in developing more sophisticated scientific reasoning. By extending assessment beyond the final response to include the path students take, AI empowers teachers to more deeply understand and guide inquiry-based learning.

However, even with these advances, important challenges remain. Even when AI scoring tools are available and capable of evaluating students' multimodal and interactive responses, many science teachers encounter practical barriers to using these tools confidently and effectively (Holstein et al., 2019). A key challenge is the lack of explainability: most AI systems do not clearly show how they arrived at a particular score or piece of feedback. This makes it difficult for teachers to interpret results, explain them to students, or use them to guide instructional next steps. Teachers need tools that are transparent and trustworthy, where they can understand the reasoning behind the AI's output—especially when student learning is on the line. Another important issue is that many AI systems, especially LLMs like ChatGPT, require prompt engineering—the practice of crafting well-structured input prompts to guide the AI's response (Lee & Zhai, 2025). However, most teachers have not been trained in how to write effective prompts or adapt them to specific learning goals. Without support, they may receive vague, unhelpful, or inconsistent results. To bridge this gap, PD should include guidance on how to design and refine prompts (e.g., providing examples or rubrics in the prompt) and how to interpret AI responses critically (Zhang et al., 2023). Ideally, scoring tools should also include built-in scaffolds and prompt templates that align with classroom tasks, making it easier for teachers to use the technology without becoming AI experts. By prioritizing explainability and supporting effective prompt design, we can empower science teachers to use AI tools not just as black boxes, but as collaborative teaching aids that enhance student learning and formative assessment.

Vignette 7.3: Automatic Scoring of Three-Dimensional Learning

In Mr. Thompson's eighth-grade science class, students are exploring properties of matter through a phenomenon-based task called "Gas-Filled Balloons." The task presents a data table and image of four balloons filled with different gases, each labeled with volume, density, and flammability. Students are asked to determine which gases might be the same substance and explain their reasoning using evidence from the data. This open-ended assessment is designed to elicit students' understanding of the disciplinary core idea about the structure and properties of matter (PS1.A), the crosscutting concept of patterns, and the science practice of constructing explanations from evidence. As students submit their responses, the PASTA platform (https://ai4stem.org/pasta-teacher-platform/) automatically scores each one using an NGSS-aligned rubric (see Fig. 7.4). The rubric evaluates responses across three levels of explanation quality:

- **A1:** Correct identification that gases A and D could be the same
- **A2:** Partial justification using one data type (e.g., flammability or density)
- **A3:** Full justification using multiple data types across table columns

Immediately after submission, the system generates a diagnostic report that groups students by their level of reasoning (Fig. 7.4). Mr. Thompson opens the Holistic Report view, which shows the distribution of students across the three explanation levels (A1–A3). He notices that while many students arrived at the correct answer, fewer provided full justifications using both density and flammability data—suggesting a need to revisit how students synthesize evidence from multiple variables.

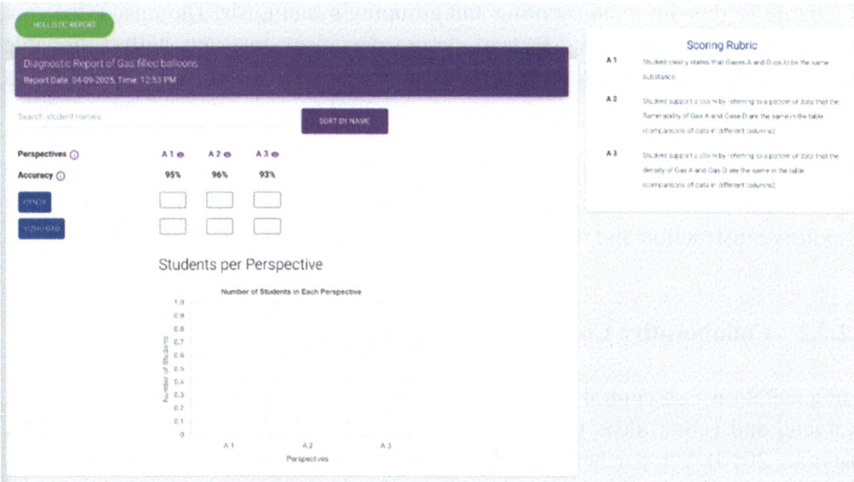

Fig. 7.4 NGSS-aligned holistic scoring report for the "gas-filled balloons" task

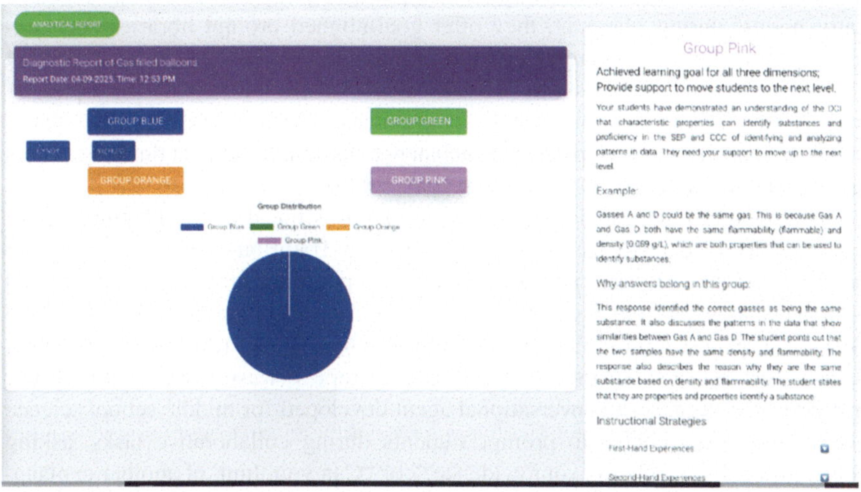

Fig. 7.5 Instructional grouping and diagnostic feedback in the PASTA analytical report

Switching to the Analytical Report (Fig. 7.5), Mr. Thompson sees that students are grouped into four instructional clusters: Group Pink, Green, Orange, and Blue. Each group is paired with a concise interpretation of students' current understanding and next-step instructional strategies. For example, Group Pink includes students who successfully integrated all three NGSS dimensions. Their sample responses are highlighted with explanations of why they are high-quality (e.g., citing both density and flammability to support their claims), and teachers are encouraged to provide enrichment or opportunities for peer coaching.

Thanks to this automatic scoring and grouping system, Mr. Thompson does not need to spend hours grading. Instead, he receives clear, instructionally actionable feedback that helps him plan targeted support and small group work for the next day. More importantly, the AI-powered scoring tool provides transparent rationales for each score, making it easier for him to give meaningful feedback to students and help them reflect on their thinking. With PASTA, Mr. Thompson can track student engagement with three-dimensional science learning in real time—empowering responsive instruction and deeper classroom dialogue.

7.2.3.2 Collaborative Conversational Agents

Argumentation is a central practice in science through which students construct, evaluate, and refine ideas based on evidence and reasoning (Driver et al., 2000; Osborne, 2010). Yet, teachers face persistent barriers—ensuring every student has a voice, monitoring evolving lines of reasoning, and delivering timely formative feedback—especially in large or time-pressed classes (Berland & Reiser, 2009; Chen et al., 2016; Forman & Ford, 2019; Mercer et al., 2019). Collaborative conversational agents (CCAs) lower these barriers by providing low-threshold, high-leverage entry points: during planning, they offer goal-aligned prompt libraries and talk-move "starters" teachers can slot into existing lessons; during enactment, they elicit claims, press for evidence, surface counter-arguments, and distribute talk equitably in real time; during reflection, they generate concise discourse analytics (e.g., claim-evidence-reasoning traces, participation balance) that make student thinking visible and targetable for next steps (de Araujo et al., 2025).

In terms of CCAs, one important advancement is the shift from scripted, rule-based agents to adaptive agents powered by LLMs. Traditional CCAs followed pre-programmed decision trees and often failed to respond meaningfully to unexpected or nuanced student input. By contrast, recent CCAs built on foundation models like GPT-4 can generate more flexible and contextually relevant responses, supporting open-ended dialogue across a range of science topics (Kasneci et al., 2023). For instance, ArguAgent—a conversational agent developed for middle school science classrooms—uses GPT-4 to prompt students during collaborative tasks, asking questions such as "What is your evidence?" or "Can you think of another explanation?" while adapting follow-up questions based on student responses (de Araujo et al., 2025). These agents promote metacognitive reflection, help students revise their reasoning in real time, and even facilitate peer-to-peer critique when integrated into group activities. In addition, they generate rich discourse data that teachers can use to assess reasoning processes that are often invisible in traditional assessments.

Recent studies have also begun exploring multimodal CCAs that combine NLP with visual or simulation-based tools. For example, Zapata-Rivera et al. (2017) integrated a conversational agent into a simulation-based inquiry task, where the agent guided students through variable manipulation, hypothesis generation, and explanation construction. In these environments, CCAs not only support real-time scientific reasoning but also help students coordinate across representations—such as graphs, tables, and visual models—during inquiry. By sustaining interaction throughout the inquiry process, these agents act more like collaborative learning partners than static assessors (Hou et al., 2025).

Despite their promise, several key challenges remain in the implementation of CCAs in science classrooms. One major concern is dialogue quality: even with advanced LLMs, CCAs can still generate responses that are generic, vague, or off-topic—particularly when student input is fragmented, unclear, or grammatically incorrect (Kuhail et al., 2023). Ensuring that agent responses are both instructionally relevant and pedagogically aligned remains a complex design challenge. Another pressing issue involves the impact of CCAs on student learning. For example, de Araujo et al. (2025) did not find significant improvements in students' conceptual understanding after using CCAs, raising questions about how these tools can be structured to support deeper learning outcomes rather than surface-level engagement. A final challenge is considering in what contexts and for what purpose chatbots are appropriate for use, especially with younger students as they are still learning how to separate make-believe from fact and exposure to chatbots could potentially blur that boundary.

This points to the need for distinguishing between different levels of goals that CCAs might support in science classrooms: (1) conversation efficacy—facilitating smooth, balanced, and on-task dialogue; (2) learning to argue—helping students develop argumentation skills such as making claims, supporting them with evidence, and responding to peers; and (3) arguing to learn—using argumentation as a mechanism for conceptual change and deeper understanding. Each of these goals requires different design considerations, particularly in terms of prompting strategies, response timing, and the depth of agent questioning. Yet, research has not clearly articulated how prompt designs map onto these different learning objectives. In the absence of this guidance, teachers may hesitate to rely on CCAs, especially when the agent's reasoning or pedagogical strategies are opaque or inconsistent (Holstein et al., 2019).

Moreover, designing effective prompts and interaction flows for CCAs requires not only familiarity with subject-matter content but also a strong grasp of learning progressions, discourse practices, and scaffolding techniques. Without adequate PD, teachers may underuse these tools or misapply them in ways that inadvertently suppress rather than support student thinking (Zhang et al., 2023). To fulfill their potential, CCAs must be accompanied by teacher-facing resources, such as goal-aligned prompt libraries, explanation features, and analytics dashboards that make student thinking visible and guide teacher interpretation and action.

Vignette 7.4: Supporting Science Talk with Clair

In Ms. Rivera's 7th-grade life science class, pairs are investigating how the digestive system breaks down and absorbs nutrients. Before the lesson, Ms. Rivera selects two goals in the *Clair* dashboard—balanced turn-taking and evidence-linked explanations—and loads a prompt bank keyed to target ideas (enzymes, villi, surface area).

Working online, Sofia and Mike stall after a hesitant start ("I think the first one is saliva." "I am not sure."). The dashboard flags 30 seconds of silence and a turn imbalance. With one tap, Ms. Rivera triggers a targeted nudge, and *Clair* posts:" It has been quiet—could you share what you think happens next, and one piece of evidence?" Mike types "Q3… Enzymes help nutrients to be absorbed." Seeing a likely misconception (confusing digestion with absorption), Ms. Rivera uses "whisper mode" to guide the next move: *Ask for an example enzyme and where absorption primarily occurs. Clair* follows up: "Sofia, could you build on Mike's idea by naming an enzyme and the place where most absorption happens?" Sofia replies, "Amylase in saliva breaks starch into sugars… and absorption mostly happens in the small intestine with the villi."

As the dialogue warms, Ms. Rivera switches *Clair* to a counter-check prompt: "What evidence shows absorption rather than digestion?", and the pair cites a diagram of villi increasing surface area. The dashboard now shows balanced turns and a complete claim, evidence, and reasoning. Ms. Rivera bookmarks the snippet for a quick share-out and circulates to other groups, confident that *Clair*'s prompts kept this pair on task while surfacing and repairing a key misconception—work that is difficult to sustain without constant facilitation.

In this vignette, Ms. Rivera sets the goals, selects prompts, tailors interventions, monitors real-time analytics, and curates artifacts for discussion. *Clair* amplifies her reach by handling routine prompts and visibility of talk patterns; Ms. Rivera ensures conceptual accuracy, aligns moves to objectives, and turns moments of confusion into teachable insights (Fig. 7.6).

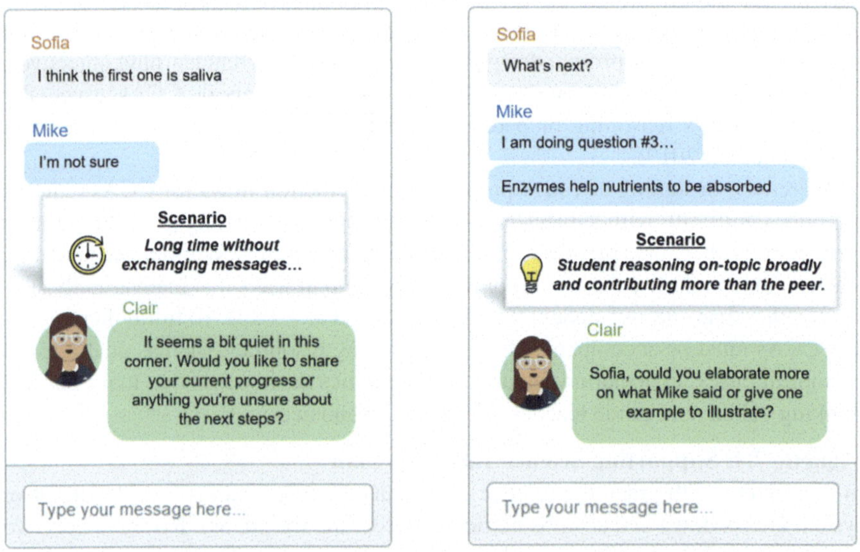

Fig. 7.6 Examples of interventions from *Clair* in a dialogue. Note. From "Investigating the impact of a collaborative conversational agent on dialogue productivity and knowledge acquisition," by A. de Araujo et al., 2025, *International Journal of Artificial Intelligence in Education*, https://doi.org/10.1007/s40593-025-00469-7. © 2025 The Author(s)

7.3 Responsible and Ethical Uses of AI in Science Education

With AI already woven into science classrooms, the question is not whether to use it but how to use it responsibly and ethically to advance science learning. In practice, AI should act as a coach rather than an answer-giver, pressing for claims, evidence, counterexamples, and uncertainty statements; pacing support with tiered, time-released hints so as to preserve productive struggle; and prompting cross-representational moves. AI should optimize learning opportunities and processes for all learners. For example, automatic scoring can return timely, level-appropriate (e.g., learning progression-based) feedback at scale while offering teacher override when confidence is low. These scores can populate dashboards that separate processes from products, enabling tiered support. Similarly, CCAs can widen participation by monitoring turn-taking, inviting quieter students in, and adapting prompts to language proficiency or prior knowledge. Together, such tools promote engagement in NGSS practices such as scientific argumentation and investigation, especially in classrooms where time, attention, and instructional bandwidth from teachers are limited.

As the integration of AI into teaching and learning accelerates, science educators are increasingly turning to AI-powered tools to meet diverse classroom needs. From planning differentiated lessons to analyzing lab reports, these tools have the potential to significantly enhance instructional efficiency and responsiveness. However, realizing that potential requires more than access, it requires intentional training, ethical awareness, and pedagogical alignment. Teachers must be supported in learning how to customize AI tools to fit their content areas, grade levels, and student populations in ways that center equity and deepen engagement. Despite widespread fears that AI might replace teachers or diminish the human aspects of education, these tools are far better suited to assist than to replace. AI cannot replicate the empathy, curiosity, or insight that teachers bring to their classrooms. Instead, when used responsibly, AI can automate routine tasks and free teachers to focus on the parts of science teaching that matter most: facilitating investigations, nurturing student inquiry, and building connections to real-world phenomena.

For AI to serve as a meaningful instructional partner, educators must see it not as a distant or impersonal force but as a natural extension of their professional toolkit. Building teacher confidence and adaptability is essential, and this begins with identifying entry points where AI applications clearly align with everyday teaching goals. Moreover, any AI system used in schools must uphold strict ethical and legal standards. Tools embedded in classrooms must comply with regulations such as the Family Educational Rights and Privacy Act (FERPA) to ensure student data remains secure. Teachers also need to be able to recognize and challenge potential biases in AI outputs that could affect learning outcomes. In science education, where inquiry, evidence, and ethical reasoning are foundational, we must ensure that our use of AI reflects those same principles. Responsible integration of AI offers not just efficiency gains but a pathway toward more personalized, inclusive, and intellectually vibrant science classrooms; classrooms that are led, always, by the professional judgment and values of teachers.

7.4 Conclusion

The integration of AI into K-12 science education offers transformative possibilities by addressing key entry points for science educators—streamlining lesson planning, enabling personalized instruction, and providing teachers with timely, actionable insights into student learning. When incorporated into PD, AI can also deepen science teachers' pedagogical content knowledge, strengthen their ability to design high-quality instruction, and expand their capacity to respond to diverse learners in real time. Yet, realizing this potential depends on more than simply providing access to AI tools. It requires deliberate, equity-driven implementation; robust, ongoing professional learning; and an unwavering commitment to ethical use. As science learning is inherently multimodal and encompasses words, graphs, diagrams, tables, and even gestures (Tang & Moje, 2010), AI must be designed to recognize and respond to these varied forms of expression. Conversational and collaborative AI systems hold promise for offering personalized, multimodal feedback to students and teachers alike, but only if teachers are equipped with the skills to design effective prompts, interpret AI-generated analytics, and connect these insights to meaningful next steps in instruction (Zhang et al., 2023). Without intentional professional learning and teacher-facing supports such as goal-aligned prompt libraries, explanation tools, and dashboards that make student thinking visible, AI tools risk being underused or misapplied, inadvertently limiting the very inquiry and critical thinking they aim to foster. Ultimately, integrating AI into K-12 science classrooms is not about replacing the human work of teaching but amplifying it. By coupling AI's efficiency and analytical power with teachers' deep knowledge of their students, science educators can create richer, more responsive learning experiences. With thoughtful planning, strong partnerships, and a commitment to ethical and equitable use, AI can become a catalyst for preparing both teachers and students to thrive in a technologically sophisticated, rapidly evolving world.

Disclosure of AI Use The authors acknowledge the use of generative artificial intelligence (GenAI) tools, including ChatGPT (OpenAI), for language editing and to enhance the clarity of the manuscript. The authors take full responsibility for all content, interpretations, and conclusions presented in this work.

Acknowledgement This material is based upon work supported by the National Science Foundation under Grant No. 2332964. Any opinions, findings, and conclusions or recommendations expressed in this material are those of the author(s) and do not necessarily reflect the views of the National Science Foundation.

References

Asano, Y., Beigman Klebanov, B., & Mikeska, J.N. (in review). What you ask is what you get? Exploring task formulation strategies to evaluate the coherence of classroom discussions with LLMs [Presentation]. In *AIED conference, Palermo, Italy.*

Beigman Klebanov, B., Suhan, M., & Mikeska, J. N. (2025, October 27-29). Towards evaluating teacher discourse without task-specific fine-tuning data. In Proceedings of the

Artificial Intelligence in Measurement and Education Conference (AIME-Con): Full Papers (pp. 192–200). https://aclanthology.org/2025.aimecon-main.21.pdf
Benedict, A., Holdheide, L., Brownell, M., & Foley, A. M. (2016). *Learning to teach: Practice-based preparation in teacher education*. Special Issues Brief.
Benedict-Chambers, A., Fick, S. J., & Arias, A. M. (2020). Preservice teachers' noticing of instances for revision during rehearsals: A comparison across three university contexts. *Journal of Science Teacher Education, 31*(4), 435–459.
Berland, L. K., & Reiser, B. J. (2009). Making sense of argumentation and explanation. *Science Education, 93*(1), 26–55.
Black, P., & Wiliam, D. (1998). Assessment and classroom learning. *Assessment in Education: Principles, Policy & Practice, 5*(1), 7–74.
Bondie, R., Mancenido, Z., & Dede, C. (2021). Interaction principles for digital puppeteering to promote teacher learning. *Journal of Research on Technology in Education, 53*(1), 107–123.
Bougdira, A., & Al Murshidi, G. (2025). Generative AI as a tool for personalized STEAM K-12 education. In *Handbook of research on innovative approaches to teaching and learning in STEAM education* (pp. 312–329). IGI Global. https://www.igi-global.com/chapter/generative-ai-as-a-tool-for-personalized-steam-k-12-education/372371
Boyd, D., Lankford, H., Loeb, S., & Wyckoff, J. (2005). Explaining the short careers of high-achieving teachers in schools with low-performing students. *American Economic Review, 95*(2), 166–171. http://www.aeaweb.org/articles.php?doi=10.1257/000282805774669628
Celik, I., Dindar, M., Muukkonen, H., & Järvelä, S. (2022). The promises and challenges of artificial intelligence for teachers: A systematic review of research. *TechTrends, 66*(4), 616–630.
Chen, J., Park, S., & Hand, B. (2016). Examining the use of talk and writing for students' development of scientific conceptual knowledge through constructing scientific explanations. *Learning and Instruction, 42*, 13–21.
Cowan, J., Goldhaber, D., Hayes, K., & Theobald, R. (2016). Missing elements in the discussion of teacher shortages. *Educational Researcher, 45*(8), 460–462.
Davis, E. A., Kloser, M., Wells, A., Windschitl, M., Carlson, J., & Marino, J.-C. (2017). Teaching the practice of leading sense-making discussions in science: Science teacher educators using rehearsals. *Journal of Science Teacher Education, 28*(3), 275–293. https://doi.org/10.1080/1046560X.2017.1302729
Davis, E. A., & Palincsar, A. S. (2023). Engagement in high-leverage science teaching practices among novice elementary teachers. *Science Education, 107*(2), 291–332.
de Araujo, A., Papadopoulos, P. M., McKenney, S., & de Jong, T. (2025). Investigating the impact of a collaborative conversational agent on dialogue productivity and knowledge acquisition. *International Journal of Artificial Intelligence in Education, 35*, 2254–2280.
Demszky, D., Liu, J., Hill, H. C., Sanghi, S., & Chung, A. (2025). Automated feedback improves teachers' questioning quality in brick-and-mortar classrooms: Opportunities for further enhancement. *Computers & Education, 227*, 105183.
Dieker, L. A., Rodriguez, J. A., Lignugaris/Kraft, B., Hynes, M. C., & Hughes, C. E. (2014). The potential of simulated environments in teacher education: Current and future possibilities. *Teacher Education and Special Education, 37*(1), 21–33. https://doi.org/10.1177/0888406413512683
Dillenbourg, P., & Hillaire, G. (2023). AI-based orchestration tools for classroom teaching. In *Advances in learning analytics and AI in education*. Springer.
Driver, R., Newton, P., & Osborne, J. (2000). Establishing the norms of scientific argumentation in classrooms. *Science Education, 84*(3), 287–312.
Erdemir, N., & Yeşilçınar, S. (2021). Reflective practices in micro teaching from the perspective of preservice teachers: Teacher feedback, peer feedback and self-reflection. *Reflective Practice, 22*(6), 766–781.
Forman, E. A., & Ford, M. J. (2019). Dialogic argumentation and students' conceptual understanding of science. *Studies in Science Education, 55*(1), 61–97.
Fuller, E. J., & Pendola, A. (2019). *Teacher preparation and teacher retention: Examining the relationship for beginning STEM teachers*. AAAS ARISE commissioned paper series.

Gobert, J. D., Sao Pedro, M., Baker, R., Toto, E., & Montalvo, O. (2012). Leveraging educational data mining for real-time performance assessment of scientific inquiry skills within microworlds. *Journal of Educational Data Mining, 4*(1), 111–143. https://jedm.educationaldatamining.org/index.php/JEDM/article/view/24

Gobert, J. D., Sao Pedro, M. A., & Betts, C. G. (2023). An AI-based teacher dashboard to support students' inquiry: Design principles, features, and technological specifications. In N. G. Lederman, D. L. Zeidler, & J. S. Lederman (Eds.), *Handbook of research on science education* (pp. 1011–1044). Routledge.

Grossman, P., Hammerness, K., & McDonald, M. (2009). Redefining teaching, re-imagining teacher education. *Teachers and Teaching: Theory and Practice, 15*(2), 273–289. https://doi.org/10.1080/13540600902875340

Gunawan, K. D. H., Kaniawati, I., & Setiawan, W. (2021, November). The responses to artificial intelligence in teacher integrated science learning training program. *Journal of Physics: Conference Series, 2098*(1), 012034. IOP.

Halder, S., & Mikeska, J. N. (2025, June 26). Exploration of recent innovations in digital teaching simulations to support science and mathematics teacher learning [Presentation]. In *International conference on STEM education: Innovation, integration, and impact*. University of Dhaka.

He, P., Shin, N., Zhai, X., & Krajcik, J. (2024). A design framework for integrating artificial intelligence to support teachers' timely use of knowledge-in-use assessments. In X. Zhai & J. Krajcik (Eds.), *Uses of artificial intelligence In STEM education* (pp. 52–94). Oxford Academic.

Hellrung, K., & Hartig, J. (2013). Understanding and using feedback—A review of empirical studies concerning feedback from external evaluations to teachers. *Educational Research Review, 9*, 174–190.

Holstein, K., McLaren, B. M., & Aleven, V. (2020). Designing for real-time classroom orchestration: The Lumilo mixed-reality teacher awareness tool. *International Journal of Artificial Intelligence in Education, 30*, 55–95. https://doi.org/10.1007/s40593-019-00186-9

Holstein, K., Wortman Vaughan, J., Daumé III, H., Dudik, M., & Wallach, H. (2019). Improving fairness in machine learning systems: What do industry practitioners need? In *Proceedings of the 2021 CHI conference on human factors in computing systems* (pp. 1–16). https://doi.org/10.1145/3290605.3300830

Hou, X., Forsyth, C., Andrews-Todd, J., Rice, J., Cai, Z., Jiang, Y., ... & Graesser, A. (2025, July). An LLM-enhanced multi-agent architecture for conversation-based assessment. In *International conference on artificial intelligence in education* (pp. 119–134). Cham: Springer Nature Switzerland.

Ilagan, M., Beigman Klebanov, B., & Mikeska, J.N. (2024). Automated evaluation of teacher encouragement of student-to-student interactions in a simulated classroom discussion. In *Proceedings of the 19th workshop on innovation use of NLP for building educational applications (BEA 2024)* (pp. 182–198). Association for Computational Linguistics. https://aclanthology.org/2024.bea-1.16/

Ingersoll, R. M., & May, H. (2011). The minority teacher shortage: Fact or fable? *Phi Delta Kappan, 93*(1), 62–65.

Ingersoll, R. M., & May, H. (2012). The magnitude, destinations, and determinants of mathematics and science teacher turnover. *Educational Evaluation and Policy Analysis, 34*(4), 435–464.

Jin, F. J. Y., Dai, W., Maheshi, B., Martinez-Maldonado, R., Gašević, D., & Tsai, Y. S. (2025). Feedback in K-12 and higher education: Educators' perspectives. *Teaching and Teacher Education, 156*, 104933.

John, G. A., Jr. (2025). AI in education: A systematic literature review of emerging trends, benefits, and challenges. *Seminars in Medical Writing and Education, 4*(2025), 795–812. https://ideas.repec.org/a/dbk/medicw/v4y2025ip795id795.html

Jong, M. S. Y., Yue, M., & Dai, Y. (2022). Pedagogical design of K-12 artificial intelligence education: A systematic review. *Sustainability, 14*(23), 15620. https://www.mdpi.com/2071-1050/14/23/15620

Jorgenson, H., Mikeska, J. N., & Cisterna, D. (2024). Professional learning leaders' perspectives on how automated and personalized feedback reports can support teacher learning. *Contemporary*

Issues in Technology and Teacher Education. George Washington University [Manuscript submitted for publication].

Jorgenson, H., Mikeska, J. & Cisterna, D. (2025). Professional learning leaders' perspectives on how automated and personalized feedback reports can support teacher learning. In R. Jake Cohen (Ed.), Proceedings of Society for Information Technology & Teacher Education International Conference (pp. 6138-6147). Orlando, FL, USA: Association for the Advancement of Computing in Education (AACE). Retrieved May 2, 2025 from https://www.learntechlib.org/primary/p/226103/

Käser, T., & Alexandron, G. (2023). Simulated learners in educational technology: A systematic literature review and a Turing-like test. *International Journal of Artificial Intelligence in Education*, 1–41.

Kasneci, E., Sessler, K., Küchemann, S., Bannert, M., Dementieva, D., Fischer, F., Gasser, U., Groh, G., Günnemann, S., Hüllermeier, E., Krusche, S., Kutyniok, G., Michaeli, T., Nerdel, C., Pfeffer, J., Poquet, O., Sailer, M., Schmidt, A., Seidel, T., ... Kasneci, G. (2023). ChatGPT for good? On opportunities and challenges of large language models for education. *Learning and Individual Differences, 103*, 102274. https://doi.org/10.1016/j.lindif.2023.102274

Kong, S. C., & Yang, Y. (2024). A human-centered learning and teaching framework using generative artificial intelligence for self-regulated learning development through domain knowledge acquisition. *IEEE Transactions on Learning Technologies*. https://ieeexplore.ieee.org/abstract/document/10507034

Krasniqi, D., & Ismajli, H. (2025). The role of feedback frequency on teachers' professional development and self-efficacy. *Educational Process: International Journal, 15*, e2025102.

Kuhail, M. A., Alturki, N., Alramlawi, S., & Alhejori, K. (2023). Interacting with educational chatbots: A systematic review. *Education and Information Technologies, 28*(1), 973–1018.

Latif, E., & Zhai, X. (2024). Fine-tuning ChatGPT for automatic scoring. *Computers and Education: Artificial Intelligence, 6*, 100210.

Lee, G., & Zhai, X. (2025). NERIF: GPT-4V for automatic scoring of drawn models. *Journal of Science Education and Technology*, 1–18. https://doi.org/10.1007/s10956-025-10262-9

Lee, G. G., Mun, S., Shin, M. K., & Zhai, X. (2025). Collaborative learning with artificial intelligence speakers: pre-service elementary science teachers' responses to the prototype. *Science & Education, 34*(2), 847–875.

Lin, P., & Van Brummelen, J. (2021). Engaging teachers to co-design integrated AI curriculum for K-12 classrooms. In *Proceedings of the 2021 CHI conference on human factors in computing systems*. https://dl.acm.org/doi/abs/10.1145/3411764.3445377

Liu, O. L., Brew, C., Blackmore, J., Gerard, L., Madhok, J., & Linn, M. C. (2014). Automated scoring of constructedresponse science items: Prospects and obstacles. *Educational measurement: issues and practice, 33(2), 19–28.* https://doi.org/10.1111/emip.12028

Liu, Z., He, X., Liu, L., Liu, T., & Zhai, X. (2023, June). Context matters: A strategy to pre-train language model for science education. In *International conference on artificial intelligence in education* (pp. 666–674). Springer Nature.

Masters, H. (2020). Using teaching rehearsals to prepare preservice teachers for explanation-driven science instruction. *Journal of Science Teacher Education, 31*(4), 414–434.

Mercer, N., Hennessy, S., & Warwick, P. (2019). Dialogue, thinking together and digital technology in the classroom: Some educational implications of a continuing line of inquiry. *International Journal of Educational Research, 97*, 187–199.

Mikeska, J.N., Beigman Klebanov, B., Marigo, A., Tierney, J., Maxwell, T., & Nazaretsky, T. (2024). Exploring the potential of automated and personalized feedback to support science teachers in learning how to attend to student ideas equitably and responsively. In A. M. Olney, I. A. Chounta, Z. Liu, O. C. Santos, I. I. Bittencourt (Eds.), *Artificial intelligence in education. AIED 2024. Lecture notes in computer science* (Vol. 14830). Springer. https://doi.org/10.1007/978-3-031-64299-9_19

Mikeska, J. N., Cross Francis, D. I., Lottero-Perdue, P. S., Park Rogers, M. A., Shekell, C., Bharaj, P. K., Howell, H., Maltese, A., Thompson, M., & Reich, J. (2025). Promoting preservice teachers' facilitation of argumentation in mathematics and science through digital simulations. *Teaching and Teacher Education, 154*. https://doi.org/10.1016/j.tate.2024.104858

Mikeska, J. N., Halder, S., Maxwell, T., Behl, K., & Bhatia, A. (2025, September 26). Exploring the affordances and challenges of generative AI teaching simulations to support science teacher learning [Presentation]. In *Northeast ASTE regional conference, Union, NJ.*

Mikeska, J. N., Tierney, J., Baehr, B., Maxwell, T., Marigo, A., Suhan, M., & Beigman Klebanov, B. (2024, September 22–25). Exploring the potential of artificial intelligence to generate feedback to support teacher learning [Paper presentation]. In *International Association for Educational Assessment (IAEA) conference, Philadelphia, PA.*

Mollick, E., & Mollick, L. (2023). *Assigning AI: Seven approaches for students, with prompts* (The Warton School Research Paper). https://ssrn.com/abstract=4475995

National Academies of Sciences, Engineering, and Medicine (NASEM). (2021). *Science and engineering in preschool through elementary grades: The brilliance of children and the strengths of educators*. https://edpreplab.org/sites/default/files/2022-01/NAP-PreK-5-STEM.pdf

National Academies of Sciences, Engineering, and Medicine (NASEM). (2023). *Advancing anti-racism, diversity, equity, and inclusion in STEMM organizations: Beyond broadening participation*. The National Academies Press. https://doi.org/10.17226/26803

National Research Council. (2012). *A framework for K-12 science education: Practices, crosscutting concepts, and core ideas*. National Academies Press. https://doi.org/10.17226/13165

Nazaretsky, T., Mikeska, J. N., & Beigman Klebanov, B. (2023). Empowering teacher learning with AI: Automated evaluation of teacher attention to student ideas during argumentation-focused discussion. In *The proceedings of learning analytics and knowledge conference, Arlington, TX*. https://dl.acm.org/doi/pdf/10.1145/3576050.3576067

NGSS Lead States. (2013). *Next generation science standards: For states, by states*. National Academies Press. https://www.nextgenscience.org/

OECD. (2024). *OECD AI Principles overview*. OECD. https://oecd.ai/en/ai-principles

Osborne, J. (2010). Arguing to learn in science: The role of collaborative, critical discourse. *Science, 328*(5977), 463–466.

Ouyang, F., Jiao, P., McLaren, B. M., & Alavi, A. H. (Eds.). (2022). *Artificial intelligence in STEM education: The paradigmatic shifts in research, education, and technology*. CRC Press.

Ovando, M. N. (2005). Building instructional leaders' capacity to deliver constructive feedback to teachers. *Journal of Personnel Evaluation in Education, 18*(3), 171–183.

Pellas, N. (2025). The impact of AI-generated instructional videos on problem-based learning in science teacher education. *Education Sciences, 15*(1), 102.

Popenici, S. A., & Kerr, S. (2017). Exploring the impact of artificial intelligence on teaching and learning in higher education. *Research and Practice in Technology Enhanced Learning, 12*(1), 22.

Quellmalz, E. S., & Silberglitt, M. D. (2017). SimScientists: Affordances of science simulations for formative and summative assessment.

Radmehr, B., Shved, E., Güreş, F. B., Singla, A., & Käser, T. (2025, July). ClickSight: Interpreting student clickstreams to reveal insights on learning strategies via LLMs. In *International conference on artificial intelligence in education* (pp. 94–102). Springer Nature.

Rahman, A., & Shuvo, T. K. (2025). Impact of artificial intelligence on learning outcomes: A study in the context of Bangladesh. *The Science Post, 4*(1), 22–36. https://www.thesciencepostjournal.com/index.php/tsp/article/view/74

Ray, P. (2023). ChatGPT: A comprehensive review on background, applications, key challenges, bias, ethics, limitations and future scope. *Internet of Things and Cyber-Physical Systems, 3*, 121–154.

Reich, J. (2022). Teaching drills: Advancing practice-based teacher education through short, low-stakes, high-frequency practice. *Journal of Technology and Teacher Education, 30*(2), 217–228.

Rodgers, W. J., Kennedy, M. J., VanUitert, V. J., & Myers, A. M. (2019). Delivering performance feedback to teachers using technology-based observation and coaching tools. *Intervention in School and Clinic, 55*(2), 103–112.

Rodriguez, J., McKinney, T., Powell, S., Walker, Z., & Vince Garland, K. (2020). Was this feedback useful? Examining the observation and feedback process for pre-service teachers. *Teaching Education, 31*(2), 144–161.

Scheeler, M. C., Ruhl, K. L., & McAfee, J. K. (2004). Providing performance feedback to teachers: A review. *Teacher Education and Special Education, 27*(4), 396–407.

Spasopoulos, T., Sotiropoulos, D., & Kalogiannakis, M. (2025). Generative AI in pre-service science teacher education: A systematic review. *Advances in Mobile Learning Educational Research, 5*(2), 1501–1523.

Straub, C., Dieker, L., Hynes, M., & Hughes, C. (2015). *Using virtual rehearsal in TLE TeachLivE™ mixed reality classroom simulator to determine the effects on the performance of science teachers: A follow-up study*. 2015 TeachLivE™ National Research Project: Year 2 Findings.

Suresh, A., Jacobs, J., Lai, V., Tan, C., Ward, W., Martin, J. H., & Sumner, T. (2021). Using transformers to provide teachers with personalized feedback on their classroom discourse: The talkmoves application. *arXiv Preprint*. https://doi.org/10.48550/arXiv.2105.07949

Tang, K.-S., & Moje, E. B. (2010). Relating multimodal representations to science literacy: An exploration of adolescents' meaning making of science texts. *Journal of Research in Science Teaching, 47*(9), 1026–1059. https://doi.org/10.1002/tea.20354

Urmeneta, A., & Romero, M. (2024). *Creative applications of artificial intelligence in education*. Springer Nature.

VanLehn, K. (2020). The relative effectiveness of human tutoring, intelligent tutoring systems, and other tutoring systems. *Educational Psychologist, 55*(4), 219–238.

Walter, Y. (2024). Embracing the future of artificial intelligence in the classroom: The relevance of AI literacy, prompt engineering, and critical thinking in modern education. *International Journal of Educational Technology in Higher Education, 21*(1), 15.

Wang, S., Wang, F., Zhu, Z., Wang, J., Tran, T., & Du, Z. (2024). Artificial intelligence in education: A systematic literature review. *Expert Systems with Applications, 252*, 124167.

Wang, X., Zheng, Y., & Rosé, C. P. (2024). Teachers' use of chat-based AI assistants to support differentiated instruction. *AI in Education Research Review, 12*(2).

Watkins, C. (2004). Feedback between teachers. In *Feedback for learning* (pp. 64–80). Routledge.

Wiuggansa, J. (2024). Mastering AI 2025 guide for teachers: Empowering educators to transform classrooms with AI and ethics.

Wolf, L. (2015). *The STEM teacher drought: Cracks and disparities in California's math and science teacher pipeline*. Education Trust-West.

Wu, X., He, X., Liu, T., Liu, N., & Zhai, X. (2023, June). Matching exemplar as next sentence prediction (MeNSP): Zero-shot prompt learning for automatic scoring in science education. In *International conference on artificial intelligence in education* (pp. 401–413). Springer Nature.

Zapata-Rivera, D., Liu, L., Chen, L., Hao, J., & von Davier, A. A. (2016). Assessing science inquiry skills in an immersive, conversation-based scenario. In *Big data and learning analytics in higher education: Current theory and practice* (pp. 237–252). Springer International Publishing.

Zawacki-Richter, O., Marín, V. I., Bond, M., & Gouverneur, F. (2019). Systematic review of research on artificial intelligence applications in higher education—Where are the educators? *International Journal of Educational Technology in Higher Education, 16*(1), 39.

Zhang, J., Zou, Y., & McKinney de Royston, M. (2023). Preparing teachers for AI-integrated classrooms: A review of teacher professional development approaches. *Computers & Education: Artificial Intelligence, 4*, 100121. https://doi.org/10.1016/j.caeai.2023.100121

Part IV
Learning Materials

Chapter 8
Transforming Science Learning Materials in the Era of Artificial Intelligence

Xiaoming Zhai and Kent Crippen

8.1 Introduction

The integration of artificial intelligence (AI) into science education is profoundly reshaping the landscape of science learning materials. AI technologies are increasingly embedded in new forms of learning materials via various educational approaches—adaptive platforms (Tan et al., 2025), intelligent tutoring systems (Létourneau et al., 2025), multimodal simulations (Almasri, 2022), and generative tools (El Fathi et al., 2025)—fundamentally altering how students' access, process, and engage with scientific knowledge. These transformed learning materials not only support personalized learning trajectories but also reflect and simulate authentic scientific practice, that includes data modeling, hypothesis testing, and evidence--based reasoning (Erduran & Levrini, 2024; Herdiska & Zhai, 2024; Xu & Ouyang, 2022). In parallel with broader societal and scientific transformations, the integration of AI in learning materials (i.e., AI support) signals a pedagogical shift: moving beyond the static delivery of facts toward environments where complex data sets can be explored, hypotheses can be generated and tested in real-time, and patterns can be interpreted with AI-supported modeling tools. Not surprisingly, science learning materials have not evolved in tandem to reflect these new affordances and to prepare students to work in AI-supported scientific contexts (Zhai & Krajcik, 2025).

Conventional textbooks and instructional designs frequently neglect the dynamic, context-sensitive, and multimodal characteristics that can be significantly enhanced by AI. They present knowledge as fixed, overlook the influence of learners' diverse

X. Zhai (✉)
AI4STEM Education Center, University of Georgia, Athens, GA, USA
e-mail: Xiaoming.zhai@uga.edu

K. Crippen
School of Teaching and Learning, University of Florida, Gainesville, FL, USA

X. Zhai, K. J. Crippen (eds.), *Advancing AI in Science Education*, Advances in Technology-Rich Science Education 3,
https://doi.org/10.1007/978-3-032-16871-9_8

perspectives, and offer limited interactivity that cannot adapt to students' questions or reasoning. These limitations restrict opportunities for authentic inquiry and evidence-based practice, areas where AI-supported materials can offer significant improvement. As scientific knowledge and practice become increasingly data-intensive and collaborative (Dong et al., 2017; Tenopir et al., 2011), there is a pressing need to redesign science curricula and materials to reflect these developments. Modern scientific inquiry frequently involves AI-supported tools for data collection, analysis, and modeling, techniques now central to disciplines ranging from climate science to genomics (Gil et al., 2014). For instance, in climate science, deep learning has been applied to predict extreme weather events and enhance climate model predictions by analyzing massive datasets from satellites and sensors (Reichstein et al., 2019). Similarly, in genomics, AI has accelerated the identification of gene–disease associations and advanced personalized medicine by enabling high-throughput analysis of sequencing data and predictive modeling of protein structures (Eraslan et al., 2019). Together, these cases illustrate how AI-supported practice has reshaped not only the scale of scientific data analysis but also the collaborative infrastructures that support discovery across domains.

To remain authentic and relevant, science learning materials must mirror these shifts by embedding AI-supported methods into instructional tasks and representations. Doing so can help students understand the nature of contemporary science, which is increasingly computational, algorithmic, and reliant on AI collaboration (Kitano, 2016). Without integrating these methods, science education risks presenting an outdated and incomplete view of how science is practiced. Moreover, students' roles are changing—they are not merely recipients of knowledge but active participants and co-constructors of meaning alongside intelligent systems, engaging in iterative processes of knowledge construction supported by AI tools that model, prompt, and adapt to learners' responses (Chen, 2025). This calls for a reimagining of how science is taught and learned, emphasizing flexibility, real-time responsiveness, and alignment of learning materials with the epistemic nature of contemporary science.

Moreover, AI is playing an increasingly important role in developing and designing learning materials with curriculum experts. Generative AI tools, such as ChatGPT and Claude, are being used collaboratively by educators to co-develop lesson plans, design formative assessments, and generate contextualized instructional content across subject areas and grade levels (Zhai, 2023). This partnership enhances the efficiency and scalability of curriculum development, allowing instructional designers to produce tailored materials that are adaptive to diverse learner needs. In addition, AI can rapidly analyze curriculum standards, student performance data, and scientific literature to recommend content sequences or highlight emerging concepts relevant to instruction (Su & Zhong, 2022). Such integration signifies a paradigmatic shift in the instructional design process, where human-AI co-authorship enables rapid prototyping and iterative refinement of educational materials. This collaborative process not only augments human creativity and productivity but also ensures a more responsive curriculum that can evolve with scientific advancements and societal needs. By leveraging large-scale datasets and natural language generation, AI enables educators to design tasks that are more closely aligned with real-world scientific inquiry (Crippen & Archambault, 2012), thereby fostering students' critical thinking and problem-solving skills.

Despite its promise, the integration of AI into science learning materials raises critical ethical and pedagogical concerns that must be urgently addressed. Unresolved issues, including algorithmic bias, misinformation, lack of transparency, and data privacy risks, pose significant threats to equitable and inclusive science learning (Akgun & Greenhow, 2022; Latif & Zhai, 2025). These risks are particularly pronounced when AI systems are used to create or curate content without adequate human oversight or validation. Without clear standards or safeguards, there is a danger of perpetuating existing educational inequities or misrepresenting scientific knowledge, especially for marginalized communities. Moreover, the opaqueness of many AI models complicates teachers' and learners' ability to critically evaluate AI-generated information, potentially undermining student and teacher agency in the classroom. These concerns highlight a pressing gap in the current educational landscape: the need to systematically examine and establish responsible and ethical norms, frameworks, and practices that ensure AI technologies support the transformation of science learning materials rather than compromise the goals of science education. The absence of such considerations could result in science learning materials that are technologically advanced but ethically fragile and pedagogically misaligned.

Given these gaps, this chapter examines the transformative potential of AI integration in science learning materials, articulating the changes brought by AI—both the opportunities and challenges associated with its integration in science learning materials. Drawing on current research, design examples, and educational frameworks, we examine key domains where AI is reshaping science learning materials, including embedded scientific practice (Bewersdorff et al., 2025), personalized learning (Yılmaz, 2024), interactive simulations (Erümit & Sarıalioğlu, 2025), multimodal content (Almasri 2022), accessibility (Lee et al., 2025), and content creation (Bewersdorff et al., 2025). While these domains can be considered separately, they are deeply interrelated: for instance, personalized learning depends on multimodal content and accessible design to be truly adaptive, while interactive simulations and embedded scientific practice together foster inquiry-rich experiences that also generate new opportunities for content creation. In practice, advances in one domain often catalyse developments in another, forming a network of mutually reinforcing transformations that together redefine what it means to design and use science learning materials. We engage with the responsible (how) and ethical (why) dimensions of these transformations, proposing solutions to guide the use of AI. By offering an analytical and normative lens, this chapter aims to support educators, developers, and decision-makers in designing inclusive, adaptive, and ethically informed science learning materials for the AI era.

8.2 Changes Brought by AI

8.2.1 AI Embedded in Scientific Practice

Contemporary science increasingly relies on AI for core practice, such as large--scale data analysis, predictive modeling, and hypothesis generation (see Chap. 2 of this book). These transformations reflect fundamental changes in the epistemic foundations of scientific research, where AI is not merely a tool but an active assistant or collaborator in scientific discovery (Bianchini et al., 2022). As scientists integrate AI in fields like climate science and genomics, where complex simulations and pattern recognition are essential (Kitano, 2016), it becomes imperative that science education reflects these changes. Science learning materials must evolve to reflect such authentic practice, enabling students to develop an understanding of how modern science is conducted and how AI is reshaping its methods, goals, and ethical considerations (Vartiainen et al., 2021).

A salient example of this shift can be observed in the case of AlphaFold, an AI system developed by DeepMind to predict protein structures with remarkable accuracy (Jumper et al., 2021). AlphaFold employs deep learning algorithms, specifically a neural network architecture trained on extensive protein sequence and structure databases, to predict three-dimensional protein configurations from amino acid sequences. Its methodology integrates multiple sequence alignments and attention mechanisms to infer spatial relationships among residues, enabling predictions that rival the accuracy of experiments. This tool has transformed genomics and structural biology by resolving the structures of proteins that were previously difficult or impossible to characterize using traditional techniques such as X-ray crystallography or cryo-electron microscopy. As a result, AlphaFold has been integrated into research workflows to accelerate drug discovery, understand genetic diseases, and interpret evolutionary relationships among organisms. In response, educational initiatives have begun integrating AlphaFold-inspired activities into biology curricula (Samson, 2025). These activities include AI-supported simulations where students explore protein folding, test hypotheses, and interpret structural data using tools similar to those employed by professional scientists.

In science classrooms, researchers have explored the integration of AI in scientific inquiry activities. For example, embedded AI tools such as Google's Teachable Machine can be used to guide students in training the algorithm to identify specific features of plants, thus mimicking scientists' work (see Fig. 8.1 for an example). Working in small groups, students first collect and label their own sets of plant images to serve as training data. They consider which features, such as leaf shape, flower color, or stem structure, would best distinguish one plant type from another, mirroring how scientists identify and operationalize variables. They upload their datasets, organizing them into labeled classes, and training the algorithm to recognize patterns across their images. Students then test the algorithm with new images, compare its classifications with their own judgments, and refine the datasets and labels to improve accuracy. Through this process, they engage in core scientific

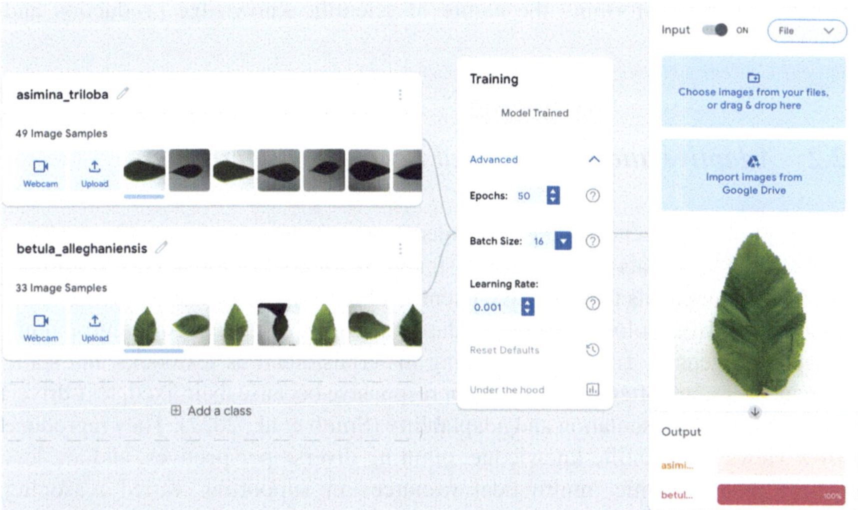

Fig. 8.1 An example of students using Google Teachable Machine to train a model to differentiate two types of leaves

practice, including the design of investigations, analyzing and interpreting data, constructing and revising models, and validating findings against empirical evidence. Such instructional experiences provide students with a window into real--world scientific processes enhanced by AI, thereby bridging the gap between classroom learning and contemporary scientific practice. In addition to Google's Teachable Machine, teacher-friendly applications such as Microsoft Lobe (https://www.lobe.ai), IBM Watson Studio (https://dataplatform.cloud.ibm.com), MIT App Inventor (https://appinventor.mit.edu), and Machine Learning for Kids (https://machinelearningforkids.co.uk) are also available to achieve this goal.

The Google Teachable Machine case exemplifies the potential for reimagining science learning materials, considering contemporary scientific practice, which increasingly relies on AI. Traditional science learning materials, often presented as linear, static, and text-driven, consistently underperform when compared to dynamic, interactive alternatives. For example, students using simulations or sequenced frames outperformed those relying on conventional static diagrams in kinematics study (Mešić et al., 2015). Moreover, frameworks like Duschl's Evidence-Explanation Continuum make explicit how scientific reasoning evolves through iterative, data-rich cycles of measurement, modeling, and explanation, cycles that static texts rarely support (Duschl et al., 2021). As exemplified by tools like AlphaFold, AI enables scientists to make discoveries that depend on large-scale data analysis, complex simulations, and predictive modeling—capabilities that are beyond the scope of traditional science instruction. Yet, most current learning materials do not provide students with opportunities to engage in AI-supported practice. Without meaningful updates to reflect how science is now practiced, students are left with an outdated view of inquiry and discovery, missing key epistemic insights into how AI is transforming the nature of scientific knowledge production and application.

8.2.2 *Adaptive and Personalized Learning Experiences*

The advent of AI presents an unprecedented opportunity to create new forms of science learning materials specifically designed to support adaptive (i.e., real-time, responsive adjustments to pace and content) and personalized learning (i.e., a broad combination of technology, including adaptive options, with pedagogy, curriculum, and student agency). Traditional learning materials, such as textbooks and static online modules, are often not inclusive or responsive because their fixed, text-driven structures limit representation and adaptability (Smith et al., 2022). They reproduce narrow views of scientific knowledge, omitting diverse perspectives, and are less effective than dynamic, multimodal resources in supporting varied reasoning approaches (Stelzer et al., 2009). Moreover, research on adaptive assessments demonstrates how static formats fail to accommodate learners' prior knowledge, underscoring the need for more responsive designs (Zhai et al., 2021).

AI-supported adaptive learning systems can dynamically generate and adjust learning content based on real-time data collected from student interactions (Kabudi et al., 2021; Létourneau et al. 2025; Wang et al., 2024). These systems analyze learner inputs, such as quiz responses, engagement metrics, and behavioral data, to develop detailed learner profiles that guide individualized instruction. Through this process, AI can deliver learning materials that adjust to each student's unique pace, scaffolding, modality, and complexity, aligning with their individual progression. Such learning environments are not only more responsive but also more inclusive, particularly for students from linguistically and culturally diverse backgrounds or those with learning differences (Nyaaba et al., 2024).

Several emerging applications of AI demonstrate the possibilities of creating novel forms of science learning materials, particularly for adaptive and personalized learning. In biology, systems such as the Learning Intelligent Tutoring System (LITS) offer interactive tutorials on cellular functions that adjust the depth and style of explanation in real-time. For example, Schmucker et al. (2024) introduced a LLM-based conversational tutoring system, which provides adaptive support to learners. The LITS system automatically generates a tutoring script from lesson texts using LLMs, which specify review questions, solutions, and learning expectations in a format that can be easily edited by instructors. The system is orchestrated through two conversational agents—Ruffle, a student, and Riley, a professor—who guide the learner in a learning-by-teaching scenario. Ruffle prompts the learner to explain concepts, while Riley provides scaffolding, corrective feedback, and clarification when misconceptions are detected. This dual-agent design enables adaptive dialogues that follow the inner- and outer-loop structures typical of intelligent tutoring systems, but with significantly reduced authoring costs compared to traditional CTSs. Empirical evaluations with biology lessons on cell organelles demonstrated that learners rated the system as more engaging, coherent, and helpful than simpler question-answer (QA) chatbots, and that many completed the entire conversational workflow. Although no significant differences in short-term test performance were observed compared to reading conditions, interaction analyses revealed that learners who engaged more deeply in conversations achieved higher learning gains, highlighting the potential of LLM-driven systems to foster adaptive, personalized science learning at scale.

The shift toward adaptive AI-supported learning materials represents more than a technological innovation—it demands a pedagogical transformation. Such materials can offer differentiated entry points into scientific content, enabling all students to engage meaningfully with core ideas and practice. Unlike conventional resources, which often assume a uniform baseline of understanding, AI-supported materials can be fluid, multimodal, and responsive to formative data. They have the potential to foster not only content mastery but also learner autonomy, self-regulation, and metacognitive reflection. To realize this potential, educators and developers must collaborate to design materials that utilize AI not only to customize learning paths but also to deepen scientific inquiry, simulate authentic practice, and promote educational accessibility. In doing so, the development of new AI-supported learning materials becomes essential—not just for personalization, but for transforming science education into a more adaptive, responsive, and student-centered enterprise.

8.2.3 Contextualized and Interactive Simulations

AI enables the development of new forms of science learning materials by powering contextualized and interactive simulations that reflect real-world phenomena (Almasri, 2022) Going far beyond traditional, static digital labs, these AI-supported simulations may dynamically adjust parameters, prompts, and investigative pathways based on students' real-time input, inquiry strategies, and learning context (Adair et al., 2023; Ben-Zion et al., 2025). This shift transforms students' roles and experience with simulations, from passive users to creators and designers. This new form of AI-supported simulations also enables highly personalized, scenario-based environments that can replicate the complexity and iterative nature of modern scientific research.

Figure 8.2 shows an example of transforming a sketch into interactive simulations when students learn projectile motion. In this example, a student drew a symbolic ball and wall with variables such as speed, gravitational coefficient, and the distance from the ball to the wall, and then supplied the sketch to ChatGPT (Fig. 8.2, left). ChatGPT then generates the code for the HTML file, which is opened in a browser. After opening the HTML file, students can study projectile motion using the simulation by manipulating variables such as the speed of release, running the simulation, and observing the resulting path of the ball as it interacts with the wall under the force of gravity (Fig. 8.2, right). In this example, AI simulations not only enhance engagement but also serve as personalized learning materials that involve

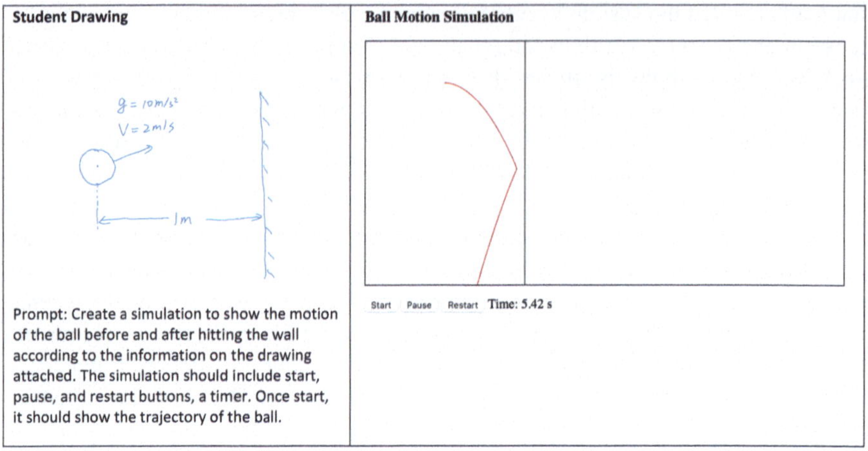

Fig. 8.2 Sketch to simulation. Students use ChatGPT to create an interactive simulation based on a hand sketch (This example idea credited to Jennifer Kleim, a Ph.D. student of the first author.)

students in creating, using, and testing simulations of observable phenomena for science learning. However, it also raises important questions about the accuracy of the underlying simulation. Thoughtful pedagogy such as engaged critiquing is needed to leverage these student-AI generated simulations for epistemic purposes.

Besides the simple on-the-fly simulations, AI can generate and model complex scientific phenomena that are otherwise inaccessible without significant efforts in coding and testing. For example, Ben-Zion et al. (2024) discuss leveraging generative AI to rapidly develop physics simulations, allowing educators to create personalized virtual labs such as a pendulum simulation (see Fig. 8.3). Unlike PhET simulations (Finkelstein, 2006), which are professionally developed, thoroughly validated, and designed for broad classroom use, such AI-generated simulations are highly customizable, allowing educators and even students to generate and refine simulations tailored to specific concepts or edge cases not covered by existing tools. While PhET offers stable, classroom-ready models, AI simulations uniquely position learners as co-designers, putting them in charge of the iterative processes of creation, validation, and refinement, as the practice of science. Moreover, students can explore complex physical phenomena, such as quantum mechanics and wave functions, in a controlled, risk-free environment, thereby overcoming traditional barriers related to safety and resource limitations (Ben-Zion et al., 2024). In biology, similar simulations can model disease transmission with adjustable mutation rates and vaccine efficacy, enabling students to explore causal reasoning and systems thinking. These simulations are not merely interactive tools; they are intelligent learning materials that evolve in response to student inputs, fostering a deeper understanding through contextualized, personalized scenarios.

Moreover, AI simulations also introduce new material formats through the inclusion of adaptive avatars and conversational agents. These elements support collaborative inquiry by facilitating dialogue, offering counterarguments, and prompting

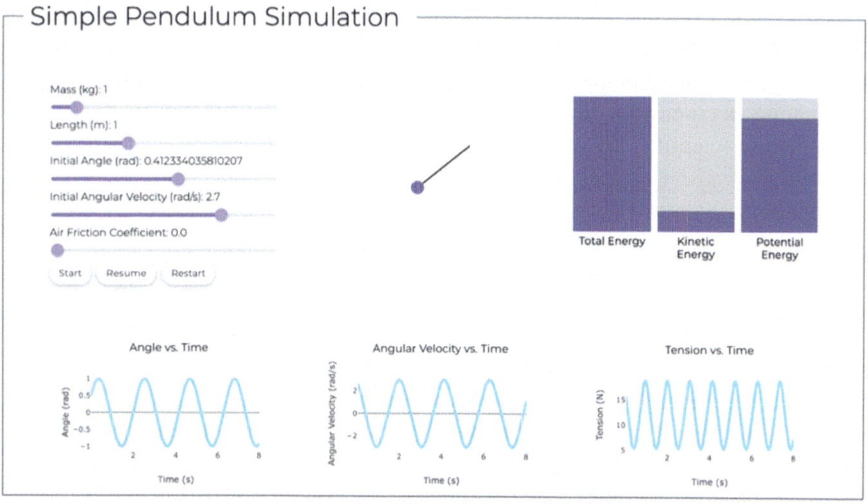

Fig. 8.3 Pendulum simulation featuring large-angle dynamics, friction effects, pendulum tension graph, and energy distribution visualization. Reprint from Ben-Zion et al. (2024), arXiv:2412.07482 under CC BY 4.0 (https://creativecommons.org/licenses/by/4.0/). No changes were made

metacognitive reflection (Makransky et al., 2019). A recent review study suggests that AI-supported virtual agents can engage students in decision-making tasks that mirror authentic challenges, yielding medium overall effects on learning ($g = 0.43$) when compared to simulations without AI-supported agents (Dai et al., 2024). In science classrooms, AI-supported virtual agents function as adaptive guides, providing feedback, hints, and personalized support to help students navigate complex inquiry tasks and decision-making processes (Thompson & McGill, 2017; Ward et al., 2013). As such, these new forms of learning materials embed students within scientific discourse communities, encouraging not only content mastery but also the development of communication and argumentation skills essential to science literacy.

To capitalize on these advancements, the design of AI-supported simulations must be rooted in robust pedagogical frameworks. Educators must ensure that these materials go beyond delivering automated answers to cultivating critical inquiry, reflection, and agency. Especially given that hallucinations represent a persistent issue impacting the validity and reliability of any output from AI (Ji et al., 2022). This can be partially achieved by using inquiry-based and modeling practices to structure the use of simulations. Rather than positioning AI as a source of solutions, teachers can frame it as a partner in investigation, prompting students to evaluate, refine, and extend its outputs. Embedding opportunities for reflection, such as asking students to explain why a simulation behaved in a certain way or to critique the validity of AI-generated results, encourages deeper reasoning. Agency can be further supported by allowing students to choose parameters, datasets, or approaches, positioning them as decision-makers in the learning process. In this way, AI becomes a catalyst for evidence-based thinking and metacognitive growth, rather than a shortcut to answers. When properly integrated, AI-supported simulations represent a significant evolution in science learning materials: from passive, uniform resources to dynamic, personalized, and context-sensitive environments that prepare students for authentic scientific engagement.

8.2.4 Integration of Multimodal and Dynamic Content

AI supports automatic generation and analysis of multimodal materials (e.g., text, image, video, data), leading to a fundamental transformation of science learning materials. Traditionally limited to static, text-centric formats, science materials are now evolving into dynamic, interactive, and richly multimodal resources (Stelzer et al., 2009; Zhai et al., 2021). AI can automatically generate visualizations from raw data, produce narrated video summaries of complex experiments, or create interactive diagrams that adapt in real time to student interactions (Bewersdorff et al., 2025). These affordances enable science content to be represented and for those representations to change dynamically in ways that are more aligned with how knowledge is constructed and communicated in professional scientific practice (Lee et al., 2023). As a result, learners benefit from more engaging and accessible materials that support diverse preferences, learning profiles, and disciplinary literacy skills.

Generative AI can also be used to create visualized learning materials, such as analogies to science concepts, to facilitate learning. For example, electric voltage and current have been challenging for middle school students to learn due to their abstract nature. To help students understand the concepts, Cao et al. (2023) used generative AI to generate multimodal learning materials, images, and videos of water pressure/current that analogize electric voltage and current, reducing the learning difficulty. Chemistry concepts, such as electrolysis and gas emission, can be difficult for students to visualize because they involve abstract or microscopic processes. To address this, Akaygun and Kilic (2025) had preservice teachers use generative AI to create text- and image-based visual analogies. The AI-generated materials were perceived as attractive and engaging, facilitating students' understanding of the macroscopic aspects of chemical processes. However, teachers also noted limitations in scientific plausibility and tool constraints, underscoring the need for teacher guidance in their use.

Another major shift enabled by AI is the capacity to generate up-to-date, contextually relevant content that reflects the rapidly evolving nature of scientific knowledge. Science learning content needs to reflect the contemporary and current development of science, but curating and transforming these new developments into accessible learning materials has traditionally been infeasible. With the development of AI, tools such as Elicit and Semantic Scholar can now scan and synthesize emerging research, providing real-time summaries and curated insights that can be directly integrated into classroom science learning materials. These systems can suggest research questions, generate comparisons, summaries, or explanations.

Notebook LM, a current product from Google, allows students to upload class materials, such as lab guides, readings, published papers, or data sets, and then ask questions, receive summaries, or generate explanations grounded in those sources. Its purpose is to help students connect ideas across documents, making complex science content more accessible while keeping inquiry focused on teacher-curated materials. This interaction positions students as investigators, using AI to highlight connections, identify gaps in their understanding, and test their interpretations against multiple sources, including primary, peer-reviewed, empirical research publications. Teachers could further scaffold reflection by asking students to critique the AI's outputs, reinforcing skills in evidence evaluation and scientific reasoning while preventing over-reliance on automated responses. This type of use enables science learning resources to remain current and epistemologically authentic, exposing students to the provisional and dynamic character of scientific understanding. Using these AI-supported science learning materials, students can enhance their awareness of the evolving nature of science, highlighting how such materials support deeper epistemic cognition.

However, the shift toward AI-supported materials demands critical attention to their educational validity and use. There is an inherent risk of over-reliance on algorithmic outputs, which may oversimplify complex scientific concepts, present the concepts inaccurately, or unintentionally perpetuate bias. Therefore, the transformation of science learning materials must be guided by intentional design, with educators equipped to interpret, adapt, and contextualize AI-generated content. As AI continues to reshape how learning materials are created and used, it is essential that its role be understood as augmenting—and not replacing—the pedagogical expertise of teachers and the epistemic engagement of students (Wang et al., 2025).

8.2.5 Language and Accessibility Support

AI-supported translation, summarization, and speech-to-text features are transforming the accessibility of science learning materials, particularly for multilingual and neurodiverse learners. These tools enable the real-time adaptation of complex scientific texts into multiple languages or simplified summaries, making the content more accessible to students with varied linguistic backgrounds and reading abilities. For example, platforms such as Microsoft's Reading Progress were integrated into digital learning environments to promote English Language Learners' (ELLs') reading aloud, thereby enhancing access to learning materials for ELLs and students with auditory processing challenges (Jose, 2025).

Moreover, AI technologies support neurodiverse learners by facilitating multimodal access to information (Panjwani-Charania & Zhai, 2024). Speech-to-text tools can convert students' spoken responses into written form, supporting those with dyslexia, dysgraphia, or fine motor difficulties. Likewise, summarization algorithms help reduce cognitive load by distilling key concepts and eliminating extraneous information from scientific texts. A study by Thaqi et al. (2024) using an

AI-supported reading assistant to help learners, particularly with lower learning proficiency or dyslexia, address challenges such as unfamiliar vocabulary and complex sentence structures by providing contextual explanations, rephrased content, and multilingual support. This approach highlights AI's potential to enhance the reading experience and support science learning for students with lower reading proficiency or neurodiverse learners.

AI can also be used to revise text, adaptable to students' level of language proficiency. For example, the study by Bewersdorff et al. (2025) illustrates how students attempt to understand the function of an insect's compound eye using an image sourced from Wikipedia. To support students at both grades 5 and 12, they leveraged generative AI by enabling adaptive interaction with visual materials and enriching the accompanying textual content, taking into account students' prior knowledge and background. Figure 8.4 illustrates how generative AI analyzes the image and delivers information tailored to learners' levels of competence.

In addition, AI can also be used to promote inclusive participation by allowing learners to interact with materials in the modalities that best suit their cognitive strengths and preferences. For instance, a neurodiverse student may prefer auditory explanations over written texts, while a multilingual student might benefit from side-by-side translations. When AI tools are integrated into learning platforms, they create opportunities for personalized and dignified engagement with science content, reflecting a more equitable vision of education.

However, the deployment of AI for accessibility must be underpinned by careful design considerations. Developers and educators must ensure that automated translations accurately convey the meaning of scientific terminology and that summarization tools preserve critical disciplinary knowledge. This requires aligning language models with domain-specific vocabularies and validation processes to check the fidelity of outputs, ensuring that nuanced concepts such as "force" or "energy" are not oversimplified or mistranslated. Yet, accomplishing this in real-time is difficult, as it demands continuous monitoring, expert oversight, and safeguards that extend beyond what individual teachers can realistically manage, which underscores the shared responsibility of designers, institutions, and policymakers in supporting responsible and ethical AI use in science education (see Chap. 3 in this book). For example, ethical concerns regarding data privacy, algorithmic fairness, and cultural sensitivity must be addressed to ensure these tools genuinely serve the needs of diverse learners. In this way, AI can become a powerful ally in the pursuit of universally designed science learning materials that respect and uplift all students.

8.2.6 AI as a Co-Author

Educators and curriculum developers are increasingly turning to generative AI tools to co-create lesson plans, assessments, simulations, and activities, marking a significant shift in the development of instructional materials. This shift is not merely a matter of efficiency—it represents a transformation in the role of teachers and

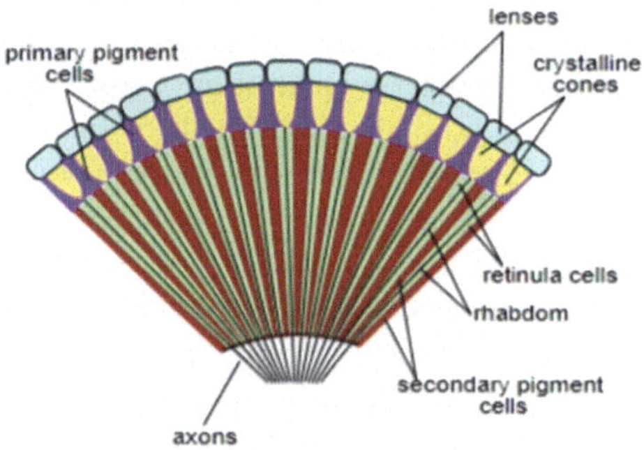

You are a science teacher. Give information about the provided image for 1. a 5th grader 2. a 12th grader. Write no more than 3 sentences each.

ChatGPT

1. For a 5th grader:

   This picture shows the inside of an insect's eye, which is a bit like a honeycomb made of many tiny lenses. Each lens helps the insect to see a small part of the world, and all of them together give the insect a big picture of what's around it. The colors show different parts that work together so the insect can see light and shapes.

2. For a 12th grader:

   The image illustrates a section of the compound eye of an insect, which is comprised of numerous ommatidia, each functioning as a separate visual unit. The ommatidia are made up of lenses, crystalline cones, and photoreceptor cells that detect light and color, with pigment cells that help to sharpen the image. This arrangement allows insects to have a wide field of vision and to detect rapid movements, though at a lower resolution than our eyes.

Fig. 8.4 Explanatory information generated by GPT-4V for 5th and 12th-grade students regarding a graphical representation of an insect's eyes. Reprinted from Bewersdorff, A., Hartmann, C., Hornberger, M., Seßler, K., Bannert, M., Kasneci, E., Kasneci, G., Zhai, X., & Nerdel, C. (2024). Taking the next step with generative artificial intelligence: The transformative role of multimodal large language models in science education. Learning and Individual Differences, 118, 102601. https://doi.org/10.1016/j.lindif.2024.102601—licensed under Creative Commons Attribution 4.0 (CC BY 4.0) (https://creativecommons.org/licenses/by/4.0/). No changes were made

developers as co-designers working alongside intelligent systems. Tools like ChatGPT, Claude, and Copilot are enabling educators to rapidly generate diverse instructional materials tailored to varying cognitive levels and subject domains. For instance, Lee and Zhai (2024) engaged pre-service science teachers in using GPT-based systems to produce lesson plans for science teaching, allowing educators to quickly iterate and customize learning materials based on students' needs.

Pre-service teachers demonstrated moderately high competence in integrating ChatGPT into lesson planning, with particularly strong alignment to instructional strategies and curriculum goals. They envisioned diverse applications across various scientific domains, utilizing ChatGPT for questioning, individualized support, and formative assessment. However, they struggled with selecting appropriate functions, sometimes misapplied the tool (e.g., relying on hallucinated resources), and raised concerns about misinformation and student overdependence. These findings highlight both the promise of AI integration and the need for teacher guidance and systemic safeguards.

One of the most profound changes brought by AI co-authorship is the ability to produce contextually relevant materials on demand. This responsiveness allows educational content to remain current with scientific advancements and sociocultural shifts. For example, Nyaaba and Zhai (2025) introduced the *Culturally Responsive Lesson Planner (CRLP)*, a theory-grounded AI tool designed to embed cultural and linguistic context into science lesson plans. Built upon Culturally Responsive Pedagogy (CRP), Funds of Knowledge, and the GenAI-Culturally Responsive Science Assessment (GenAI-CRSciA) framework (Nyaaba et al., 2024), the CRLP employs an Interactive Semi-Automated (ISA) prompting system that guides teachers through a step-by-step dialogue to supply essential cultural parameters such as local language, indigenous knowledge, and community practices. Rather than producing generic outputs, the system requires teachers to provide contextual details that the AI then weaves throughout objectives, activities, assessments, and extension tasks. When tested in Ghana's Ashanti Region, the CRLP generated Grade 7 science lessons on states of matter that integrated Asante Twi terminology (e.g., *ɛpono* for wooden furniture, *dadeɛ nkuto* for metal objects, *abɔɔ* for stones) and locally relevant cultural practices like fufu preparation and palm oil processing. Comparative evaluations by expert reviewers showed that CRLP-produced lessons contained more than double the cultural elements of standard GPT-4o outputs (48.5 vs. 21) and achieved higher ratings in accuracy (1.8 vs. 1.2) and curriculum relevance (2.0 vs. 1.3). Importantly, the CRLP also included contextualized teaching resources, bilingual vocabulary, and culturally responsive assessments, while inviting teacher input and modification before finalizing the lesson plan. Beyond improving cultural authenticity, the iterative dialogue with the CRLP fostered teacher agency and AI literacy, positioning GenAI not as a static generator but as a collaborative co-author that adapts content to local contexts in real time. Taken together, the CRLP exemplifies how AI co-authorship can mitigate Western-centric biases and enable the rapid creation of educational materials that are not only scientifically accurate but also culturally meaningful across diverse global contexts.

Generative AI can also democratize the design process, enabling educators with limited time or technical resources to participate in the creation of high-quality content. Studies by Kim and Wargo (2025) show that teachers in rural or underfunded schools reported improved confidence and perceived usefulness when using AI tools to generate science learning activities. This shift reduces reliance on prepackaged curriculum and empowers teachers to act as adaptive, responsive designers of learning experiences. Moreover, AI systems can provide scaffolding during content creation by suggesting revisions, checking factual accuracy, or enhancing clarity, effectively serving as collaborative writing partners.

However, the integration of generative AI into curriculum development also raises critical questions about authorship, quality assurance, and pedagogical integrity. It is essential that AI-supported materials undergo rigorous review by educators to ensure alignment with learning objectives, standards, and ethical norms. While this requirement creates a tension with the benefits of real-time adaptivity described previously, it also highlights a productive balance: rapid responsiveness must be coupled with deliberate oversight to ensure that adaptivity enhances, rather than undermines, responsible and ethical practice. Without careful oversight, there is a risk of reinforcing biases, inaccuracies, or superficial content. As such, professional development and institutional guidelines are needed to support educators in leveraging generative AI responsibly and effectively. When appropriately integrated, AI co-authorship holds the potential to enhance creativity, efficiency, and accessibility in science education.

8.3 Responsible and Ethical Uses of AI in Response to the Changes

While the integration of AI into the development and design of science learning materials offers unprecedented opportunities, it is equally important to embed responsible and ethical principles transparently into both their construction and deployment (as articulated in the REP framework, Chap. 3 of this book). Such efforts are essential to ensure that AI-supported educational innovations not only enhance learning outcomes but also promote accessibility, inclusion, and human-centered values in science education. The thoughtful implementation of AI should be guided by frameworks that prioritize student empowerment, educator agency, and social justice. In the following sections, we elaborate on the key dimensions of ethical considerations that must shape the responsible use of science learning materials in light of the changes brought by AI.

8.3.1 Promote Transparency and Critical AI Literacy in Scientific Practice

As students increasingly encounter AI while engaging in scientific practice—such as creating models, analyzing large datasets, conducting virtual experiments, and generating hypotheses—it becomes crucial to ensure that these experiences are grounded in *transparency and explainability*. Students must be provided with opportunities to build foundational AI literacy that encompasses not only a technical understanding of how AI algorithms operate, but also a reflective awareness of how training data, model architectures, and design assumptions influence AI outputs, potentially introducing epistemic bias, privileging certain forms of knowledge or perspectives (see Chap. 4 of this book). *Transparency* and *human oversight* are necessary. AI systems used in scientific contexts should be accompanied by documentation and user-facing explanations that enable students to critically evaluate the sources, processes, and reliability of AI-generated content, yet recent research demonstrates that even many applied researchers disagree on or misunderstand explainability outputs—calling for careful design and scaffolding rather than assuming explanations mean understanding (Krishna et al., 2023). For example, Zhai, Shi, and Nehm (2021) show that even in machine learning-based science assessments, human oversight is essential because students and educators must interpret outputs critically and reconcile them with disciplinary reasoning.

Educators must scaffold learning experiences that encourage students to critique AI outputs, identify errors or inconsistencies, and make informed decisions about when and how to use AI tools. Additionally, institutions must ensure that the integration of AI in science learning aligns with ethical review standards, minimizes biases in representations and predictions, and upholds the integrity of scientific inquiry by treating AI as a tool to augment, not replace, human reasoning.

8.3.2 Design Accessible, Transparent, and Privacy-Conscious Personalized Learning

AI-supported personalization presents compelling opportunities to deliver content, pacing, and feedback tailored to learners' individual needs, skills, and interests. However, this personalization must be designed with *accessibility* at its core. Ethically adaptive systems should be sensitive to cultural, linguistic, and socioeconomic differences among learners, ensuring that recommendations or adaptations do not inadvertently perpetuate systemic inequities. For example, developers must audit AI models for bias in data collection and algorithmic decision-making to uphold *fairness*, while policymakers and educators must advocate for *transparency* in algorithmic design and outputs. Privacy and data protection are also critical, as adaptive systems often rely on the ongoing collection of sensitive behavioral and academic data. Clear consent mechanisms, data anonymization strategies, and compliance with data protection laws (e.g., FERPA, GDPR) must be enforced. Importantly, adaptive learning environments should maintain the agency of students and educators, allowing them to adjust or override AI-generated recommendations. Transparency reports, user dashboards, and opt-out features are crucial for promoting trust and digital autonomy. Research on intelligent tutoring and personalization underscores both the promise and risks of such systems: they can scaffold inquiry and self-directed learning, but they also risk amplifying inequities if not carefully designed and monitored (Holmes et al., 2019).

8.3.3 Ensure Scientific Integrity and Human Oversight in AI Simulations

AI has opened the door to richly contextualized and responsive simulations that enable students to engage with complex scientific phenomena in ways that were previously inaccessible due to safety, cost, or logistical constraints. For instance, students can now manipulate variables in a virtual chemical reaction, observe astrophysical events over time, or conduct experiments with AI-generated feedback. These tools hold the potential to foster deep conceptual understanding and active learning, but only if they are grounded in scientifically accurate models and designed with pedagogical intention. Developers must work closely with scientists, science educators, and curriculum designers to ensure that simulations accurately reflect representations and support key learning objectives. Educators should be provided with professional development to effectively integrate these tools, including strategies for guiding inquiry, diagnosing misconceptions, and encouraging students to reflect on the scientific reasoning behind observed outcomes. Moreover, students should be supported in understanding the limitations of simulations and the assumptions underlying the models they explore, promoting a nuanced and critical engagement with digital tools. This is evident when comparing robust, research-validated platforms, such as PhET (Finkelstein, 2006), with newly emerging AI-generated simulations (Ben-Zion et al., 2025), which can be highly flexible but require additional validation and oversight.

8.3.4 Curate Accurate and Pedagogically Coherent Multimodal Content

AI technologies now enable the generation and analysis of complex, multimodal learning content that includes not only written texts but also audio explanations, diagrams, infographics, interactive graphs, and videos. Such content has the potential to make abstract scientific concepts more tangible and engaging, particularly for students who benefit from visual or auditory learning modalities. However, this abundance of dynamic content must be curated with a focus on scientific accuracy, cognitive load, and instructional coherence. Developers should be held accountable for validating the sources, claims, and data underpinning AI-generated materials to promote transparency. Cross-disciplinary teams of content experts, educators, and technologists should collaborate to design and review AI-generated learning resources. Furthermore, multimodal systems should incorporate features that enable differentiation and Universal Design for Learning (UDL), ensuring that content is accessible, inclusive, and adaptable to diverse learner needs. Educators also require training and support to critically evaluate and effectively integrate multimodal AI content into their lesson planning and formative assessment practice. Studies have shown, for instance, that multimodal and interactive formats outperform static text in supporting conceptual understanding in physics (Stelzer et al., 2009) and that multimodal cues, such as facial expressions, can help diagnose and address student misconceptions in kinematics (Liaw et al., 2021).

8.3.5 *Advance Inclusive Access Through Ethical Language and Accessibility Tools*

One of AI's most transformative potentials in attending to science learning materials lies in its ability to bridge linguistic and accessibility gaps. Tools that offer real-time language translation, text simplification, voice-to-text transcription, and multimodal summarization can democratize access to scientific knowledge for students who are multilingual, neurodivergent, or have disabilities. However, ethical implementation requires more than just technical inclusion. These tools should be developed in consultation with diverse user communities to ensure they are culturally responsive and affirming of students' linguistic identities (i.e., fairness). Rather than merely translating dominant forms of scientific discourse, AI systems should also validate and incorporate multiple ways of knowing and expressing scientific understanding. Interface design must emphasize usability across devices, connectivity levels, and contexts. Educators must also be equipped with pedagogical strategies for using these tools to foster agency and participation among all students. Accessibility, explainability, and transparency should be built into the DNA of AI learning platforms, not added as an afterthought. For example, Watters et al. (2021) demonstrated how AI-supported tools can improve access for visually impaired students in science classrooms, while Smith, Avraamidou, and Adams (2022) highlight the importance of culturally sustaining pedagogies to ensure that inclusive access extends beyond language or disability accommodations.

8.3.6 *Maintain Human Oversight and Clarify AI's Role in Educational Co-Creation*

The rise of generative AI introduces new opportunities for educators to co-create content such as assessment questions, lesson plans, explanatory texts, or even simulations. Yet, this co-authoring role demands a clear framework for accountability, transparency, and the ethical use of resources. Educators must remain the final arbiters of the quality, alignment, and appropriateness of AI-supported materials, ensuring that these resources meet educational goals and uphold professional standards and scientific integrity. Materials should clearly indicate the involvement of AI in their creation, allowing students and other stakeholders to engage with them critically and transparently. Moreover, the anthropomorphising of AI tools (i.e., describing or treating AI systems as if they possess human-like qualities)—in public discourse and classroom practice alike—must be actively countered. As Nash (2024) cautions, referring to AI as if it "knows," "thinks," or "believes" obscures its fundamental nature as a statistical inference tool. Educators should help students develop epistemic clarity, emphasizing that AI lacks consciousness, understanding, or intent. Teaching these distinctions fosters digital literacy and safeguards students' critical reasoning skills. When used responsibly, AI can support co-creation, ideation, and iteration, but it must always be contextualized within human values of autonomy, dignity, and inclusive participation. In this vision, AI becomes a collaborative instrument for enhancing inquiry and deepening engagement—a means of expanding, rather than narrowing, the role of human judgment in science education. This was illustrated in a study of pre-service teachers using ChatGPT for lesson planning, where participants generated creative science activities but also over-relied on AI outputs, underscoring the importance of human oversight (Herdiska & Zhai, 2024).

8.4 Conclusion

The integration of AI into science learning materials represents a paradigm shift in science education, redefining not only what students learn but also how they engage with scientific practice and construct scientific knowledge. This chapter has synthesized emerging developments and empirical examples to illustrate how AI is fundamentally transforming the design, representation, and accessibility of science learning materials across multiple dimensions. From embedding authentic scientific practice and enabling adaptive learning experiences to generating multimodal content and expanding accessibility through language and cognitive support, AI technologies offer unprecedented opportunities to enrich and personalize science education. Moreover, the co-creative role of educators and intelligent systems signals a new era of instructional material development, marked by responsiveness, cultural relevance, and iterative design.

These transformations, however, present challenges. The affordances of AI must be balanced with ethical and pedagogical imperatives that ensure transparency, inclusivity, and scientific integrity. Without thoughtful integration, AI-generated materials risk perpetuating epistemic bias (i.e., privileging certain ways of knowing, reasoning, or representing science while marginalizing others), diminishing learner agency, or undermining the pedagogical coherence of science instruction. As such, the development and deployment of AI-supported learning materials must be guided by robust frameworks that center human judgment, critical inquiry, and equitable access.

The implications for science education are profound. The shifting nature of science, as increasingly computational, data-driven, and AI-supported, necessitates corresponding shifts in how science is taught and learned. Science educators, curriculum designers, and policymakers must collaborate to engage with the opportunities and risks posed by AI, creating learning environments that are both innovative and ethically grounded. Professional development efforts should prioritize building educators' capacity to evaluate, adapt, and design AI-supported materials. In parallel, students must be equipped not only to use AI tools but also to critically interrogate their outputs and understand their epistemic boundaries.

In sum, the future of science education in the AI era hinges on the field's ability to leverage AI's transformative potential while upholding the foundational goals of scientific literacy, critical thinking, and social responsibility. AI should not replace human-centered teaching and learning but rather serve as a catalyst for reimagining science education in ways that are more authentic, inclusive, and responsive to the complexities of the contemporary world.

Declaration of AI Use Parts of this manuscript were prepared with the assistance of generative AI tools for language editing or paragraphing of the authors' writing. The authors reviewed, revised, and verified all AI-assisted content for accuracy, appropriateness, and originality. No confidential or sensitive data were entered into AI systems. The final responsibility for the content rests solely with the authors.

Acknowledgement This material is based upon work supported by the National Science Foundation under Grant No. 2332964. Any opinions, findings, and conclusions or recommendations expressed in this material are those of the author(s) and do not necessarily reflect the views of the National Science Foundation.

References

Adair, A., Pedro, M. S., Gobert, J., & Segan, E. (2023). Real-time AI-driven assessment and scaffolding that improves students' mathematical modeling during science investigations. In N. Wang, G. Rebolledo-Mendez, N. Matsuda, O. C. Santos, & V. Dimitrova (Eds.), *Artificial intelligence in education: 24th international conference, AIED 2023, Tokyo, Japan, July 3–7, 2023, proceedings* (Vol. 13916, pp. 202–216). Springer. https://doi.org/10.1007/978-3-031-36272-9_17

Akaygun, S., & Kilic, I. (2025). Generative artificial intelligence (GenAI) as the artist of chemistry visuals: Chemistry preservice teachers' reflections on visuals created by GenAI. *Journal of Chemical Education, 102*(7), 2549–2564. https://doi.org/10.1021/acs.jchemed.4c00775

Akgun, S., & Greenhow, C. (2022). Artificial intelligence in education: Addressing ethical challenges in K-12 settings. *AI and Ethics, 2*(3), 431–440. https://www.ncbi.nlm.nih.gov/pmc/articles/PMC8455229/pdf/43681_2021_Article_96.pdf

Almasri, F. (2022). Simulations to teach science subjects: Connections among students' engagement, self-confidence, satisfaction, and learning styles. *Education and Information Technologies, 27*(5), 7161–7181. https://doi.org/10.1007/s10639-022-10940-w

Ben-Zion, Y., Zarzecki, R. E., Glazer, J., & Finkelstein, N. D. (2024). Leveraging AI for rapid generation of physics simulations in education: Building Your own virtual lab. *arXiv Preprint*. https://doi.org/10.48550/arXiv.2412.07482

Bewersdorff, A., Hartmann, C., Hornberger, M., Seßler, K., Bannert, M., Kasneci, E., Kasneci, G., Zhai, X., & Nerdel, C. (2025, February 1). Taking the next step with generative artificial intelligence: The transformative role of multimodal large language models in science education. *Learning and Individual Differences, 118*(102601), 1–13. https://doi.org/10.1016/j.lindif.2024.102601

Bianchini, S., Müller, M., & Pelletier, P. (2022, December 1). Artificial intelligence in science: An emerging general method of invention. *Research Policy, 51*(10), 104604. https://doi.org/10.1016/j.respol.2022.104604

Cao, C., Ding, Z., Jiao J., & Zhai, X. (2023). Demystifying STEM concepts through Generative AI: A multimodal exploration of analogical reasoning. In *IJCAI2023 multi reasoning (multimodal reasoning: Techniques, applications, and challenges).*

Chen, B. (2025). Beyond tools: Generative AI as epistemic infrastructure in education. *arXiv Preprint*. https://doi.org/10.48550/arXiv.2504.06928

Crippen, K. J., & Archambault, L. (2012). Scaffolded inquiry-based instruction with technology: A signature pedagogy for STEM education. *Computers in the Schools, 29*(1–2), 157–173. https://doi.org/10.1080/07380569.2012.658733

Dai, C.-P., Ke, F., Pan, Y., Moon, J., & Liu, Z. (2024). Effects of artificial intelligence-powered virtual agents on learning outcomes in computer-based simulations: A meta-analysis. *Educational Psychology Review, 36*(1), 31.

Dong, Y., Ma, H., Shen, Z., & Wang, K. (2017). A century of science: Globalization of scientific collaborations, citations, and innovations. *ArXiv*. https://doi.org/10.48550/arxiv.1704.05150

Duschl, R., Avraamidou, L., & Azevedo, N. H. (2021). Data-texts in the sciences: The evidence-explanation continuum. *Science & Education, 30*(5), 1159–1181. https://doi.org/10.1007/s11191-021-00225-y

El Fathi, T., Saad, A., Larhzil, H., Lamri, D., & Al Ibrahmi, E. M. (2025). Integrating generative AI into STEM education: Enhancing conceptual understanding, addressing misconceptions, and assessing student acceptance. *Disciplinary and Interdisciplinary Science Education Research, 7*(1), 6. https://doi.org/10.1186/s43031-025-00125-z

Eraslan, G., Avsec, Ž., Gagneur, J., & Theis, F. J. (2019). Deep learning: New computational modelling techniques for genomics. *Nature Reviews. Genetics, 20*(7), 389–403. https://doi.org/10.1038/s41576-019-0122-6

Erduran, S., & Levrini, O. (2024). The impact of artificial intelligence on scientific practices: An emergent area of research for science education. *International Journal of Science Education*, 1–8. https://doi.org/10.1080/09500693.2024.2306604

Erümit, A. K., & Sarıalioğlu, R. Ö. (2025). Artificial intelligence in science and chemistry education: A systematic review. *Discover Education, 4*(1), 178. https://doi.org/10.1007/s44217-025-00622-3

Finkelstein, N. (2006). High-tech tools for teaching physics: The Physics Education Technology project (W. Adams, C. Keller, K. Perkins, & C. Wieman, Trans.). *MERLOT Journal of Online Learning and Teaching, 2*(3), 110–121.

Gil, Y., Greaves, M., Hendler, J., & Hirsh, H. (2014). Amplify scientific discovery with artificial intelligence. *Science, 346*(6206), 171–172.

Herdiska, A., & Zhai, X. (2024). Artificial intelligence-based scientific inquiry. In X. Zhai & J. Krajcik (Eds.), *Uses of artificial intelligence in STEM education* (pp. 179–197). Oxford University Press.

Holmes, W., Bialik, M., & Fadel, C. (2019). *Artificial intelligence in education: Promises and implications for teaching and learning* (p. 242). Center for Curriculum Redesign.

Ji, Z., Lee, N., Frieske, R., Yu, T., Su, D., Xu, Y., Ishii, E., Bang, Y., Madotto, A., & Fung, P. (2022). Survey of hallucination in natural language generation. *ACM Computing Surveys, 55*(12). https://doi.org/10.1145/3571730

Jose, J. (2025). The impact of integrating Microsoft Teams—Reading progress as an artificial intelligence (AI) platform for promoting learners' reading aloud skills. *Education and Information Technologies, 30*(6), 7077–7115.

Jumper, J., Evans, R., Pritzel, A., Green, T., Figurnov, M., Ronneberger, O., Tunyasuvunakool, K., Bates, R., Žídek, A., & Potapenko, A. (2021). Highly accurate protein structure prediction with AlphaFold. *Nature, 596*(7873), 583–589. https://www.nature.com/articles/s41586-021-03819-2.pdf

Kabudi, T., Pappas, I., & Olsen, D. H. (2021). AI-enabled adaptive learning systems: A systematic mapping of the literature. *Computers and Education: Artificial Intelligence, 2*, 100017, 1–12.

Kim, J., & Wargo, E. (2025). Empowering educational leaders for AI integration in rural STEM education: Challenges and strategies. *Frontiers in Education, 10*, 1567698.

Kitano, H. (2016). Artificial intelligence to win the Nobel Prize and beyond: Creating the engine for scientific discovery. *AI Magazine, 37*(1), 39–49.

Krishna, S., Han, T., Gu, A., Jabbari, S., Wu, Z. S., & Lakkaraju, H. (2023). The disagreement problem in explainable machine learning: A practitioner's perspective. *Research Square*. https://doi.org/10.21203/rs.3.rs-2963888/v1

Latif, E., & Zhai, X. (2025). Privacy-preserved automated scoring using federated learning for educational research. In *International conference of artificial intelligence in education, Palermo, Italy*.

Lee, G., Shi, L., Latif, E., Gao, Y., Bewersdorf, A., Nyaaba, M., Guo, S., Wu, Z., Liu, Z., Wang, H., Mai, G., Liu, T., & Zhai, X. (2023). Multimodality of AI for education: Towards artificial general intelligence. *IEEE Transactions on Learning Technologies, 18*, 666–683. https://doi.org/10.1109/TLT.2025.3574466

Lee, G., Yun, M., Zhai, X., & Crippen, K. (2025). Artificial intelligence in science education research: Current states and challenges. *Journal of Science Education and Technology*, 1–18. https://doi.org/10.1007/s10956-025-10239-8

Lee, G., & Zhai, X. (2024). Using ChatGPT for science learning: A study on pre-service teachers' lesson planning. *IEEE Transactions on Learning Technologies, 17*, 1683–1700. https://doi.org/10.1109/TLT.2024.3401457

Létourneau, A., Deslandes Martineau, M., Charland, P., Karran, J. A., Boasen, J., & Léger, P. M. (2025). A systematic review of AI-driven intelligent tutoring systems (ITS) in K-12 education. *NPJ Science of Learning, 10*(1), 29, 1-13. https://doi.org/10.1038/s41539-025-00320-7

Liaw, H., Yu, Y.-R., Chou, C.-C., & Chiu, M.-H. (2021). Relationships between facial expressions, prior knowledge, and multiple representations: A case of conceptual change for kinematics instruction. *Journal of Science Education and Technology, 30*(2), 227–238. https://doi.org/10.1007/s10956-020-09863-3

Makransky, G., Terkildsen, T. S., & Mayer, R. E. (2019). Adding immersive virtual reality to a science lab simulation causes more presence but less learning. *Learning and Instruction, 60*, 225–236.

Mešić, V., Dervić, D., Gazibegović-Busuladžić, A., Salibašić, D., & Erceg, N. (2015). Comparing the impact of dynamic and static media on students' learning of one-dimensional kinematics. *Eurasia Journal of Mathematics, Science and Technology Education, 11*(5). https://doi.org/10.12973/eurasia.2015.1385a

Nyaaba, M., & Zhai, X. (2025). Developing a theory-grounded AI tool for the generation of culturally responsive lesson plans. *Computers and Education: Artificial Intelligence, 9*, 100474.

Nyaaba, M., Zhai, X., & Faison, M. Z. (2024). Generative AI for culturally responsive assessment in science: A conceptual framework. *Education Sciences, 14*(12), 1–22.

Panjwani-Charania, S., & Zhai, X. (2024). AI for students with learning disabilities: A systematic review. In Z. Xiaoming & J. Krajcik (Eds.), *Uses of artificial intelligence in STEM education* (pp. 469–493). Oxford University Press.

Reichstein, M., Camps-Valls, G., Stevens, B., Jung, M., Denzler, J., Carvalhais, N., & Prabhat. (2019). Deep learning and process understanding for data-driven Earth system science. *Nature, 566*(7743), 195–204. https://doi.org/10.1038/s41586-019-0912-1

Samson, A. -H. (2025). Protein folding: An A.I. teaching module. https://www.dawsoncollege.qc.ca/ai/portfolios/protein-folding-an-a-i-teaching-module/?utm_source=chatgpt.com

Schmucker, R., Xia, M., Azaria, A., & Mitchell, T. (2024, July). Ruffle & Riley: Insights from designing and evaluating a large language model-based conversational tutoring system. In *International conference on artificial intelligence in education* (pp. 75–90). Springer Nature.

Smith, T., Avraamidou, L., & Adams, J. D. (2022). Culturally relevant/responsive and sustaining pedagogies in science education: Theoretical perspectives and curriculum implications. *Cultural Studies of Science Education, 17*(3), 637–660. https://doi.org/10.1007/s11422-021-10082-4

Stelzer, T., Gladding, G., Mestre, J. P., & Brookes, D. T. (2009). Comparing the efficacy of multimedia modules with traditional textbooks for learning introductory physics content. *American Journal of Physics, 77*(2), 184–190. https://doi.org/10.1119/1.302820

Su, J., & Zhong, Y. (2022). Artificial Intelligence (AI) in early childhood education: Curriculum design and future directions. *Computers and Education: Artificial Intelligence, 3*, 100072.

Tan, L. Y., Hu, S., Yeo, D. J., & Cheong, K. H. (2025). Artificial intelligence-enabled adaptive learning platforms: A review. *Computers and Education: Artificial Intelligence, 9*, 100429. https://doi.org/10.1016/j.caeai.2025.100429

Tenopir, C., Allard, S., Douglass, K., Aydinoglu, A. U., Wu, L., Read, E., Manoff, M., & Frame, M. (2011). Data sharing by scientists: Practices and perceptions. *PLoS One, 6*(6), e21101. https://doi.org/10.1371/journal.pone.0021101

Thaqi, E., Mantawy, M. O., & Kasneci, E. (2024). SARA: Smart AI reading assistant for reading comprehension. In M. Khamis, Y. Sugano, & L. Sidenmark (Eds.), *Proceedings of the 2024 symposium on eye tracking research and applications* (pp. 1–3). ACM. https://doi.org/10.1145/3649902.3655661

Thompson, N., & McGill, T. J. (2017). Genetics with Jean: The design, development and evaluation of an affective tutoring system. *Educational Technology Research and Development, 65*(2), 279–299. https://doi.org/10.1007/s11423-016-9470-5

Vartiainen, H., Toivonen, T., Jormanainen, I., Kahila, J., Tedre, M., & Valtonen, T. (2021). Machine learning for middle schoolers: Learning through data-driven design. *International Journal of Child-Computer Interaction, 29*, 100281.

Wang, X., Zainuddin, Z., & Leng, C. (2025). Generative artificial intelligence in pedagogical practices: A systematic review of empirical studies (2022–2024). *Cogent Education, 12*(1). 1–21. https://doi.org/10.1080/2331186X.2025.2485499

Ward, W., Cole, R., Bolaños, D., Buchenroth-Martin, C., Svirsky, E., & Weston, T. (2013). My science tutor: A conversational multimedia virtual tutor. *Journal of Educational Psychology, 105*(4), 1115–1125. https://doi.org/10.1037/a0031589

Watters, J. D., Hill, A., Weinrich, M., Supalo, C., & Jiang, F. (2021). An artificial intelligence tool for accessible science education. *Journal of Science Education for Students with Disabilities, 24*(1), 1–14. https://doi.org/10.14448/jsesd.13.0010

Xu, W., & Ouyang, F. (2022). The application of AI technologies in STEM education: A systematic review from 2011 to 2021. *International Journal of STEM Education, 9*(1), 59. https://doi.org/10.1186/s40594-022-00377-5

Yılmaz, Ö. (2024). Personalised learning and artificial intelligence in science education: Current state and future perspectives. *Educational Technology Quarterly, 2024*(3), 255–274. https://doi.org/10.55056/etq.744

Zhai, X. (2023). ChatGPT for next generation science learning. *XRDS: Crossroads*. https://doi.org/10.1145/3589649

Zhai, X., & Krajcik, J. (2025). *Uses of artificial intelligence in STEM education*. Oxford University Press.

Zhai, X., Shi, L., & Nehm, R. H. (2021). A meta-analysis of machine learning-based science assessments: Factors impacting machine-human score agreements. *Journal of Science Education and Technology, 30*(3), 361–379. https://doi.org/10.1007/s10956-020-09875-z

Part V
Assessment

Chapter 9
Addressing Challenges of Science Assessment with Artificial Intelligence

Lei Liu, Kevin Haudek, Xiaoming Zhai, Ross H. Nehm, and Knut Neumann

9.1 Introduction

The landscape of science assessment is undergoing profound changes with the rapid integration of Artificial Intelligence (AI) into science education. The *K-12 Framework for Science Education* (2012) has already redefined science learning through three-dimensional (3D) learning (NRC, 2012), which requires the integration of disciplinary core ideas (DCIs), crosscutting concepts (CCCs), and science and engineering practices (SEPs) to foster a deep understanding of science. This

L. Liu (✉)
Educational Testing Service, Princeton, NJ, USA
e-mail: lliu001@ets.org

K. Haudek
Biochemistry and Molecular Biology Department, Michigan State University, East Lansing, MI, USA

CREATE for STEM Institute, Michigan State University, East Lansing, MI, USA
e-mail: haudekke@msu.edu

X. Zhai
AI4STEM Education Center, University of Georgia, Athens, GA, USA
e-mail: xiaoming.zhai@uga.edu

R. H. Nehm
Ecology and Evolution, Applied Mathematics and Statistics, Stony Brook University, Stony Brook, NY, USA
e-mail: ross.nehm@stonybrook.edu

K. Neumann
Department of Physics Education, IPN-Leibniz-Institute for Science and Mathematics Education, Kiel, Germany
e-mail: neumann@leibniz-ipn.de

X. Zhai, K. J. Crippen (eds.), *Advancing AI in Science Education*, Advances in Technology-Rich Science Education 3,
https://doi.org/10.1007/978-3-032-16871-9_9

shift demands science assessments to move beyond content-focused assessments toward more authentic, performance-based, and process-oriented evaluations of student learning. Therefore, traditional assessments, such as multiple-choice items and isolated problem-solving tasks, struggle to capture the complexity of 3D learning and the iterative nature of scientific inquiry (NRC, 2014).

At the same time, AI is fundamentally reshaping the nature of science knowledge, problem solving, and human-AI collaboration in science learning, which requires additional attention to science assessment, including its design, interpretation, and implementation. With Generative AI's capacity to generate scientific explanations, analyze data, and model complex phenomena, science assessments now need to capture not only how well students understand scientific principles but also how they critically engage with AI systems and their outputs, assess the trustworthiness of AI-driven conclusions, and apply ethical reasoning in AI-assisted scientific problem solving (Osborne, 2023; Ng et al., 2021). These developments introduce changes necessary to science assessment practices in maintaining its validity, reliability, relevance, and fairness.

This chapter examines how AI is reshaping science assessments, focusing on key changes in assessment constructs, measurement approaches, data interpretation, and responsible use of AI in assessments. We explore strategies for designing AI-resilient science assessments, where students demonstrate authentic scientific reasoning rather than passively relying on AI-generated responses. Additionally, we address how AI can be leveraged to support dynamic formative assessments to provide real-time feedback, track learning progress, and support personalized science learning experiences. Furthermore, this chapter highlights responsible use of AI in science assessments to ensure fairness and mitigate potential biases while maintaining the rigor of measurement and applicability in science classrooms.

9.2 Changes to Assessment Brought About by AI

9.2.1 Redefining Science Assessment Constructs

AI is fundamentally transforming the way people learn, engage with, and apply scientific knowledge, which indicates a need to redefine science assessment constructs (i.e., the specific set of knowledge, skills and practices that an assessment is designed to measure). The integration of AI into science education is changing the cognitive and social landscape of how science is learned, practiced, and applied. This transformation calls for redefining science assessment constructs to reflect the evolving realities of learning and working in AI-augmented environments, which requires capturing how students reason scientifically with AI, collaborate effectively in human-AI partnerships, critically evaluate machine-generated outputs, and navigate ethical dilemmas inherent in AI-supported scientific inquiry.

Traditionally, science assessments focused on evaluating content proficiency, often through decontextualized tasks like multiple-choice or short constructed-response items. These formats assume that knowledge resides primarily in the learner's mind and that reasoning unfolds in a linear, siloed process. However, this traditional approach is at odds with the vision of the *K-12 Framework*, which emphasizes integrating three key dimensions of science learning and posits that assessments should measure these dimensions to determine whether meaningful science learning has occurred. AI is reshaping the epistemic structure of science learning by functioning as a real-world analytical, interpretive, and collaborative partner. Therefore, the construct of scientific reasoning must be updated to include how learners interact with and through AI systems to engage in scientific inquiry. Additionally, the use of AI tools is fundamentally changing how scientists engage in the practice of science, from generating hypotheses to data analysis and constructing explanations based on patterns or anomalies identified through AI-assisted detection (Erduran & Levrini, 2024). These changing scientific practices demand rethinking our expectations of students' engagement in science within an AI-supported world.

For example, AI tools can assist students in formulating hypotheses, simulating experimental outcomes, and generating explanations based on large datasets. In this context, collaboration in science is no longer limited to peer-to-peer or student-teacher interactions. Instead, students are increasingly expected to engage in human-AI partnerships, where reasoning is distributed across humans and machines (Lee et al., 2025). This change brought by AI challenges current assessment practices that primarily measure either individual cognition or traditional group work, and stresses the need to design tasks that assess how students integrate AI contributions, interrogate AI outputs, and maintain epistemic agency in AI-mediated science learning. Such shifts suggest that new science assessment constructs should include:

- AI-human collaborative reasoning, where students are evaluated on their ability to work with, critique, and refine AI-generated hypotheses or simulations.
- Critical data literacy, involving the ability to discern trustworthy AI outputs from biases or flawed analyses (Ng et al., 2021).
- Ethical AI reasoning,where students must justify not just scientific sensemaking, but also the ethical implications of those solutions in AI-enhanced contexts.

Such shifts in what we value developing in students, and thus measure via assessment, align with changing ways scientists incorporate AI into their scientific endeavors (see Chap. 2, Nehm & Kubsch).

Science literacy has undergone substantial conceptual shifts over time, evolving from a focus on content knowledge to a broader competency that includes scientific reasoning, engagement with socio-scientific issues (Feinstein, 2011; Laugksch, 2000; Ryder, 2001), and critical evaluation of information (Osborne, 2023). These shifts have been driven by the increasing complexity of scientific knowledge and its intersections with society. Now, with the emergence of AI in science and education, science literacy is undergoing another transformation that requires new competencies in AI fluency, epistemic vigilance, and ethical decision making in AI-supported scientific processes (see Chap. 4, Liu & Tripp). One major addition to science literacy in the AI era is the need to understand how AI models function in scientific learning and research. Individuals now need to grasp how AI systems generate insights, what data they rely on, and their potential biases embedded in AI-supported analyses (Ng et al., 2021). Building on Osborne's emphasis on critical evaluation of information, AI introduces a new challenge: assessing the trustworthiness of AI-generated results. This shift extends traditional notions of science literacy by requiring learners to critically assess not just human-generated knowledge (Osborne, 2023), but also machine-generated outputs. In addition, as Feinstein (2011) highlights the role of science literacy in real-world decision-making, AI expands this need by introducing ethical considerations in AI-generated scientific solutions. Students and professionals must now evaluate whether AI-supported outcomes in fields like climate science, healthcare, and engineering are ethical, fair, and aligned with societal values. All these emerging changes require the goal of science education not only to focus on making sense of natural phenomena but also to develop understanding of how AI systems model and generate scientific knowledge, to question the limitations and biases of AI-generated analyses, and to evaluate the social ethical, and environmental implications of AI-augmented scientific decisions (Cope et al., 2021; Krenn et al., 2022).

9.2.2 Developing AI-Resilient, Performance-Based Science Assessments

AI's capability to instantly generate accurate answers to many traditional science assessment items, such as formula-based problems, short explanations, and even longer written responses, challenges the validity of science assessments that focus on content recall (Ng et al., 2021). These formats no longer provide reliable measures of students' own understanding since students can simply outsource the task to AI. This phenomenon creates a growing need to develop AI-resilient assessment models that require students to demonstrate reasoning processes, problem-solving strategies, and scientific thinking that cannot easily be mimicked or faked by an AI system.

As traditional assessments become more vulnerable to AI-enabled shortcuts, the field must revisit and expand the concept of assessment validity. AI-resilient tasks must be validated not just in terms of their psychometric properties, but also in terms of their ability to measure non-algorithmic thinking, resist over-reliance on AI-generated responses, verify the agency and authorship of the student's contributions, and reflect authentic scientific inquiry.

Performance-based assessments offer a solution. These tasks require students to engage in complex reasoning in interactive or authentic contexts to make judgments and decisions based on evidence, including opportunities to demonstrate how they collaborate productively with AI systems as part of the problem-solving process.

AI-resilient assessments must assess not just what students know, but also how they design and conduct investigations, interpret data including AI-generated data, identify flaws or limitations in AI predictions, justify explanations or scientific arguments using multiple sources of evidence or use scientific reasoning to address socio-scientific problems. As Osborne (2023) noted, science education in the twenty-first century must prioritize engaging learners in real-world applications of science, especially as technology blurs the line between knowing and finding out. In the contexts of performance-based assessments such as scenario-driven tasks and scientific modeling, these tasks provide a richer and more authentic space to assess how students apply science knowledge, evaluate scientific claims, and navigate AI-informed decision making.

9.2.3 Ensuring Validity and Reliability of Automatically Generated Scores

As AI reshapes the goals and formats of science assessments, another pressing challenge emerges: how can we ensure that AI-based evaluations of these complex assessments are valid, fair, and interpretable? This becomes especially urgent as the design of AI-resilient science assessments grows more sophisticated. These assessments often include open-ended or ill-structured tasks that reflect the three--dimensional nature of science learning envisioned by the *K-12 Framework*. Considering scalability and efficiency, the burden of evaluation should increasingly rely on AI-based scoring systems.

While AI offers the promise of scoring consistency and rapid turnaround, it also brings new risks that go beyond accuracy. One major concern is whether these systems can validly and fairly interpret the complexity of students' scientific reasoning, especially when students draw on diverse backgrounds, reasoning patterns, and linguistic expressions. For instance, AI models trained on limited or biased data may overemphasize superficial features and underrepresent deeper conceptual understanding. This issue of construct underrepresentation undermines the very goal of AI-resilient design to measure authentic scientific reasoning in real-world contexts (Shute & Rahimi, 2021; Zhai et al., 2021).

Another concern centers on fairness issues. Students from linguistically diverse or marginalized backgrounds may respond differently than the training data anticipates, resulting in systematic under-scoring or misinterpretation of their work (Baker & Hawn, 2022). As assessments become more complex and personalized, maintaining comparability and fairness across varied student populations becomes more challenging. Adding to this challenge is the trend for AI systems to become more autonomous, to both evaluate and respond to students' input with decreasing amounts of human oversight.

Even when AI scoring is technically robust, questions of interpretability and transparency remain. Educators, school administrators, and students must be able to understand not only what the AI generated insights are, but also why these insights were generated. Without this understanding, AI scores risk becoming a black-box of outputs that may lose teacher trust, limit instructional utility, and reduce students' ownership of their learning (Gardner et al., 2021; Williamson et al., 2012). For science teachers especially, trust in AI scores hinges on whether they can see and understand the reasoning behind a score to effectively use them to compare student work and make informed instructional decisions. In the absence of transparent explanations, there is a risk that educators will either over-trust AI outputs or reject them completely.

Several principles might be helpful to guide future AI-based scoring or evaluation. The first principle is to carefully consider the AI-based approach and use of scores for an intended task. Since different AI approaches have different strengths and challenges, especially as they relate to accuracy and reliability, it is critical to choose an approach that can best support the intended outcome and use of AI evaluation. The second principle is to create human-in-the-loop systems to involve educators or researchers with relevant expertise in verifying scores, particularly when stakes are higher or responses are atypical. Finally, AI scoring systems must be regularly evaluated against diverse student data for bias auditing and validation across populations.

9.2.4 Transforming Classroom Assessment Practices

The integration of AI into educational environments has significantly transformed classroom science assessment practices, introducing new ways to evaluate students' learning processes, conceptual understanding, and epistemic engagement. This transformation is not merely a technological shift but also a pedagogical one, where AI functions as a formative partner in assessing, diagnosing, and guiding student learning. Classroom-based assessments—traditionally teacher-administered quizzes, tests, and performance tasks—are increasingly being supplemented or replaced by AI-supported tools that allow for real-time, nuanced, and continuous assessment. As AI becomes more embedded in learning environments, its capacity to collect, analyze, and interpret diverse data types enables more responsive and individualized assessments that align with constructivist and sociocultural theories of learning (Luckin et al., 2016; Siemens & Long, 2011).

One of the most notable changes is the transformation of formative assessment practices through AI-supported learning environments. Intelligent tutoring systems (ITS) like ASSISTments and Carnegie Learning's MATHia, as well as platforms such as Knewton and SOWISO, have demonstrated the ability to provide instant feedback to students while simultaneously generating diagnostic insights for teachers (Koedinger et al., 2013; Pane et al., 2014). These systems leverage large-scale data analytics and machine learning (ML) to track student progress over time, detect misconceptions, and adapt instruction accordingly. For instance, an ITS used in a physics classroom may monitor how students manipulate force vectors in a simulation and use this data to diagnose their understanding of Newton's laws of motion. Teachers can access visual dashboards summarizing trends in student thinking, allowing them to adjust their instruction in real-time to better meet individual and group learning needs. The formative nature of these tools supports the implementation of responsive pedagogy, where assessment is seamlessly embedded within instruction rather than treated as an afterthought (Black & Wiliam, 2009; Shute & Rahimi, 2021).

Moreover, AI technologies facilitate the use of multimodal assessment, enabling educators to capture and interpret learning data from a wide array of sources beyond traditional written or selected-response formats. This includes gesture, speech, interaction logs, eye tracking, and modeling behaviors—all of which can be processed by AI to infer students' reasoning processes and engagement levels (Azevedo et al., 2019; Lee et al., 2023). For example, in science inquiry tasks conducted in virtual labs or augmented reality environments, AI can analyze students' verbal explanations and hand movements as they construct and test hypotheses, providing deeper insights into their scientific practices and conceptual understanding (Bewersdorff et al., 2025; Lee & Zhai, 2025). Such assessments are particularly valuable for evaluating, for example, Next Generation Science Standards performance expectations (NGSS Lead States, 2013) that emphasize scientific modeling, explanation, and argumentation, which are difficult to assess through static formats. Tools like nQuire and Go-Lab enable learners to engage in science investigations while AI systems track and assess their inquiry processes, offering personalized prompts or feedback to support metacognition and reflection (de Jong et al., 2014).

Another transformative shift involves the redistribution of assessment agency. With AI tools becoming more student-facing, learners are increasingly positioned as active participants in their own assessment processes. This is particularly evident in platforms that support self-assessment, automated feedback, and peer evaluation, often guided by AI-generated rubrics or exemplar-based comparisons (Luaces et al., 2018). For instance, AI writing assistants like Grammarly or Quill Bot can provide automated suggestions on students' scientific writing, while also prompting students to evaluate and revise their work based on the feedback. More advanced systems, such as those utilizing large language models (LLMs), are capable of not only generating feedback but also engaging students in dialogic revision processes, prompting them to reflect on the quality of their arguments and evidence. These systems extend the assessment beyond correctness toward epistemic engagement, pushing students to justify claims, revise based on feedback, and critically evaluate the feedback itself (Lee et al., 2025). This shift fosters greater learner autonomy and supports the development of metacognitive and epistemic skills essential for scientific literacy in an AI-mediated world (Wiliam, 2011).

In parallel, the role of the teacher in assessment is undergoing significant redefinition. Rather than being the sole assessor, teachers now work alongside AI as co-evaluators, interpreters, and learning designers. While AI tools can generate vast quantities of assessment data, it is ultimately the teacher who contextualizes, interprets, and acts upon these insights. Research indicates that teachers often view AI tools as helpful in highlighting learning patterns or identifying students who need additional support, but they also express reservations about the accuracy and appropriateness of AI-generated feedback, particularly for complex reasoning or culturally nuanced responses (Popenici & Kerr, 2017; Zhai et al., 2023). Effective use of AI in the classroom thus requires professional development that supports teachers' data literacy, assessment literacy, and understanding of AI limitations. Co-design approaches that involve teachers in the development and customization of AI-based assessments have been found to increase trust, usability, and instructional alignment (Holstein et al., 2019; Tsai et al., 2022). These developments suggest that rather than replacing teachers, AI reconfigures their roles toward more sophisticated decision-making, formative interpretation, and instructional mediation.

Furthermore, AI enables the convergence of instruction and assessment, breaking down the artificial divide between teaching and testing. AI-powered platforms can integrate learning activities with embedded assessments that continuously collect evidence of learning as students engage with tasks. In dynamic science simulations, such as PhET Interactive Simulations or Labster, AI systems can record students' decisions, infer misconceptions, and offer tailored scaffolds that function simultaneously as instruction and assessment (Wieman et al., 2010). This approach aligns with "stealth assessment," which involves unobtrusive, ongoing evaluation embedded in authentic tasks (Shute & Ventura, 2013). For example, while exploring climate models, students may receive real-time prompts from an AI agent asking them to explain the causes of observed trends. Their responses are analyzed to evaluate conceptual coherence, causal reasoning, and argumentation skills—all without disrupting the learning experience. This integration allows teachers to access evidence-rich profiles of students' scientific understanding over time, supporting continuous formative feedback and more meaningful summative evaluations.

Finally, the increasing presence of AI in classrooms underscores the urgent need to address equity and bias in assessment practices. While AI tools have the potential to support inclusive and differentiated learning, they may also reinforce systemic biases if not carefully audited and validated (Baker & Hawn, 2022; Binns, 2018). For instance, speech-recognition systems may misinterpret the responses of students who speak non-standard dialects or whose first language is not English, leading to inaccurate assessments of their knowledge and reasoning. Similarly, AI scoring engines trained on narrow datasets may privilege certain forms of expression or reasoning, disadvantaging students from underrepresented or marginalized groups. Ensuring equitable assessment practices requires diverse training data, culturally responsive algorithms, transparent scoring criteria, and human oversight—particularly in high-stakes settings (Kizilcec & Lee, 2022; Williamson et al., 2012). Furthermore, participatory design approaches involving students, educators, and communities can help to surface potential biases and align AI tools with culturally sustaining pedagogies. Without intentional efforts to design for fairness, the risk is that AI will not merely mirror existing inequities but may magnify them under the guise of objectivity.

In sum, AI is revolutionizing classroom science assessment by enabling real-time formative feedback, multimodal evidence capture, learner-driven evaluation, and seamless integration of assessment and instruction. These changes have the potential to enrich science education by making assessment more personalized, authentic, and instructionally embedded. However, realizing this potential requires careful attention to validity, teacher capacity, ethical design, and equity—ensuring that AI serves not only as a tool for efficiency but also as a catalyst for deeper and more inclusive learning.

9.3 Responsible and Ethical Uses (REU) of AI to Address the Changes

9.3.1 REU #1: The Use of AI in Science Learning Necessitates Redefining Science Assessments to Measure Scientific Reasoning, AI-Human Collaboration, Critical Engagement with AI-Generated Data, and Ethical Decision-Making in AI-Augmented Scientific Inquiry

Alignment with REP Principles This REU advances *Beneficial & Meaningful Use* by ensuring assessments capture authentic, AI-supported science practices; *Robustness & Scientific Integrity* by grounding tasks in valid, reliable frameworks; *Transparency & Explainability* by requiring students to understand and critique AI reasoning.

The rapid integration of AI into science education is altering the ways students learn, reason, and engage with scientific knowledge. AI is reshaping scientific inquiry into a more distributed, collaborative, and critically mediated process. As a result, science assessments need to evolve as assessments should reflect how students actually learn (NRC, 2014). What we measure in science assessments should now reflect these shifts in how AI mediates how students make sense of phenomena. This will require developing new science assessments that evaluate these skills and performances. One such process to develop these new assessments is the Evidence Centered Design (ECD) framework, which aims to ensure the validity of assessment scores by linking constructs to assessment tasks and student evidence (Mislevy et al., 2006). Central to the ECD process is defining a student model (what is to be measured), an evidence model (what student behaviors or outputs show proficiency) and a task model (what assessment tasks elicit that evidence). Further work has extended the ECD framework to demonstrate how science assessments can be developed for science learning that integrates multiple dimensions of content and practice (Harris et al., 2019). Thus, transforming science assessment for a future with AI does not just involve eliciting more or different types of evidence, but also fundamentally involves considering what proficiencies students need to develop to engage in future science endeavors, which will be supported by AI.

One significant change in science learning is that students now often engage in AI-supported inquiry and reasoning. Instead of relying solely on human-generated explanations, students interact with AI systems during their inquiry process (Lee et al., 2025). Scientific reasoning involves not only simple applications of procedures, it also requires critically evaluating and refining AI-generated scientific outputs. This shift challenges science assessments that measure isolated recall or procedural fluency. Rather, science assessments must capture how students critique AI-supported reasoning, validate AI-generated claims, and integrate human and machine insights into scientifically coherent explanations. For example, Liu and her colleagues (2024) developed an AI-supported system to diagnose and support student reasoning patterns in NGSS-aligned science tasks. Their system uses ML models to analyze constructed responses, focusing not simply on correctness but on identifying distinct reasoning profiles based on students' use of evidence and reasoning to support claims they made. Instead of simply scoring student responses, the system examines the structure and quality of students' scientific practices of constructing explanations and arguments to support fine-grained diagnosis of their reasoning development (Liu et al., 2024). Similarly, Haudek and Zhai (2024) demonstrated that ML scoring models can reliably evaluate complex aspects of students' scientific arguments, such as the presence of causal claims, reasoning links, and evidentiary support. Their study focused on how assessment constructs, such as the degree of reasoning complexity required, affect the accuracy and reliability of automated scoring models. Their findings suggest that ML-based assessment systems can validly capture critical elements of students' scientific reasoning beyond simple factual recall.

In addition, AI-driven personalization is transforming how students learn science. In AI-enhanced learning environments, adaptive systems personalize not only the pace of instruction but also the choices of learning pathways based on each students' prior knowledge, interests, and cultures (Shute & Rahimi, 2021). Rather than progressing through a uniform curriculum, AI may support students to experience individualized scientific explorations that are more relevant and motivating to them. This change presents important implications for the design of science assessments, which must allow for personalization of task contexts to enable students to investigate different phenomena or solve problems more relevant to them (Ryoo & Linn, 2015). Research (He et al., 2024) has shown that AI can support teachers in dynamically adapting knowledge-in-use assessments to better match students' emerging needs, allowing for more responsive science instruction and evaluation. Furthermore, recent studies emphasize that personalization must go beyond tailoring content to meaningfully include students' cultural experiences and identities. Nyaaba et al. (2024) propose that generative AI can be used to design culturally responsive science assessments that provide contextually relevant and motivating tasks for diverse learners. Similarly, Penuel et al. (2025) argue that considering student interest as a design principle for assessments can expand opportunities for engagement and deeper learning in science. Moreover, research by Kang, Talafian, and Tschida (2023) illustrates how using unconventional and flexible forms of assessment in secondary science classrooms (e.g., student-driven investigations, and project-based tasks) can expand students' opportunities to demonstrate complex reasoning and problem solving. These findings all advocate for students' agency and ownership in personalized assessment environments (Kallick & Zmuda, 2017).

As AI becomes increasingly integrated into science learning, human-AI collaboration should be explicitly considered in the design of future science assessments. Students nowadays may collaborate with AI systems to analyze complex data and construct scientific explanations. This evolution mirrors trends observed in professional scientific and medical fields, where human-AI partnerships have demonstrated superior outcomes compared to either human or AI performance alone (Reverberi et al., 2022). In a recent medical decision-making study, Reverberi and colleagues found that humans working collaboratively with AI achieved more accurate and reliable diagnostic judgments than humans or AI agents working independently. This suggests that the skill of critically engaging with AI output, such as integration, questioning, and refining AI-generated insights, is becoming a core scientific competency. According to von Davier and Burstein (2024), future assessment ecosystems must adopt a human-centered AI perspective, where success is judged not only by technological performance but by how effectively human learners can leverage AI to enhance reasoning, decision-making, and learning. This shift requires assessing new constructs, such as how students validate AI-supported claims, recognize biases in AI-generated outputs, and make ethical decisions when integrating AI insights into scientific sensemaking.

Additionally, research in human-AI interaction emphasizes the importance of developing students' metacognitive awareness of AI's capabilities and limitations. Maadi et al. (2021) argue that effective human-AI collaboration depends on users' ability to understand when and how to trust AI, and when to insert human oversight. In the context of science assessment, this suggests that tasks should not simply ask students to accept AI-generated answers but should require them to critically evaluate, revise, or extend AI outputs based on scientific evidence and reasoning.

9.3.2 REU#2: Future Science Tasks Should Be Performance Based to Enable AI-Resilient Assessments

Alignment with REP Principles This REU supports *Fairness* by enabling all learners to demonstrate competencies in diverse contexts; *Human Oversight & Autonomy* by preserving teachers' interpretive roles; *Robustness & Scientific Integrity* by designing performance-based tasks that yield discipline-aligned evidence of student learning.

As AI tools increasingly generate accurate responses to traditional assessment items, science educators must design tasks that assess scientific reasoning, decision-making, and authentic scientific practices—skills that AI cannot easily replicate. Performance-based and competency-focused assessments offer a promising way forward by requiring students to engage in real-world problem-solving, evaluate AI outputs, and construct evidence-based explanations, for example. These assessments are essential given the additional challenges of supporting deep, meaningful science learning towards knowledge-in-use.

Recent studies have explored strategies to address the challenges of designing NGSS-aligned assessments. For example, the Cognitively Based Assessment of, for, and as Learning (CBAL) in Science initiative aimed to create competency models and learning progressions that support the development of assessments aligned with NGSS (Liu et al., 2013). The project focused on designing tasks that assess students' competencies to apply scientific knowledge and engage in evidence-based reasoning through multiple scientific practices such as modeling, constructing arguments and explanations. The Next Generation Science Assessment project focused on creating assessment tasks that align with NGSS, targeting both upper-elementary and middle school students, which provided educators with tools to assess complex science learning and formative assessment strategies to inform instructional decision-making (Harris et al., 2019).

Science learning is an iterative, constructive and developmental process rather than a one-time event. Traditional assessments often provide only snapshots of student conceptual understanding, making it difficult to track learning over time. Many assessments fail to capture students' ability and processes to revise and refine their ideas, which is critical to scientific reasoning (Harris et al., 2019). Therefore, there is a high demand for appropriately designed tools to capture the iterative and progressive learning to track students' progress and identify persistent naive ideas or incremental conceptual change (Krajcik & Shin, 2023). While these challenges are often discussed in the context of science education, they are equally relevant across the broader STEM disciplines, where learning is similarly cumulative, interconnected, and iterative. Addressing them requires assessment approaches that can span multiple STEM domains while remaining sensitive to the unique practices of each field.

The increasing use of digital tools in STEM education, together with the rapidly increasing potential of AI enables a new, almost revolutionary approach to assessment of STEM learning: an on-the-fly assessment, sometimes also referred to as stealth assessment (Shute & Ventura, 2013). That is, assessment that uses data that students produce as they learn (e.g., text created in response to tasks, but also mouse clicks or keystrokes), to assess their learning, track developmental progression in their learning over time, and predict the extent to which their progression indeed leads to a greater competence in STEM proficiencies in the long run (Kubsch et al., 2022). The idea to use data that students produce as they learn is not new. In fact, ITSs (for an overview see Graesser et al., 2018) or Exploratory Learning Environments (ELEs) (Amershi & Conati, 2010) have used such data to model student learning, and predict long(er) term outcomes for student learning for quite some time. Findings from these and other studies suggest that including and integrating multimodal data can significantly improve the assessment of student learning (e.g. Gorham et al., 2019; for an overview see Ochoa, 2017).

In the context of regular STEM instruction, however, there is substantially less work on how to use data that students produce as they learn for tracking their learning. For one, researchers have struggled to interpret the data emerging from complex classroom situations. In fact, much of the work in the context of ITSs or ELEs has worked off of models of how students learn about STEM which model the domain as a collection of knowledge and skills to be acquired, but overlook how students' understanding progresses under curriculum-guided instruction. This is counter to recent perspectives that students should progress in acquiring and using knowledge and skills, demonstrating increasing integration of knowledge and skills to explain phenomena or solve problems (NRC, 2012). To do that, the analysis of students' learning must be rooted in domain-specific models of students' learning over time (i.e., learning progressions; Jin et al., 2024). That is, instruction must be designed to support learning based on a learning progression, such that the assessment of students' learning can be aligned with the learning progression.

Designing instruction based on a learning progression often requires more detailed guidance than the broader levels that progressions typically provide. Drawing from underlying learning theories, such as knowledge integration theory (Clark & Linn, 2013), can offer finer-grained descriptions needed to design tasks that track and support students' progress within an instructional unit. For example, Neumann et al. (2013) proposed a learning progression, grounded in knowledge integration theory, that outlined how students move from understanding energy forms to reasoning about energy transfer and transformation. Building on this work, Wyrwich et al. (2025) demonstrated how such a progression can guide the design of instructional activities that support students in making these conceptual shifts. With a clear vision of what students are to learn in an epistemic activity (e.g., creating and using a model of energy transformation processes taking place as a pendulum swings), we can delineate learning performances—that is, the knowledge and skills students are expected to develop and combine to meet the goals of the activity. Using principles from ECD, we can specify what counts as evidence that students have demonstrated particular knowledge, skills, or their integration (Harris et al., 2019). Gombert et al. (2024) have demonstrated how this approach enables accurate assessment of students' knowledge, skills, and their ability to integrate them into a learning performance.

Kubsch et al. (2022) have described applying ECD by delineating the domain-specific model into underlying knowledge and skills, specifying learning performances, and articulating evidentiary statements that describe what to accept as evidence throughout the process of developing curriculum materials. Kubsch et al. (2024) demonstrated how this procedure can be adapted to the AI-based analysis of existing curriculum materials, as long as these materials incorporate digital tools. In subsequent studies, the authors have provided evidence that following this procedure supports tracking student learning over time and predicting long(er) term outcomes such as the development of competence (Kubsch et al., 2025; Wyrwich et al., 2025). Further work also demonstrated that feeding information about students' learning performances back to teachers together with the ability to provide feedback can improve student learning (Karademir, Borgards, et al., 2024; Karademir, Di Mitri, et al., 2024). Thus, a well-designed system that incorporates instruction, assessment and AI evaluation and is founded on developmental models of learning provides enormous potential.

Overall, however, the field has only started to tap into the potential of using the wealth of data that is produced as students work with digital technologies as part of regular classroom teaching. And this potential cannot be underestimated. We acknowledge the change that AI is bringing to assessment of students learning in regular STEM classrooms. With AI this assessment may render explicit formative and summative assessment unnecessary, providing the teacher with much better, more fine-grained information, about each student on-the-fly without the substantial effort of scoring of students' products. This also provides an opportunity to better understand how students develop expertise in science domains, and offer more fine-grained and relevant instructional supports to students at developmentally appropriate levels. As the use of AI in assessment continues to develop, the procedure employed by Kubsch et al. (2022, 2024) could inform the use of generative AI as part of these designed learning systems, which would in turn allow for using the power of generative AI for creating additional tasks for students demonstrating limited growth or students who have already mastered the level of competence students are envisioned to master; effectively allowing for a different, more individualized but eventually better STEM education (see Chap. 6, Neumann, Zhai & Kubsch).

9.3.3 REU #3: Validity and Reliability Related to Automated Evaluation of Complex Assessments Remains at the Center of AI Fairness, Beneficial Use and Robustness

Alignment with REP Principles This REU addresses *Fairness* by identifying and mitigating subgroup bias; *Transparency & Explainability* by making scoring criteria and AI reasoning interpretable; *Human Oversight & Autonomy* by embedding human oversight into scoring processes, and *Robustness & Scientific Integrity* by ensuring automated evaluation meet rigorous validity and reliability standards.

AI-based automated scoring systems in science education are validated by gathering evidence that their scores are both accurate (matching human judgments) and construct-valid (reflecting the intended skills or knowledge). There are a number of foundational reports that posit approaches and criteria for establishing and maintaining validity in automatic scoring practice which will remain important as AI-based scoring expands (Williamson et al., 2012; Yan et al., 2020). In practice, this means developers of AI-based scoring systems must show that such a system evaluates responses in a way that fits the intended use and interpretation of the assessment. Important criteria of validity arguments include high agreement with expert human raters, consistency across diverse groups of test-takers, and logical alignment between the scoring algorithm's features and the construct being measured. Measures to validate these systems usually involve comparing AI-generated scores to human scores as a benchmark for accuracy. Studies often use metrics like percent agreement, Cohen's kappa, F1scores and/or correlation coefficients to quantify how closely the AI matches human scoring (Zhai et al., 2021). High inter-rater reliability between the AI and humans provides some evidence that the automated system is scoring accurately.

Other perspectives have posited that assessment validity is a holistic concept—beyond correlating with human scores, one must ensure the AI is measuring the right competencies and not introducing bias (Wood et al., 2021). Thus, validation of these systems should also examine how AI arrives at its scores and whether those scores truly reflect the underlying assessment construct (Kaldaras et al., 2024). For example, a number of studies have recommended documenting some of the technical aspects of algorithm use and ensuring it mirrors the criteria defined in coding rubrics aligned with the assessment (Kaldaras & Haudek, 2022; Zhai et al., 2020). However, matching rubrics may still fall short of fully capturing construct validity, particularly if the system fails to reflect the reasoning captured through real-world performances. As Beggrow et al. (2014) demonstrated, even when automated scoring aligned with rubric-based ratings, it may diverge significantly from insights gained through clinical interviews, which more authentically capture students' scientific thinking. This indicates the importance of incorporating external validation sources, such as interviews or task-based observations, to ensure AI scoring systems support meaningful inferences about student learning.

There is also an increasing focus on responsible use of AI in assessment. Burstein and LaFlair (2024) propose a unified framework that merges validity theory with AI ethics, underscoring that validity, reliability, and fairness remain the foundational criteria for evaluating AI scoring systems. This multi-faceted approach to developing validity arguments and maintaining consistency is essential for gaining acceptance from educators, testing agencies, and other stakeholders in education. Here we consider several approaches to ensuring and maintaining validity and reliability of automated scoring systems for science assessments, including consideration of scoring approaches, including human oversight in the scoring process and centering fairness in the development of these systems.

The AI approach employed in automated assessment may impact what is recognized as evidence of student reasoning, therefore affecting the reliability and accuracy of scoring. Machine learning (ML) and LLMs offer distinct approaches to automated scoring of student-generated text, particularly in jargon-rich and linguistically ambiguous science domains (e.g., biology) (Pan & Nehm 2025). Each method has strengths and weaknesses in areas including training data, scoring accuracy, reliability, and technological resources. Training data critically influences scoring accuracy. ML models in science education are trained on highly specialized corpora containing student-generated responses, closely aligning training and testing data. Such text corpora (e.g., children writing about phenomena) are rare in LLM training. Indeed, LLMs are typically trained on broader but general-purpose datasets, and they often lack direct exposure to the specialized scientific language and unique misconceptions found in student-generated text. The mismatch between training and testing data poses challenges for the scoring accuracy of LLMs (Huebner et al., 2021; McCoy et al., 2024; Wei et al., 2021). Recent research indicates improvements in LLM performance may occur through domain-specific fine-tuning, such as SciEdBERT developed by Liu et al. (2023).

Linguistic complexities inherent to specific scientific domains (e.g., evolutionary biology) also present unique challenges to automated scoring systems. Ambiguities, undefined jargon, and common misconceptions permeate both public and academic discourse (Nehm, 2010; Rector et al., 2013). Issues such as lexical ambiguity (e.g., how to interpret use of the term 'adapt'; the shifting use of appropriate and inappropriate teleological language) further complicate automated assessment (Rector et al., 2013). Moreover, linguistic patterns and jargon change as students develop deeper disciplinary proficiency, adding further complexity to training and testing efforts (Wang et al., 2025). These nuances pose particular difficulties for LLMs, which rely heavily on general linguistic patterns.

In terms of scoring accuracy and reliability, ML methods may achieve higher accuracy rates, comparable to human expert scorers, and deliver deterministic results that facilitate longitudinal studies and robust comparisons (Moharreri et al., 2014; Nehm, 2025). However, few studies have directly compared LLM vs. ML performance (Pan & Nehm, 2025). LLMs such as GPT-4o exhibit robust but slightly lower accuracy and a probabilistic output structure that may complicate reliability and replicability over time (Pack et al., 2024).

Finally, the choice between ML and LLM scoring involves careful consideration of resource allocation during assessment development and validation. While ML requires substantial initial investment in scoring data for training, the deployment and ongoing scoring process incur minimal additional costs. ML scoring generally demands significant initial rubric development and data scoring but may leverage existing institutional data (e.g., large graded corpora) and also produce valuable insights into student learning difficulties during iterative rubric refinement (Nehm, 2025; Sripathi et al., 2023). LLMs, on the other hand, require significant time invested in prompt engineering or corpus augmentation. These efforts are complicated by the fact that LLM scoring decisions are often challenging to understand even in cases of "explainable AI" (Cohn et al., 2024; Dann, 2014; Yan et al., 2024).

The computational infrastructure (e.g., ML, LLM) used to generate assessment scores emerges as an important but often neglected consideration in the development and validation of AI-enabled assessments. Corpora properties, scoring replicability, and system interpretability are impacted by computational approach. Future work should establish guidelines for maximizing the strengths of a computational approach while minimizing its ethical and interpretive risks.

Moving from consideration of AI methods for automated scoring, developers need to consider the role of human oversight in the development and deployment process. Although human expertise is critical in developing automated scoring systems, providing human oversight throughout the process of AI-based evaluation and checking for reliability over periods of time is essential in maintaining trust in AI systems. A human-in-the-loop (HITL) approach to AI integrates human expertise into automated processes to ensure that AI systems remain transparent, adaptable, and pedagogically sound (Memarian & Doleck, 2024; U.S. Department of Education, 2023). Recent studies demonstrate that HITL approaches enhance science education by combining AI's computational power with important pedagogical insights from humans. For example, researchers achieved success in automating science assessment by keeping teachers "in the loop" during AI-based grading. Cohn et al. (2024) used a chain-of-thought prompting technique with an LLM to evaluate middle-school science responses while allowing researchers to iteratively correct the model's reasoning using active learning techniques. In most cases, the AI model with corrected reasoning showed improved performance on score prediction as experts help to guide its focus. Similarly, Funayama et al. (2023) tested an automated short-answer scoring system in which cases assigned low confidence scores by the system can be returned to human graders. This system maintained high scoring quality, and could potentially reduce science teacher workload in assessment by examining only select, borderline cases of scoring. In a personalized tutoring context, Thomas et al. (2024) compared a hybrid human-AI tutoring model to traditional adaptive tutoring software in secondary math classes. The study reported learning gains when AI-based tutors were paired with human tutors at specific points throughout the lessons, with lower-achieving students benefiting the most from this hybrid approach.

Fairness is another crucial aspect of validity for automated systems: evaluators compare the AI's scoring performance across subgroups (e.g. different genders, language backgrounds) to ensure the system does not systematically favor or penalize any group. Although there are multiple definitions of fairness and/or bias in education (Baker & Hawn, 2022; Binns, 2018), here we focus on the ability of AI models to score or evaluate student responses from different sub-groups of target students with no significant difference in performance metrics. That is, bias due to Evaluation/Testing Bias during Model Learning (Baker & Hawn, 2022; Krist & Kubsch, 2023). For scoring of student responses, misclassified responses by AI may result in a range of effects from having students assigned an incorrect score on a high-stakes assessment (e.g. essay grading) to having an AI-based system provide a non-relevant learning resource based on the outcome of a formative assessment embedded in a curriculum. That is, the use of the AI scores is paramount to understanding the risk to students and other stakeholders. Obviously, if AI scores are biased towards or against some sub-group of students, this is deeply problematic as some students will be inadvertently disadvantaged, experience more misclassifications of their work and have a false sense of their learning achievement.

Some AI systems only respond to student input (e.g. chatbots), without formally scoring or classifying student work. However, due to the probabilistic nature of generative models, the same (or very similar input) from different students may result in different output returns for different students. This is particularly concerning given the potential for students to interact individually with autonomous chatbots as part of online tutoring systems. Thus, different students could receive different types of feedback, or differently stated feedback from the chatbot, despite having very similar understanding of a phenomenon. Thus, there may be some advantage to using predictive or deterministic ML models with prescribed scripts for returning student-facing feedback. Conversely, more research examining how to develop and constrain generative AI agents for assigned tasks in science education (e.g. feedback delivery, suggesting variables to manipulate in experiments, scaffolding grade-appropriate data analysis) would be beneficial.

As shown by Baker & Hawn (2022) in a review of existing studies, there are relatively few that have examined for bias based on native language and a number of other important sub-populations for education. This remains a challenge to be addressed as AI-based assessment tasks become more complex and include more text-based responses as part of performance assessments. A recent study showed that different human coding approaches used to train AI models for student scientific arguments lead to different degrees of bias in AI predicted scores for English learners as compared to non-English learners (Wilson et al., 2024). Further, this bias in evaluating scientific arguments from different sub-groups of students can be "hidden" by examining the AI performance measures over an entire assessment as compared to performance on individual items (Bracey et al., 2023).

As science assessment may increasingly require multi-modal inputs (e.g. drawings, verbal discourse), AI-based evaluation models need to maintain performance measures across student subpopulations for the essential input. For example, Zhai, He and Krajcik (2022) noted that an AI algorithm developed to evaluate students' drawn models struggled to correctly evaluate student models with "unexpected" representations (i.e. model elements not used by the majority of students). Relatedly, Li, Haudek and Krajcik (2025) found that AI-predicted scores of student-drawn models differed from human scores when model elements were hand-drawn as opposed to using predefined icons as part of the model. It is critical that such AI-based scoring models evaluate the underlying construct of the scientific modeling practice, which can be expressed in a variety of ways, and not just image elements of the majority group. Researchers should ensure that both training and testing data are collected from students in classrooms, at developmentally appropriate stages in learning, as opposed to solely relying on data gleaned from open source databases or web-based information.

9.3.4 REU #4: Responsible Use of AI as Part of Classroom Science Assessments Requires Both Teachers and Students to Understand the Benefits and Limitations of AI

Alignment with REP Principles This REU advances *Human Oversight & Autonomy* by equipping educators and students to make informed decisions about AI use; *Transparency & Explainability* by clarifying how AI tools operate; *Privacy & Data Security* by safeguarding student information; and *Respect for Human Values* by promoting ethical and fair assessment practices.

While AI offers transformative potential to enhance the validity, responsiveness, and personalization of science assessments, its use must be guided by a clear understanding of its benefits, limitations, and broader implications. Teachers and students—along with curriculum designers, technology developers, and school leaders—must be equipped to critically engage with AI technologies, evaluate their outputs, and make informed decisions about how and when to rely on them.

A foundational step toward responsible use in assessment practices is fostering AI literacy among both educators and learners. This entails developing a conceptual understanding of how AI tools function, what types of data they process, how algorithms identify patterns, generate predictions or feedback, and under what conditions their outputs may be flawed, biased, or misleading. For students, this involves cultivating epistemic vigilance—being able to question and critique the validity of AI-generated feedback or recommendations (Ng et al., 2021). For example, a student engaging with an AI-supported writing tool should not merely accept suggested revisions but reflect on their appropriateness within the scientific argument being developed. Similarly, students using simulation-based science assessments must recognize that AI-generated prompts or analyses are contingent on training data and assumptions that may not fully reflect real-world complexities (Krenn et al., 2022). Embedding such reflective practices in classroom instruction not only supports deeper science learning but also empowers students to become responsible users of emerging technologies.

Teachers, in turn, require targeted professional development to responsibly integrate AI into assessment practices in pedagogically sound and ethically informed ways. This includes professional training in interpreting AI-generated dashboards, identifying algorithmic errors, and understanding the limitations of predictive analytics in diverse classroom contexts. Teachers must also learn to recognize when AI tools might inadvertently disadvantage certain students—for example, when natural language processing models fail to account for non-standard dialects, or when adaptive systems prematurely narrow students' learning paths based on initial performance (Baker & Hawn, 2022). Rather than acting as passive users of technology, teachers should be positioned as active mediators, and maintain human oversight of the resulting use of AI. This requires teachers who can evaluate AI-generated insights, contextualize them within classroom dynamics, and supplement them with their own professional judgment. Building this capacity requires not only technical knowledge but also ethical reasoning, data literacy, and cultural competence.

In practice, responsible AI use in classroom assessments calls for HITL systems—designs that retain teacher oversight and enable collaborative decision-making between humans and machines (Holstein et al., 2019). These systems can ensure that final judgments about student performance are not solely automated but involve teacher review, especially when student responses are ambiguous, atypical, or high-stakes. For instance, while an AI model might flag a student's explanation as lacking causal reasoning, it is the teacher who can interpret whether the explanation reflects a conceptual misunderstanding, linguistic barrier, or creative scientific reasoning not captured by the algorithm. HITL approaches uphold the professional agency of teachers while enhancing the scalability and efficiency of AI-supported assessment.

Transparency and explainability are also central to responsible AI use. Both teachers and students must be able to understand how AI systems arrive at specific assessments or recommendations. Without this transparency, AI tools risk becoming opaque "black boxes," undermining trust and reducing the instructional value of the feedback they generate (Gardner et al., 2021; Wu et al., 2025). Developers of AI-based assessment systems should therefore prioritize model interpretability, offering insights into the criteria used for scoring or feedback and indicating confidence levels or sources of uncertainty. For example, a system that evaluates students' scientific arguments might highlight which parts of the response contributed most to the score, or which conceptual elements were inferred as missing. Providing such explanatory feedback not only supports instructional decision-making but also encourages students to engage in self-regulated learning and argument revision.

Responsible use of AI in assessment must be situated within a broader vision of equity and inclusion in science education. AI should be leveraged not merely to optimize performance metrics but to help create more inclusive learning environments that recognize and value diverse ways of knowing, reasoning, and expressing understanding. This includes designing assessment tasks that are culturally relevant, allowing for multiple modes of representation, and ensuring that AI feedback supports growth and learning. AI systems should be used to illuminate opportunity gaps and support differentiated instruction, rather than entrenching deficit-based narratives about student performance.

Ethical considerations play a crucial role when choosing between LLMs and ML tools for automated evaluation of student work, particularly concerning issues of beneficence and privacy (Yan et al., 2024). Beneficence involves proactively minimizing potential harm arising from implementing a specific tool or action. The majority of discussions in the automated scoring literature in science education have been technically oriented, emphasizing scoring accuracy, and implicitly addressing beneficence by justifying acceptable error levels. Arguments supporting LLM scoring errors as acceptable typically compare these errors with human scoring variability, emphasize performance relative to teacher performance, or suggest that such errors are tolerable in lower-stakes or formative assessment settings. Additionally, arguments highlight the practical benefits of automated scoring, such as reducing workload for teachers, thus allowing them to dedicate more time to tasks with higher instructional impact (Zhai et al., 2020).

However, these technically focused justifications have notable shortcomings (Pan & Nehm 2025). Most have appeared in academic literature without adequately engaging broader stakeholder perspectives, such as those of parents, educators, or school boards. Consequently, definitions of what constitutes acceptable accuracy or "successful" scoring have largely been determined by researchers developing and testing these systems. This narrow approach risks overlooking whether stakeholders perceive these accuracy differences—particularly between ML and LLM scoring—as meaningful or impactful. Given the rapid integration of LLMs in educational contexts, there is a clear need to broaden ethical discussions by including diverse stakeholder groups in the establishment of acceptable accuracy thresholds and in evaluating potential harms and benefits associated with automated scoring. HITL discussions have rarely considered which humans are "in the loop."

Privacy represents another significant ethical concern differentiating ML and LLM approaches. The deployment of AI-powered assessment tools often involves the collection and processing of large amounts of student data—including interaction logs, written responses, speech patterns, and potentially biometric data. Ensuring that these data are securely stored, anonymized where appropriate, and used solely for pedagogical purposes is essential to protecting student rights and upholding ethical standards in educational technology (Smuha, 2022). Some ML tools in science education attempt to mitigate privacy risks by employing anonymized datasets, ensuring student responses are never linked to personal identifiers or retained beyond immediate scoring purposes. In contrast, proprietary LLMs often retain submitted data, potentially using student-generated content for future profit-driven activities without explicit consent. This raises serious ethical concerns regarding data ownership, informed consent, and the ability to maintain clear distinctions between human-generated and AI-generated content within educational environments.

Additionally, corporations behind popular LLMs face legal challenges due to alleged copyright infringements resulting from unauthorized use of proprietary material during their training processes (Brittain, 2025). Educational institutions employing or financially supporting these proprietary LLMs may unintentionally participate in ethically problematic or potentially unlawful activities. Despite their significance, these privacy-related ethical challenges have been largely overlooked in research evaluating LLM effectiveness.

Ethical considerations related to beneficence and privacy must be carefully evaluated when choosing between ML and LLM tools. Expanding discussions beyond technical accuracy to include extensive stakeholder participation and transparent analysis of potential harms and benefits is essential for the ethically responsible adoption and use of automated scoring systems in educational contexts.

9.4 Conclusions

The integration of AI into science education necessitates a fundamental re-evaluation of assessment practices. Traditional models that emphasize content recall and isolated cognition are insufficient to capture dynamic, collaborative, and AI-supported learning that characterizes modern scientific inquiry. As AI becomes more embedded in formal science classrooms and informal learning environments, science assessments must change to remain valid, equitable, and instructionally meaningful.

This chapter has emphasized four key changes where responsible and ethical use of AI is critical to future assessment use. First, assessment constructs must be expanded to reflect how students engage in reasoning with and through AI. Students must demonstrate their ability to critique AI-generated outputs, integrate AI-based outputs with human judgments, and apply ethical reasoning in AI-supported scientific contexts. This necessitates principled assessment design that is grounded in clearly defined constructs, evidence-centered design frameworks, and validity arguments, to ensure that the tasks truly measure the competencies we value in science education. Second, the design of performance-based, developmentally grounded assessments should draw on cognitive models of how students learn and engage with science. These performance assessments should be multi-modal to capture a range of student engagements and outputs. When supported by AI-driven analytics, these assessments can offer real time insights into student thinking, as well as track student progress over developmentally appropriate and targeted learning progressions.

Equally important are considerations of validity, fairness, and interpretability in AI-based scoring systems used with assessments. As automated assessments continue to become more prevalent, ensuring systems are appropriately calibrated to diverse student populations is critical. Consideration of the choice of AI approaches for automated evaluation in science domains, and aligned with assessment use, continues to be a critical issue. In parallel, both teachers and students must develop AI literacy in the scope of assessments and acknowledge system limitations. Teachers need support to critically interpret AI-generated data from assessments, while merging its use in classroom instruction with teacher expertise and contextual knowledge of students and classrooms. Students should learn to engage reflectively and collaboratively with AI tools as part of scientific inquiry.

Looking forward, there is likely to be increasing use of AI across all grade levels in science education, both in K-12 and post-secondary levels. The advent of AI is opening the door for a promising transformation of science education in the twenty-first century, yet as outlined here, there are a number of challenges that must be addressed to ensure the responsible and ethical use of AI. Recent initiatives in the U.S. demonstrate a growing recognition of AI's role in education. For example, recent national efforts have committed to fostering AI literacy as part of K-12 schools and teacher preparation. As these policies take shape, science education exists at a critical juncture: the opportunity to shape the future of AI-supported assessments that not only enhance science learning but also support and evaluate the skills students will need to engage in the science workforce in the future.

In sum, the development of AI-integrated assessments should be guided not only by what is possible, but by what is pedagogically and ethically sound. Future research should explore how to operationalize these principles in diverse educational settings, helping ensure that AI can serve as a catalyst for deeper science learning, as opposed to a shortcut for easy evaluation. By centering validity and classroom assessment, science educators and researchers can harness the potential of AI to support more authentic, holistic, and adaptive science assessments.

Declaration of AI Use All sections of this manuscript were drafted by the authors. AI-based writing tools were used in portions of this manuscript to assist with language editing and improve clarity of expression. These revisions were subsequently reviewed by the authors. All intellectual content and interpretations are the authors' own.

Acknowledgement This material is based upon work supported by the National Science Foundation under Grant No. 2332964. Any opinions, findings, and conclusions or recommendations expressed in this material are those of the author(s) and do not necessarily reflect the views of the National Science Foundation.

References

Amershi, S., & Conati, C. (2010). Automatic recognition of learner types in exploratory learning environments. In C. Romero, S. Ventura, M. Pechenizkiy, R. S. J. d. Baker (Eds.), *Handbook of educational data mining* (1st ed., p. 18). CRC Press.

Azevedo, R., Mudrick, N. V., Taub, M., & Bradbury, A. E. (2019). Self-regulation in computer-assisted learning systems. In J. Dunlosky & K. A. Rawson (Eds.), *The Cambridge handbook of cognition and education* (pp. 587–618). Cambridge University Press; Cambridge Core. https://doi.org/10.1017/9781108235631.024

Baker, R. S., & Hawn, A. (2022). Algorithmic bias in education. *International Journal of Artificial Intelligence in Education, 32*(4), 1052–1092. https://doi.org/10.1007/s40593-021-00285-9

Beggrow, E. P., Ha, M., Nehm, R. H., Pearl, D., & Boone, W. J. (2014). Assessing scientific practices using machine-learning methods: How closely do they match clinical interview performance? *Journal of Science Education and Technology., 23(1)*, 160–182. https://doi.org/10.1007/s10956-013-9461-9

Bewersdorff, A., Hartmann, C., Hornberger, M., Seßler, K., Bannert, M., Kasneci, E., Kasneci, G., Zhai, X., & Nerdel, C. (2025). Taking the next step with generative artificial intelligence: The transformative role of multimodal large language models in science education. *Learning and Individual Differences, 118*, 102601. https://doi.org/10.1016/j.lindif.2024.102601

Binns, R. (2018). Fairness in machine learning: Lessons from political philosophy. In S. A. Friedler & C. Wilson (Eds.), *Proceedings of the 1st conference on fairness, accountability and transparency* (Vol. 81, pp. 149–159). PMLR. https://proceedings.mlr.press/v81/binns18a.html

Black, P., & Wiliam, D. (2009). Developing the theory of formative assessment. *Educational Assessment, Evaluation and Accountability(Formerly: Journal of Personnel Evaluation in Education), 21*(1), 5. https://doi.org/10.1007/s11092-008-9068-5

Bracey, Z. B., Stuhlsatz, M. A. M., Wilson, C. D., Cheuk, T., Santiago, M. M., Osborne, J., Haudek, K., & Donovan, B. M. (2023). Investigating differential severity across linguistic subgroups in automated scoring of student argumentation. In X. Liu & W. J. Boone (Eds.), *Advances in applications of Rasch measurement in science education* (pp. 385–426). Springer. https://doi.org/10.1007/978-3-031-28776-3_15

Brittain, B. (2025). OpenAI must face part of Intercept lawsuit over AI training. *Reuters*. Retrieved May 22, 2025, from https://www.reuters.com/legal/litigation/openai-must-face-part-intercept-lawsuit-over-ai-training-2025-02-20/

Burstein, J., & LaFlair, G. T. (2024). *Where assessment validation and responsible AI meet*. https://arxiv.org/abs/2411.02577

Clark, D. B., & Linn, M. C. (2013). The knowledge integration perspective: Connections across research and education. In S. Vosniadou (Ed.), *International handbook of research on conceptual change* (pp. 520–538). Routledge.

Cohn, C., Hutchins, N., Le, T., & Biswas, G. (2024). A chain-of-thought prompting approach with LLMs for evaluating students' formative assessment responses in science. *Proceedings of the AAAI Conference on Artificial Intelligence, 38*(21), 23182–23190. https://doi.org/10.1609/aaai.v38i21.30364

Cope, B., Kalantzis, M., & Searsmith, D. (2021). Artificial intelligence for education: Knowledge and its assessment in AI-enabled learning ecologies. *Educational Philosophy and Theory, 53(12)*, 1229–1245. https://doi.org/10.1080/00131857.2020.1728732

Dann, R. (2014). Assessment as learning: Blurring the boundaries of assessment and learning for theory, policy and practice. *Assessment in Education: Principles, Policy & Practice, 21(2)*, 149–166. https://doi.org/10.1080/0969594X.2014.898128

de Jong, T., Sotiriou, S., & Gillet, D. (2014). Innovations in STEM education: The Go-Lab federation of online labs. *Smart Learning Environments, 1*, 3. https://doi.org/10.1186/s40561-014-0003-6

Erduran, S., & Levrini, O. (2024). The impact of artificial intelligence on scientific practices: An emergent area of research for science education. *International Journal of Science Education, 46*(18), 1982–1989. https://doi.org/10.1080/09500693.2024.2306604

Feinstein, N. (2011). Salvaging science literacy. *Science Education, 95*(1), 168–185. https://doi.org/10.1002/sce.20414

Funayama, H., Asazuma, Y., Matsubayashi, Y., Mizumoto, T., & Inui, K. (2023). Reducing the cost: Cross-prompt pre-finetuning for short answer scoring. In N. Wang, G. Rebolledo-Mendez, N. Matsuda, O. C. Santos, & V. Dimitrova (Eds.), *Artificial intelligence in education* (pp. 78–89). Springer Nature.

Gardner, J., O'Leary, M., & Yuan, L. (2021). Artificial intelligence in educational assessment: 'Breakthrough? Or buncombe and ballyhoo?'. *Journal of Computer Assisted Learning, 37*(5), 1207–1216. https://doi.org/10.1111/jcal.12577

Gombert, S., Fink, A., Giorgashvili, T., Jivet, I., Di Mitri, D., Yau, J., Frey, A., & Drachsler, H. (2024). From the automated assessment of student essay content to highly informative feedback: A case study. *International Journal of Artificial Intelligence in Education, 34*(4), 1378–1416. https://doi.org/10.1007/s40593-023-00387-6

Gorham, T., Jubaed, S., Sanyal, T., & Starr, E. L. (2019). Assessing the efficacy of VR for foreign language learning using multimodal learning analytics. In C. N. Giannikas, E. Kakoulli Constantinou, & S. Papadima-Sophocleous (Eds.), *Professional development in CALL: A selection of papers* (pp. 101–116). Research-Publishing.net. https://doi.org/10.14705/rpnet.2019.28.873

Graesser, A. C., Hu, X., & Sottilare, R. (2018). Intelligent tutoring systems. In F. Fischer, C. E. Hmelo-Silver, S. R. Goldman, & P. Reimann (Eds.), *International handbook of the learning sciences* (1st ed.). Routledge.

Harris, C. J., Krajcik, J. S., Pellegrino, J. W., & DeBarger, A. H. (2019). Designing knowledge-in-use assessments to promote deeper learning. *Educational Measurement: Issues and Practice, 38*(2), 53–67. https://doi.org/10.1111/emip.12253

Haudek, K. C., & Zhai, X. (2024). Examining the effect of assessment construct characteristics on machine learning scoring of scientific argumentation. *International Journal of Artificial Intelligence in Education, 34*(4), 1482–1509. https://doi.org/10.1007/s40593-023-00385-8

He, P., Shin, N., Zhai, X., & Krajcik, J. (2024). A design framework for integrating artificial intelligence to support teachers' timely use of knowledge-in-use assessments. In X. Zhai & J. Krajcik (Eds.), *Uses of artificial intelligence in STEM education* (pp. 52–94). Oxford University Press.

Holstein, K., Wortman Vaughan, J., Daumé, H., Dudik, M., & Wallach, H. (2019). Improving fairness in machine learning systems: What do industry practitioners need? In *Proceedings of the 2019 CHI conference on human factors in computing systems* (pp. 1–16). https://doi.org/10.1145/3290605.3300830

Huebner, P. A., Sulem, E., Cynthia, F., & Roth, D. (2021, November 10–11). BabyBERTa: Learning more grammar with small-scale child-directed language [Paper presentation]. In *25th conference on computational natural language learning* (pp. 624–646).

Jin, H., Yan, D., & Krajcik, J. (2024). *Handbook of research on science learning progressions* (1st ed.). Routledge.

Kaldaras, L., Akaeze, H. O., & Reckase, M. D. (2024). Developing valid assessments in the era of generative artificial intelligence. *Frontiers in Education, 9*. https://www.frontiersin.org/journals/education/articles/10.3389/feduc.2024.1399377

Kaldaras, L., & Haudek, K. C. (2022). Validation of automated scoring for learning progression-aligned Next Generation Science Standards performance assessments. *Frontiers in Education, 7*. https://doi.org/10.3389/feduc.2022.968289

Kallick, B., & Zmuda, A. (2017). *Students at the center: Personalized learning with habits of mind* (1st ed.). ASCD.

Kang, H., Talafian, H., & Tschida, P. (2023). Expanding opportunities to learn in secondary science classrooms using unconventional forms of classroom assessments. *Journal of Research in Science Teaching, 60*(5), 1053–1091. https://doi.org/10.1002/tea.21824

Karademir, O., Borgards, L., Di Mitri, D., Strauß, S., Kubsch, M., Brobeil, M., Grimm, A., Gombert, S., Rummel, N., Neumann, K., & Drachsler, H. (2024). Following the impact chain of the LA cockpit: An intervention study investigating a teacher dashboard's effect on student learning. *Journal of Learning Analytics, 11*(2), 215–228. https://doi.org/10.18608/jla.2024.8399

Karademir, O., Di Mitri, D., Schneider, J., Jivet, I., Allmang, J., Gombert, S., Kubsch, M., Neumann, K., & Drachsler, H. (2024). I don't have time! But keep me in the loop: Co-designing requirements for a learning analytics cockpit with teachers. *Journal of Computer Assisted Learning, 40(6)*, 2681–2699. https://doi.org/10.1111/jcal.12997

Kizilcec, R., & Lee, H. (2022). Algorithmic fairness in education. In W. Holmes & K. Porayska-Pomsta (Eds.), *The ethics of artificial intelligence in education* (1st ed., p. 29). Routledge.

Koedinger, K. R., Booth, J. L., & Klahr, D. (2013). Instructional complexity and the science to constrain it. *Science, 342*(6161), 935–937. https://doi.org/10.1126/science.1238056

Krajcik, J., & Shin, N. (2023). Student conceptions, conceptual change, and learning progressions. In N. G. Lederman & S. K. Abell (Eds.), *Handbook of research on science education* (Vol. 3, pp. 116–138). Routledge. https://doi.org/10.4324/9780367855758-7

Krenn, M., Pollice, R., Guo, S. Y., Aldeghi, M., Cervera-Lierta, A., Friederich, P., dos Passos Gomes, G., Häse, F., Jinich, A., Nigam, A., Yao, Z., & Aspuru-Guzik, A. (2022). On scientific understanding with artificial intelligence. *Nature Reviews Physics, 4*(12), 761–769. https://doi.org/10.1038/s42254-022-00518-3

Krist, C., & Kubsch, M. (2023). Bias, bias everywhere: A response to Li et al. And Zhai and Nehm. *Journal of Research in Science Teaching, 60*(10), 2395–2399. https://doi.org/10.1002/tea.21913

Kubsch, M., Czinczel, B., Lossjew, J., Wyrwich, T., Bednorz, D., Bernholt, S., Fiedler, D., Strauß, S., Cress, U., Drachsler, H., Neumann, K., & Rummel, N. (2022). Toward learning progression analytics—Developing learning environments for the automated analysis of learning using evidence centered design. *Frontiers in Education, 7*. https://www.frontiersin.org/journals/education/articles/10.3389/feduc.2022.981910

Kubsch, M., Grimm, A., Neumann, K., Drachsler, H., & Rummel, N. (2024). Using evidence-centered design to develop an automated system for tracking students' physics learning in a digital learning environment. In X. Zhai & J. Krajcik (Eds.), *Uses of artificial intelligence in STEM education*. Oxford University Press. https://doi.org/10.1093/oso/9780198882077.003.0011

Kubsch, M., Strauß, S., Grimm, A., Gombert, S., Drachsler, H., Neumann, K., & Rummel, N. (2025). Self-regulated learning in the digitally enhanced science classroom: Toward an early warning system. *Educational Psychology Review, 37*(2), 34. https://doi.org/10.1007/s10648-025-10011-9

Laugksch, R. C. (2000). Scientific literacy: A conceptual overview. *Science Education, 84*(1), 71–94. https://doi.org/10.1002/(SICI)1098-237X(200001)84:1<71::AID-SCE6>3.0.CO;2-C

Lee, G.-G., Mun, S., Shin, M.-K., & Zhai, X. (2025). Collaborative learning with artificial intelligence speakers. *Science & Education, 34*(2), 847–875. https://doi.org/10.1007/s11191-024-00526-y

Lee, G.-G., Shi, L., Latif, E., Gao, Y., Bewersdorff, A., Nyaaba, M., Guo, S., Wu, Z., Liu, Z., Wang, H., Mai, G., Liu, T., & Zhai, X. (2023). *Multimodality of AI for education: Towards artificial general intelligence*. https://arxiv.org/abs/2312.06037

Lee, G.-G., & Zhai, X. (2025). Realizing visual question answering for education: GPT-4V as a multimodal AI. *TechTrends, 69*(2), 271–287. https://doi.org/10.1007/s11528-024-01035-z

Li, T., Haudek, K., & Krajcik, J. (2025). Utilizing deep learning AI to analyze scientific models: Overcoming challenges. *Journal of Science Education and Technology*. https://doi.org/10.1007/s10956-025-10217-0

Liu, L., Cisterna, D., Kinsey, D., Qi, Y., & Steimel, K. (2024). 162AI-based diagnosis of student reasoning patterns in NGSS assessments. In X. Zhai & J. Krajcik (Eds.), *Uses of artificial intelligence in STEM education* (pp. 162–176). Oxford University Press. https://doi.org/10.1093/oso/9780198882077.003.0008

Liu, L., Rogat, A., & Bertling, M. (2013). A CBAL™ science model of cognition: Developing a competency model and learning progressions to support assessment development. *ETS Research Report RR-13-29*. https://doi.org/10.1002/j.2333-8504.2013.tb02336.x

Liu, Z., He, X., Liu, L., Liu, T., & Zhai, X. (2023). Context matters: A strategy to pre-train language model for science education. In N. Wang, G. Rebolledo-Mendez, V. Dimitrova, N. Matsuda, & O. C. Santos (Eds.), *Artificial intelligence in education. Posters and late breaking results, workshops and tutorials, industry and innovation tracks, practitioners, doctoral consortium and blue sky* (pp. 666–674). Springer Nature.

Luaces, O., Díez, J., & Bahamonde, A. (2018). A peer assessment method to provide feedback, consistent grading and reduce students' burden in massive teaching settings. *Computers & Education, 126*, 283–295. https://doi.org/10.1016/j.compedu.2018.07.016

Luckin, R., Holmes, W., Griffiths, M., & Forcier, L. B. (2016). *Intelligence unleashed: An argument for AI in education*. UCL Knowledge Lab / Pearson.

Maadi, M., Akbarzadeh Khorshidi, H., & Aickelin, U. (2021). A review on human–AI interaction in machine learning and insights for medical applications. *International Journal of Environmental Research and Public Health, 18*(4), 2121. https://doi.org/10.3390/ijerph18042121

McCoy, R. T., Yao, S., Friedman, D., Hardy, M. D., & Griffiths, T. L. (2024). Embers of autoregression show how large language models are shaped by the problem they are trained to solve. *Proceedings of the National Academy of Sciences of the United States of America, 121*(41), e2322420121. https://doi.org/10.1073/pnas.2322420121

Memarian, B., & Doleck, T. (2024). Human-in-the-loop in artificial intelligence in education: A review and entity-relationship (ER) analysis. *Computers in Human Behavior: Artificial Humans, 2*(1), 100053. https://doi.org/10.1016/j.chbah.2024.100053

Mislevy, R. J., Steinberg, L. S., & Almond, R. G. (2006). On the structure of educational assessments. *Measurement: Interdisciplinary Research and Perspectives, 4*(1–2), 3–62. https://doi.org/10.1207/S15366359MEA0101_02

Moharreri, K., Ha, M., & Nehm, R. H. (2014). EvoGrader: An online formative assessment tool for automatically evaluating written evolutionary explanations. *Evolution: Education and Outreach, 7*(1), 1–14. https://doi.org/10.1186/s12052-014-0015-2

National Research Council (NRC). (2012). *A framework for K-12 science education: Practices, crosscutting concepts, and core ideas*. The National Academies Press. http://www.nap.edu/openbook.php?record_id=13165

National Research Council (NRC). (2014). *Developing assessments for the next generation science standards*. The National Academies Press. https://doi.org/10.17226/18409

Nehm, R. H. (2010). Understanding undergraduates' problem-solving processes. *Journal of Microbiology & Biology Education, 11*(2), 119–122. https://doi.org/10.1128/jmbe.v11i2.203

Nehm, R. H. (2025). AI in biology education assessment: How automation can drive educational transformation. In X. Zhai et al. (Eds.), *Uses of artificial intelligence in STEM education*. Oxford University Press.

Neumann, K., Viering, T., Boone, W. J., & Fischer, H. E. (2013). Towards a learning progression of energy. *Journal of Research in Science Teaching, 50*(2), 162–188. https://doi.org/10.1002/tea.21061

Ng, D. T. K., Leung, J. K. L., Chu, S. K. W., & Qiao, M. S. (2021). Conceptualizing AI literacy: An exploratory review. *Computers and Education: Artificial Intelligence, 2*, 100041. https://doi.org/10.1016/j.caeai.2021.100041

NGSS Lead States. (2013). *Next generation science standards: For states, by states*. The National Academies Press.

Nyaaba, M., Zhai, X., & Faison, M. Z. (2024). Generative AI for culturally responsive science assessment: A conceptual framework. *Education Sciences, 14*(12). https://doi.org/10.3390/educsci14121325

Ochoa, X. (2017). Multimodal learning analytics. In C. Lang, A. F. Wise, A. Merceron, D. Gašević, & G. Siemens (Eds.), *Handbook of learning analytics* (pp. 129–141). Society for Learning Analytics Research. https://doi.org/10.18608/hla17.011

Osborne, J. (2023). Science, scientific literacy, and science education. In N. G. Lederman & S. K. Abell (Eds.), *Handbook of research on science education* (Vol. 3, 1st ed., pp. 785–816). Routledge.

Pack, A., Barrett, A., & Escalante, J. (2024). Large language models and automated essay scoring of English language learner writing: Insights into validity and reliability. *Computers and Education: Artificial Intelligence, 6*, 100234. https://doi.org/10.1016/j.caeai.2024.100234

Pan, Y., & Nehm, R. H. (2025). Large language model and traditional machine learning scoring of evolutionary explanations: Benefits and drawbacks. *Education Sciences, 15*(6), 676. https://doi.org/10.3390/educsci15060676

Pane, J. F., Griffin, B. A., McCaffrey, D. F., & Karam, R. (2014). Effectiveness of cognitive tutor algebra I at scale. *Educational Evaluation and Policy Analysis, 36*(2), 127–144. https://doi.org/10.3102/0162373713507480

Penuel, W. R., O'Connor, K., Allen, A.-R., Jacobs, J. K., & Lo, A. S. (2025). Examining science teachers' conceptions of student interest as a consideration in designing assessments. *Journal of Science Teacher Education, 36*(5), 643–663. https://doi.org/10.1080/1046560X.2024.2435747

Popenici, S. A. D., & Kerr, S. (2017). Exploring the impact of artificial intelligence on teaching and learning in higher education. *Research and Practice in Technology Enhanced Learning, 12*(1), 22. https://doi.org/10.1186/s41039-017-0062-8

Rector, M. A., Nehm, R. H., & Pearl, D. (2013). Learning the language of evolution: Lexical ambiguity and word meaning in student explanations. *Research in Science Education, 43*(3), 1107–1133. https://doi.org/10.1007/s11165-012-9296-z

Reverberi, C., Rigon, T., Solari, A., Hassan, C., Cherubini, P., Antonelli, G., Awadie, H., Bernhofer, S., Carballal, S., Dinis-Ribeiro, M., Fernández-Clotett, A., Esparrach, G. F., Gralnek, I., Higasa, Y., Hirabayashi, T., Hirai, T., Iwatate, M., Kawano, M., Mader, M., … GI Genius CADx Study Group. (2022). Experimental evidence of effective human–AI collaboration in medical decision-making. *Scientific Reports, 12*(1), 14952. https://doi.org/10.1038/s41598-022-18751-2

Ryder, J. (2001). Identifying science understanding for functional scientific literacy. *Studies in Science Education, 36*(1), 1–44. https://doi.org/10.1080/03057260108560166

Ryoo, K., & Linn, M. C. (2015). Designing and validating assessments of complex thinking in science. *Theory into Practice, 54*(3), 238–254. https://doi.org/10.1080/00405841.2015.1044374

Shute, V., & Ventura, M. (2013). *Stealth assessment: Measuring and supporting learning in video games*. MIT Press. https://doi.org/10.7551/mitpress/9589.001.0001

Shute, V. J., & Rahimi, S. (2021). Stealth assessment of creativity in a physics video game. *Computers in Human Behavior, 116*, 106647. https://doi.org/10.1016/j.chb.2020.106647

Siemens, G., & Long, P. (2011). Penetrating the fog: Analytics in learning and education. *EDUCAUSE Review, 46*(5), 30–40.

Smuha, N. (2022). Pitfalls and pathways for trustworthy artificial intelligence in education. In W. Holmes & K. Porayska-Pomsta (Eds.), *The ethics of artificial intelligence in education* (1st ed., p. 33). Routledge.

Sripathi, K. N., Moscarella, R. A., Steele, M., Yoho, R., You, H., Prevost, L. B., Urban-Lurain, M., Merrill, J., & Haudek, K. C. (2023). Machine Learning Mixed Methods Text Analysis: An Illustration From Automated Scoring Models of Student Writing in Biology Education. *Journal of Mixed Methods Research, 15586898231153946*. https://doi.org/10.1177/15586898231153946

Thomas, D. R., Lin, J., Gatz, E., Gurung, A., Gupta, S., Norberg, K., Fancsali, S. E., Aleven, V., Branstetter, L., Brunskill, E., & Koedinger, K. R. (2024). Improving student learning with hybrid human-AI tutoring: A three-study quasi-experimental investigation. In *Proceedings of the 14th learning analytics and knowledge conference* (pp. 404–415). https://doi.org/10.1145/3636555.3636896

Tsai, C.-W., Ma, Y.-W., Chang, Y.-C., & Lai, Y.-H. (2022). Integrating multiculturalism into artificial intelligence-assisted programming lessons: Examining inter-ethnicity differences in learning expectancy, motivation, and effectiveness. *Frontiers in Psychology, 13*. https://www.frontiersin.org/journals/psychology/articles/10.3389/fpsyg.2022.868698

US Department of Education, Office of Educational Technology. (2023). *Artificial intelligence and future of teaching and learning: Insights and recommendations*.

von Davier, A. A., & Burstein, J. (2024). AI in the assessment ecosystem: A human-centered AI perspective. In P. Ilic, I. Casebourne, & R. Wegerif (Eds.), *Artificial intelligence in education: The intersection of technology and pedagogy* (Intelligent systems reference library) (Vol. 261). Springer. https://doi.org/10.1007/978-3-031-71232-6_6

Wang, H., Haudek, K., Manzanares, A. D., Romulo, C., Royse, E. A., & Azzarello, C. (2025). Extending a pretrained language model (BERT) using an ontological perspective to classify students' scientific expertise level from written responses. *Disciplinary and Interdisciplinary Science Education Research, 7*, 26.

Wei, J., Garrette, D., Linzen, T., & Pavlick, E. (2021, November 7–11). Frequency effects on syntactic rule learning in transformers. In *Proceedings of the 2021 conference on empirical methods in natural language processing* (pp. 932–948). Association for Computational Linguistics.

Wieman, C. E., Adams, W. K., Loeblein, P., & Perkins, K. K. (2010). Teaching physics using PhET simulations. *The Physics Teacher, 48*(4), 225–227. https://doi.org/10.1119/1.3361987

Wiliam, D. (2011). What is assessment for learning? *Studies in Educational Evaluation, 37*(1), 3–14. https://doi.org/10.1016/j.stueduc.2011.03.001

Williamson, D. M., Xi, X., & Breyer, F. J. (2012). A framework for evaluation and use of automated scoring. *Educational Measurement: Issues and Practice, 31*(1), 2–13. https://doi.org/10.1111/j.1745-3992.2011.00223.x

Wilson, C. D., Haudek, K. C., Osborne, J. F., Buck Bracey, Z. E., Cheuk, T., Donovan, B. M., Stuhlsatz, M. A. M., Santiago, M. M., & Zhai, X. (2024). Using automated analysis to assess middle school students' competence with scientific argumentation. *Journal of Research in Science Teaching, 61*(1), 38–69. https://doi.org/10.1002/tea.21864

Wood, S., Yao, E., Haisfield, L., & Lottridge, S. (2021). *Establishing standards of best practice in automated scoring* (p. 27). ACT Research. https://files.eric.ed.gov/fulltext/ED616491.pdf

Wu, X., Xu, C., Li, X., & Zhang, Y. (2025). Enhancing transparency and explainability in large language model-driven educational assessment. *JMIR AI, 4*(1), e51834. https://doi.org/10.2196/51834

Wyrwich, T., Kubsch, M., Drachsler, H., & Neumann, K. (2025). Tracking students' progression in developing understanding of energy using AI technologies. *Physical Review Physics Education Research, 21*(1), 010152. https://doi.org/10.1103/PhysRevPhysEducRes.21.010152

Yan, D., Rupp, A., & Foltz, P. W. (2020). *Handbook of automated scoring: Theory into practice* (1st ed.). Chapman and Hall/CRC. https://doi.org/10.1201/9781351264808

Yan, L., Sha, L., Zhao, L., Li, Y., Martinez-Maldonado, R., Chen, G., Li, X., Jin, Y., & Gašević, D. (2024). Practical and ethical challenges of large language models in education: A systematic scoping review. *British Journal of Educational Technology, 55*(1), 90–112. https://doi.org/10.1111/bjet.13370

Zhai, X., He, P., & Krajcik, J. (2022). Applying machine learning to automatically assess scientific models. *Journal of Research in Science Teaching, 59*(10), 1765–1794. https://doi.org/10.1002/tea.21773

Zhai, X., Neumann, K., & Krajcik, J. (2023). Editorial: AI for tackling STEM education challenges. *Frontiers in Education, 8*, 1183030. https://doi.org/10.3389/feduc.2023.1183030

Zhai, X., Shi, L., & Nehm, R. H. (2021). A meta-analysis of machine learning-based science assessments: Factors impacting machine-human score agreements. *Journal of Science Education and Technology, 30*(3), 361–379. https://doi.org/10.1007/s10956-020-09875-z

Zhai, X., Yin, Y., Pellegrino, J. W., Haudek, K. C., & Shi, L. (2020). Applying machine learning in science assessment: A systematic review. *Studies in Science Education, 56*(1), 111–151. https://doi.org/10.1080/03057267.2020.1735757

Chapter 10
AI-Based Science Assessment: Potential Pitfalls and Responsible and Ethical Practices

Gyeonggeon Lee, Xiaoming Zhai, and Okhee Lee

10.1 Introduction

Science assessment is one of the educational fields that has most actively adopted artificial intelligence (AI) technology. The earliest AI-related studies in science education occurred in the 1980s and leveraged AI to capture learners' problem-solving process based on expert systems (e.g., Good et al., 1986). The application of AI to science assessment became more commonplace in the early 2010s, when scholars began to utilize supervised machine learning techniques to train scoring models to functionally imitate the marking of human scorers on student-constructed responses (e.g., Nehm et al., 2012). During this period, various machine learning algorithms, such as decision trees, naive Bayes, Bayes nets, and support vector machines, were employed to develop scoring models (Zhai, Yin, et al., 2020). Each algorithm analyzes the semantic features of student writing for classification and numerical prediction. After the models are tested, they may be applied in large-scale classes to provide timely scoring of students' constructed responses. These algorithms show significant potential yet need a large amount of human-labeled data on the researcher's end to train and test, which is costly and thus calls for more efficient methods.

G. Lee (✉)
Natural Sciences and Science Education Department, National Institute of Education, Nanyang Technological University, Singapore, Singapore
e-mail: gyeonggeon.lee@nie.edu.sg

X. Zhai
AI4STEM Education Center, University of Georgia, Athens, GA, USA

O. Lee
Department of Teaching and Learning, Steinhardt School of Culture, Education, and Human Development, New York University, New York, NY, USA

X. Zhai, K. J. Crippen (eds.), *Advancing AI in Science Education*, Advances in Technology-Rich Science Education 3,
https://doi.org/10.1007/978-3-032-16871-9_10

One alternative is large language models (LLMs). Different from the algorithms used in supervised machine learning-based assessment, LLMs are pretrained with large datasets by their developers before being used by researchers. LLMs thus possess a certain level of language proficiency without further training or finetuning, which significantly improves their efficiency by reducing the need for human--labeled data. For example, the most commonly known LLM, BERT (i.e., Bidirectional Encoder Representations from Transformers), was trained using Wikipedia data, which can understand the natural language content within context (Devlin et al., 2019). This development not only reduced the need for large datasets for training but, more importantly, increased scoring accuracy (Zhai, 2024a).

In recent years, AI-based science assessment as a topic in the scholarly literature has risen dramatically (see Fig. 10.1) due to two main factors. The first factor involves generative AI—a critical aspect of most LLMs that can generate text, image, videos, etc. and transforms the development of science assessment tasks. The second factor involves the growing diversity of data modality, which extends text-based assessment prompts to image, sound, and video and enables AI to analyze multimodal inputs from students (Lee, Shi, et al., 2025; Lee, Yun, et al., 2025).

Along with the increasing attention and growing potential of AI in science assessment, concerns are being raised. Among these concerns is maintaining robustness and integrity, as AI-based assessments must align with rigorous educational standards and accurately reflect students' scientific knowledge (see the Responsible and Ethical Principles Framework in Chap. 3). The frequently reported student outsourcing of assessments using AI has also drawn significant concerns. In addition, as AI systems may inadvertently perpetuate biases, concerns have been raised about

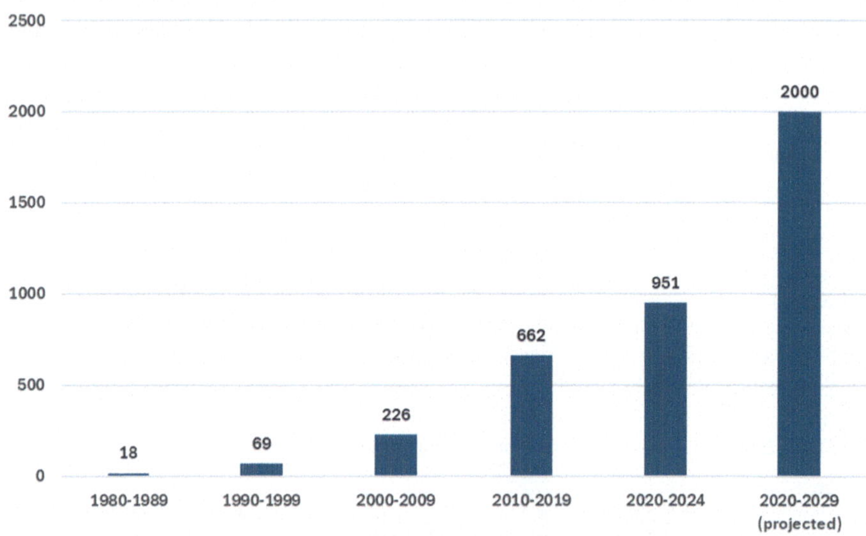

Fig. 10.1 Number of "science assessment" + "artificial intelligence" hits in Google Scholar (the projected number of 2000 for 2020–2029 was extrapolated based on 951 hits in 2020–2024, as of June 11, 2025)

their potential to disadvantage certain student groups based on factors such as gender, race, ethnicity, or socioeconomic status. Moreover, there is a need to ensure that AI deployment in science assessment consistently demonstrates respect for human values, emphasizing dignity, agency, and inclusivity for all learners. Finally, AI-based assessment draws concerns about intersecting factors such as language, culture, and accessibility, which afford or limit equitable use of AI in science assessment.

In this chapter, we present changes to science assessment brought by AI, particularly focusing on three potential pitfalls: (a) outsourcing of cognitive tasks, (b) bias and achievement disparity in AI scoring, and (c) locality. We then discuss how science teachers and researchers can overcome these potential pitfalls based on the responsible and ethical principles (REP) (see Chap. 3) for practicing AI-based science assessment. We close the chapter with a discussion of the implications of AI use in future science assessment research and practice.

10.2 AI-Based Science Assessment: Changes and Potential Pitfalls

AI is transforming assessment practices in science classrooms—targeting more complex constructs, enriching the ways evidence of students' thinking is collected, and automating the delivery and scoring of assessments (Zhai, Haudek, Shi, et al., 2020; Zhai & Wiebe, 2023). However, despite these benefits, we contend that practicing AI-based science assessment is not without challenges. Recognizing the potential pitfalls brought by AI to science assessments in the classroom is essential for developing effective solutions.

10.2.1 Outsourcing of Cognitive Tasks

Being able to solve challenging problems, including those in science, is one of the benchmark capabilities for most LLMs. Research suggests that ChatGPT and GPT-4 perform comparably with students when solving K-12 science tasks from the NAEP (National Assessment of Educational Progress) test and particularly excel at cognitively demanding questions (Zhai et al., 2025). These LLMs exhibit robust reasoning ability, demonstrating not only inferential but also pedagogical knowledge, which is noteworthy given their conversational capabilities (Lee, Lee, et al., 2025; Lee, Shi, et al., 2025). With such advanced capabilities, it is not surprising that students have turned to LLMs as tools to outsource classroom assessments, drawing concerns about academic integrity and plagiarism.

Cognitive outsourcing involves delegating information processing to external agents (e.g., AI) to save one's cognitive resources (Tao et al., 2024). However, this raises concerns about who or what is responsible for knowing. In science education, in particular, AI-involved cognitive outsourcing reduces students' learning opportunities and thus may negatively impact students' learning abilities and autonomy. In their quasi-experimental study, Xu et al. (2025) showed that a generative AI (GenAI)-enhanced learning environment decreased students' self-regulated learning ability when there was no metacognitive support. This may be due to students' decreased cognitive engagement and dependence on powerful GenAI models. Moreover, cognitive outsourcing can also compromise learning, particularly when some students simply adopt (i.e., copy) AI-generated answers without engagement or personal efforts (Sullivan et al., 2023). In such cases, AI-based assessment cannot promote authentic student learning, as students have not processed the target learning content.

Building on the concern of cognitive outsourcing, Gao et al. (2025) researched why students find it easy to offload science tasks to GenAI. They introduced a two--dimensional framework—modality and interactivity—to analyze how susceptible different types of assessment tasks are to GenAI outsourcing. According to their findings, tasks with lower interactivity and simpler modality, such as monomodal static tasks (e.g., text-based multiple-choice questions) and multimodal static tasks (e.g., drawing-based models), are particularly vulnerable. These tasks effectively assess scientific knowledge and practices but can be easily handled by powerful models like ChatGPT-4, making them easy targets for students to outsource. In contrast, tasks that are both multimodal and interactive (e.g., see Fig. 10.2, from Gao et al., 2025) require sustained engagement, dynamic responses, and often hands-on input—making them more resistant to GenAI outsourcing and thus better suited for preserving authentic student learning.

While much of the literature has focused on learners' outsourcing during science assessment, teachers are not immune to similar tendencies, which can jeopardize integrity and compromise the value of assessment. Teachers are increasingly relying on GenAI models to support various aspects of assessment development and evaluation, including generating items (Chan et al., 2025), automatic scoring (Lee et al., 2024), crafting feedback (Ba et al., 2025; Guo, Latif, et al., 2024), and critiquing items based on curriculum standards (e.g., Next Generation Science Standards [NGSS], NGSS Lead States, 2013; Nguyen & Hayward, 2025). Although these uses can increase efficiency and reduce workload, they also present potential pitfalls: Overreliance on GenAI may lead to a decline in teachers' assessment literacy, reduced critical engagement in item design, and diminished capacity to make pedagogically sound judgments. Moreover, if teachers adopt AI-generated assessment materials without thorough review or contextual adaptation, the assessments may not align with instructional goals or may fail to meet the diverse needs of learners. In this sense, teacher outsourcing, much like student outsourcing, risks weakening the integrity and educational value of science assessments.

A contrast between traditional (without AI) and AI-based (outsourcing) science assessments shows a clear picture of the potential pitfalls (see Table 10.1). Ideally, traditional science assessment involves teachers designing assessment items using their expertise with instruction and assessment. This process requires thoughtful

Fig. 10.2 Example of multimodal interactive tasks and responses from ChatGPT-4o (see Gao et al., 2025, p. 9)

alignment with curriculum, learning objectives, and students' developmental levels. Students respond to assessment items with their cognitive efforts, and their responses are intended to reflect their proficiency in the subject matter. Teachers then evaluate students' responses, often using a given scoring rubric, and subsequently provide detailed, context-aware feedback. This approach ensures that both teachers and students are informed by professional judgement and grounded in educational goals (NRC, 2001).

When AI is introduced in science assessment, each step in the process could be altered. While it is true that teachers can leverage AI to enhance work efficiency, careful consideration of outsourcing can upgrade the value and overall outcome of assessment. When teachers design items relying solely on AI, it may reduce their instructional dedication and also introduce unexpected errors or even bias. These issues could negatively impact the integrity of the assessment if efforts are not taken to fix these issues. Students, in turn, may use AI to generate responses, circumventing meaningful engagement with the content. Scoring and feedback can also be handled by AI, often without clear curriculum alignment. As a result, teachers' expertise, students' cognitive efforts, and curriculum relevance may be diminished or bypassed entirely in the AI-outsourced science assessment system. This AI-driven approach shifts control away from educators and toward algorithms, raising

Table 10.1 Comparison of traditional and AI-based science assessment

Assessment component	Traditional science assessment	AI-based (outsourced) science assessment
Educational Value	High—supports learning, diagnosis, and instructional decisions	Questionable—may obscure real learning needs and inflate student performance
Curriculum Alignment	Strong—aligned with standards, learning objectives, and content knowledge	Weak or absent—AI-generated assessments may not map to intended curricular goals
Item Design	Designed by teachers using their pedagogical and assessment expertise	Generated or assisted by AI, outsourcing teacher expertise
Teacher Involvement	High—teachers exercise judgment, align with the curriculum, and understand student needs	Minimal—teachers rely on AI tools, reducing their active role
Student Engagement	Students respond based on their own cognitive efforts and understanding	Students may use AI to generate responses, reducing authentic cognitive engagement
Reflection of Student Ability	Responses reflect student proficiency and understanding of the target content	Responses may reflect AI's capabilities more than students' actual knowledge
Scoring	Done by teachers or with rubrics; aligned with standards and informed by student context	Automated by AI; may lack contextual accuracy and curricular alignment
Feedback Quality	Constructive, targeted, and context--aware feedback based on teacher judgment	Generic or misaligned feedback from AI; may lack depth or educational insight
Risks	Time-consuming for teachers; dependent on their expertise	Risk of de-skilling teachers and students, misalignment with educational goals
Control Over Assessment	Teachers and schools control the assessment process	AI systems (and their training data) shape the assessment process

significant concerns about the educational value and integrity of assessments conducted without careful oversight.

10.2.2 Bias and Achievement Disparities in AI Scoring

Science assessment has long been concerned with validity (i.e., the extent to which an assessment measures what it is supposed to measure) to avoid impartial assessment of student performance and thinking. With the introduction of AI into the scoring process, researchers have raised concerns about validity, particularly the potential bias that may exacerbate the known inequity in education and compromise the validity of assessment uses (Baker & Hawn, 2022; Bird et al., 2025; Idowu, 2024; Zhai et al., 2021). In their study, Krist and Kubsch (2023) classified AI-related bias into societal and technical. The former refers to historical and societal injustices embedded in data or systems and perpetuated through AI use, while the latter arises from errors in measurement, sampling, or model design. For example, due to some reason, one may find that girls in a class performed significantly better than boys in science learning. This may indicate societal bias. When AI accurately scores students' performance that reflects this gender gap, AI may perpetuate but does not cause this bias. However, if the AI's scoring accuracy is significantly higher for one group compared to the other, this could introduce technical bias. Zhai and Krajcik (2025) refer to the societal AI bias as one type of *pseudo mechanical AI bias*—the bias that AI perpetuates but does not causally contribute to. In other words, societal AI bias exists regardless of whether one uses AI or not, which is often seen in education. For example, an AI system making accurate predictions about students' grades might still contribute to societal bias if the grades are used to reallocate resources away from underserved populations, exacerbating existing disparities. In this case, the bias exists regardless of whether the predictions are made by AI or humans; thus, attributing this bias to AI is pseudo. Thus, Zhai and Nehm (2023) argue that, although addressing societal bias is essential, it is not unique to the AI research community; rejecting AI simply because it may reflect or perpetuate existing societal disparities misses the opportunity to use AI as a tool to support teachers and promote equity when implemented thoughtfully.

What should really draw concerns in the community and needs more empirical research is technical AI bias. Technical bias arises from flaws in training data, algorithms, measurement, or deployment. Examples include label noise, skewed sampling, and flawed feature engineering. It is generally identifiable and fixable within the system itself (e.g., improving data quality, rebalancing classes, or adjusting algorithmic design). To deal with technical AI bias, Zhai and Krajcik (2025) suggest that one needs to first differentiate bias from pseudo bias. All bias entails three conditions: (a) *deviation*—bias represents the differences between observations and ground-truth or the actual value—the errors; (b) *systematic nature*—bias reflects systematic rather than random errors; and (c) *tendency*—bias involves a consistent inclination to favor or disfavor certain ideas or entities over others.

Some discussions about AI bias in education extend beyond the scope of the term "bias." For instance, some researchers characterize a machine scoring error in an underserved individual's assignment as bias, despite lack of a systematic pattern. Similarly, machine scoring outcomes that don't consider the societal consequences of such outcomes may fall short of constituting technical bias. To ensure conceptual clarity, any discussions about AI bias need to conform to the three conditions—*deviation, systematic nature,* and *tendency*—to ensure the discussed bias is technically real.

In science education, although concerns about AI bias are high, there have been limited empirical studies examining the substantiation of bias. The limited research that has been conducted offers partial remedies to public concerns, while also exposing notable limitations. These studies primarily employed supervised machine learning in which researchers used human-labeled student responses to train algorithms. The resulting findings have been useful and inspiring. For instance, for the concerns about linguistic bias in AI scoring systems, empirical studies in the early days of AI-based automatic scoring suggest that such errors tend to be relatively minor and on par with those found in clinical interview interpretations (Beggrow et al., 2014; Ha & Nehm, 2016). This research, however, is based on relatively small sample sizes and qualitative analyses.

A recent study (Guo et al., 2025) using large data sets provides more solid evidence about potential AI bias, particularly toward English language learners (ELLs) when training data are unbalanced and small. The study systematically examined scoring performance using BERT-based models trained on various combinations of ELL and non-ELL responses across different data scales in science assessments (see Fig. 10.3). Results showed that when the training dataset included a large number of ELL responses (around 30,000 or 1,000), the AI scoring models did not exhibit significant bias or distort disparities between ELLs and non-ELLs. However, when the number of ELL responses was small (approximately 200), models trained solely on ELL data showed significantly lower accuracy than those on mixed data. These findings underscore that AI bias against ELLs is not inherent but rather emerges when data are insufficient or misrepresentative of real-world ELL distributions. Therefore, ensuring adequate and naturally representative ELL data in training sets is crucial for fair and accurate AI-based assessment systems.

Equally important, if not more, is to examine whether AI systematically enlarged mean score gaps (MSGs) among groups. MSG indicates the difference in average scores between demographic groups (e.g., ELLs and non-ELLs), which is critical because it reveals potential disparities in how AI systems assess students of varying

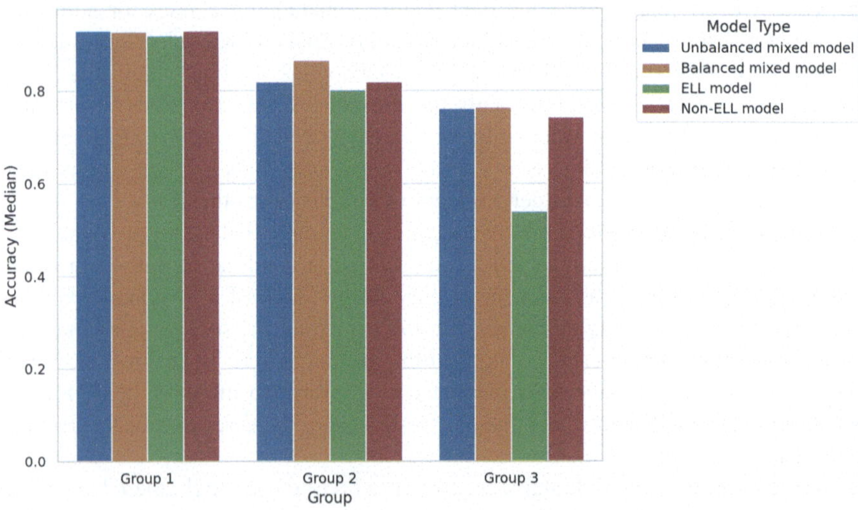

Fig. 10.3 Median accuracy of four AI scoring models across three dataset sizes in science assessments. Group 1 (≈30,000 ELLs), Group 2 (≈1,000 ELLs), and Group 3 (≈200 ELLs) show how training data scale and composition affect model performance. While all models performed similarly with large datasets, the ELL-only model showed a sharp drop in accuracy with smaller datasets, highlighting the risk of bias when training data are limited or unbalanced (Guo et al., 2025, p. 272)

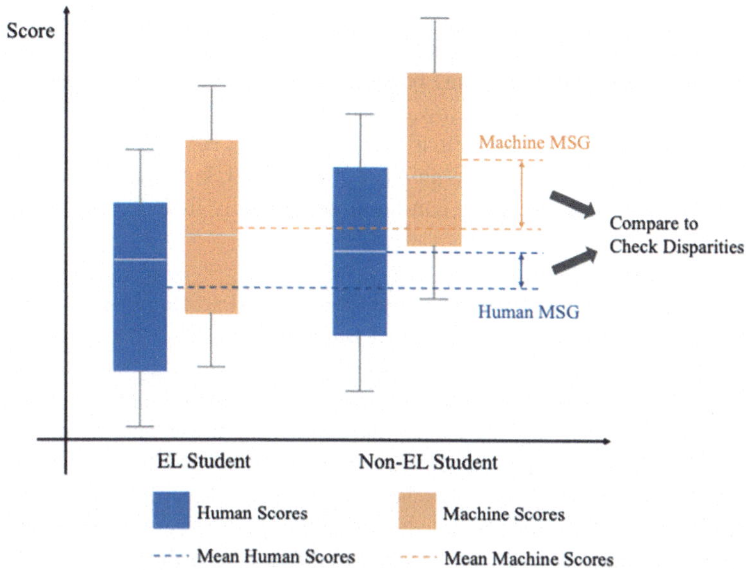

Fig. 10.4 Analyzing disparity by comparing MSGs between human and machine scores (Guo et al., 2025)

backgrounds (see Fig. 10.4). In Guo et al. (2025), MSG was used to compare the scoring patterns between human raters and AI models. The study found that when the training data were sufficiently large (e.g., ELL ≈ 30,000 or ELL ≈ 1,000), the AI-generated MSGs were not significantly different from those of human raters, indicating no distortion in disparities. However, when the dataset was small (e.g., ELL ≈ 200), significant discrepancies in MSGs emerged, particularly in models trained on balanced or ELL-only datasets. These findings suggest that while AI scoring may not inherently amplify disparities, imbalanced or limited training data can lead to misrepresentation of group performance, underscoring the importance of using adequately sized and naturally representative datasets to ensure fairness in AI-based assessment systems.

To dig into the issues of AI scoring (particularly technical bias), we suggest more empirical research at two stages:

Data Sampling Stage Biases in AI scoring can result from multiple sources at the data preparation and pre-processing stage. How data have been historically accumulated in favor of majority groups over minority groups can affect how AI scores responses from individuals in each demographic group. This is also the case for how data are collected for specific research purposes at present. For example, if more data have been collected on a particular gender, race, ethnicity, socioeconomic status, disability, or language group, the scoring machine may overrepresent responses from those groups (Idowu, 2024). In comparison, how experimental data are sampled from the collected dataset can affect whether certain demographic groups are overrepresented or underrepresented in the model. For example, even if the collected dataset represents diverse demographic groups, subsampling may distort their distribution. Data imbalance (i.e., unbalanced numbers of data points for different categories, regardless of the cause) may be another source of bias at this stage and can lead the model parameters to lean toward the majority category (Fang et al., 2023). For example, a part of a large (inter)national dataset (e.g., NAEP, PISA, and TIMSS) is commonly sampled and fed into an AI scoring model, but unintentional sources of bias in the sampling procedure can be unintentionally included in the AI scoring model.

Data Labeling and Model Training Stage Biases in AI scoring outcomes can also result from multiple sources at the model training stage. The scoring rubrics can affect whether and how we widen or close the gap between higher achieving and lower achieving profiles in score distribution. If we use holistic scoring rubrics rather than analytic scoring, this may underrepresent lower achieving students' responses that are worth partial credit (Zhang et al., 2015). On the other hand, scoring results of human experts (i.e., teachers and/or researchers) for each response is considered a ground truth for machine learning. If human scorers' inconsistent scoring behavior, influenced by construct-irrelevant item scenarios (cf. Zhai, Haudek, Stuhlsatz, et al., 2020), is reflected in scoring model training, it will inevitably lead to biased scoring models. Further, if human experts have systematic bias toward certain demographic groups, their scoring results will likely be biased toward those groups, and the machine may also show the same pattern. Statistically, the type of algorithms used to build the scoring models can also affect the significance of bias as they can overestimate or underestimate the effect size (Bird et al., 2025). Finally, the architecture of the machine learning pipeline, as a set of algorithmic layers, can affect bias.

10.2.3 Locality

Given that outsourcing and AI scoring bias are more internal aspects of potential pitfalls (i.e., developers and teachers could solve them), we also need to consider external aspects that are macroscopically characterized. For example, we conceptualize locality as a distinct environment where its components intersect. In science assessment, components, such as standards, language, culture, region, and accessibility, influence how science teachers and learners interact with AI (see Fig. 10.5). While the literature introduced below has discussed some components of the locality, our experiences as science education researchers who work with pre- and in-service science teachers from various countries have shown that these components need to be comprehensively deliberated, as their combination impacts how AI can be used for science education. Note that the five components discussed here are not an exhaustive list, but necessary considerations of locality that have been highlighted as relevant to science assessment in the literature. The extent to which each

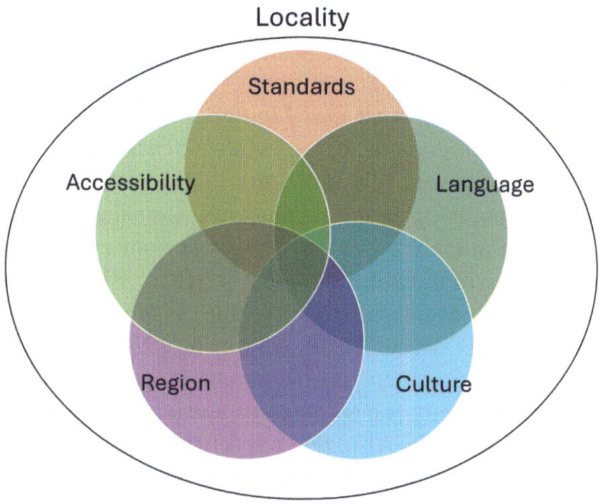

Fig. 10.5 Locality as represented by standards, language, culture, region, and accessibility

component of locality supports or deters integrating AI in science assessment depends on educational contexts.

Standards—National, Regional, and/or School Standards That Prescribe the Learning Objectives and Target Content Knowledge Aligning assessment to standards is one of the most important components in designing and administering science assessments, considering that they should work as a formative stepping-stone toward achieving learning objectives (NRC, 2001). Therefore, AI-based assessment does not replace or substitute the existing science education framework, such as the NGSS (NGSS Lead States, 2013); rather, it serves to demonstrate that students' learning progressions at each grade level or band meet those standards (Zhai et al., 2021). The Next Generation Science Assessment (NGSA) is a representative example that demonstrates how AI technology can be used to realize an existing assessment framework. The NGSA team design assessments in physical and life sciences across three dimensions of the NGSS (i.e., disciplinary core ideas, science and engineering practices, and crosscutting concepts). The interactive assessment tasks are delivered via the online portal so that teachers can use these tasks in NGSS-aligned classroom instruction. AI assessment models are tested with student responses. However, how to train AI models to understand or at least reflect alignment between standards and assessment is an issue.

Language—The Primary Language That Learners Use to Respond to Assessment Items Since LLMs are typically trained with a large English-based corpus, researchers have raised concerns about using AI for students whose native language is not English. For example, linguistic characteristics of Korean (Lee et al., 2018) and Chinese (Yang et al., 2025), coupled with a scarcity of training data in these languages, may lead to difficulty in developing AI scoring models in science assessment for speakers of these languages. Also, if scoring models are trained with a small amount of data from ELLs, their scoring accuracy can decrease (Guo et al.,

2025). Therefore, learners' primary language can influence the outcomes of AI use in science assessment.

Culture—Affordances and Limitations of Background Experiences That Affect Students' Responsive Behavior to Science Assessments NRC (2001) acknowledges that students' cultural norms and beliefs impact how they respond to assessment items. For example, if students internalize that mathematics items should be answered quickly based on their classroom culture, they may not put effort into responding to more difficult items (NRC, 2001). How much effort they commit to formal assessments in school settings is also related to their cultural backgrounds (NRC, 2001). Recently, Nyaaba et al. (2024) identified five tenets of culturally responsive assessment: (a) Indigenous language, (b) race and ethnicity, (c) religious beliefs, (d) family and community, and (e) Indigenous knowledge. If science assessment, particularly AI-based science assessment, does not consider these tenets, learners may perceive the assessment as irrelevant or inappropriate to demonstrate their understanding of the content.

Region—National- or Regional-Level Atmosphere Surrounding Technology Acceptance Regional differences can affect technology acceptance in education. Several countries in the Asia-Pacific region, such as Singapore, have implemented policy-level integration of contemporary technologies in classrooms, which also supports the use of AI in science education (UNESCO, 2024b). Similarly, Al-Zahrani and Alasmari (2024) concluded that higher education students and faculty members in Saudi Arabia held favorable attitudes toward the use of AI in education due to their social experience. However, other countries may not have such a supportive policy or environment for adopting contemporary educational technologies. If AI studies are misunderstood as pertinent only to WEIRD (White, Educated, Industrialized, Rich, and Democratic) countries, other nations may face hurdles in increasing acceptance and usage of AI in society, including the education sector (Samuel et al., 2023).

Accessibility—Individual- and Organization-Level Ease of Access to Hardware and Software Since the rise of educational technologies following the onset of the internet, accessibility has been acknowledged as a significant issue in the education field (Kim, 2009). In terms of AI usage, accessibility becomes a critical issue (EC, 2022; UNESCO, 2023, 2024a) that is compounded by the fact that certain AI technologies can be expensive. Contemporary AI models typically require an enormous number of computing resources (e.g., graphics processing unit [GPU] hardware). Furthermore, not every country allows citizens to freely access AI services. Even the free versions of ChatGPT and Google Gemini, prominent GenAI models, are not available in all countries. Thus, accessibility is a significant issue impacting AI use in science assessment.

In summary, different components of locality affect AI-integration in science assessment across different contexts. For example, the first author of this chapter studied in South Korea and has worked in the USA and Singapore. In his experience, South Korea has good accessibility to AI technology in terms of hardware and software. However, the Korean language itself inhibits the utilization of state-of-the-art AI models, which largely depend on English as a global language. In contrast, in the USA context, English as a primary language can afford AI-integration in science assessment. However, even though Singapore is an English-speaking country and has a strong educational technology driving force, the coverage of the national curriculum may differ from the data that typical AI models are trained with. As an example, the Singapore O-Level Chemistry Syllabus focuses on molecules that have covalent bonds between two or more atoms. In this case, noble gases, such as neon, helium, and argon, are not considered molecules. In contrast, ChatGPT may provide information that noble gases can be considered as molecules in the tradition of thermodynamics. This is an example of how locality can pose challenges in integrating AI into science assessment.

10.3 Responsible and Ethical Principles for Practicing AI-Based Science Assessment

The three pitfalls of AI use in science assessment—cognitive outsourcing, bias in AI scoring, and locality—call for applying REP of AI from both developers and users of AI-based science assessment. This call asks for conceptual, technological, and practical shifts from traditional science assessment practices.

10.3.1 Reconceptualizing Science Assessment Tasks and the Human-AI Relationship

Even though AI-enabled assessment outsourcing is challenging and concerning, educational institutions can proactively integrate AI into assessment practices, moving beyond a defensive stance that primarily seeks to minimize or avoid AI use (rather, the protective momentum could target technical bias). This is because students will inevitably operate within AI-enriched environments when they leave school. The competencies required for their future success will extend beyond the knowledge retention that traditional science assessments target toward the ability to leverage AI in science problem-solving. Therefore, students' increased tendency to outsource assessments highlights the necessity to reconsider the constructs targeted by traditional science assessments.

Specifically, this new challenge brought by AI demands reevaluation of science assessment tasks, particularly those addressing lower order cognitive skills, such as rote memorization and basic recall. The future workforce market urgently needs a paradigmatic shift toward developing students' higher-order thinking, such as critical thinking, creativity, and design thinking—those skills that support science and engineering practices (SEPs) of the NGSS and would not easily be substituted by AI. To achieve this goal, aligning assessment tasks with authentic AI-based scientific practices is essential, which can enhance the relevance and applicability of the evaluated competencies and benefit learners for their future lives and work, aligning with the REP framework (see the REP Framework in Chap. 3).

Additionally, it is imperative for teachers to reconsider the format and features of assessments to minimize their susceptibility to outsourcing. As indicated in Gao et al. (2025), innovations, such as multimodal tasks, and interactive tasks, such as scientific inquiry simulations on digital learning devices that shift the focus from static outcomes to dynamic processes, offer promising directions for development. Given that assessment fundamentally constitutes an evidentiary inferential process (Pellegrino et al., 2016), there is a critical need to update the nature of evidence collected, the approaches employed in collecting such evidence, and the interpretive techniques used for deriving meaningful conclusions and making instructional decisions. For example, AI-embedded analytical dashboards can support teachers in understanding learners' status and thus provide adaptive feedback (Kim, 2024). Embracing such assessment frameworks not only addresses the concerns about outsourcing but also ensures alignment with educational goals that can better prepare students to navigate and thrive within AI-based professional and societal contexts, with ethical integrity.

While recent studies indicate that AI can augment teachers' assessment practices through teacher-AI collaboration (Kim, 2024), teachers' overreliance on AI is drawing concerns due to AI's seemingly powerful performance and ease of use. We urge teachers to move beyond this reliance and actively develop their AI competency to leverage AI and incorporate AI as an agent and more than a tool, fostering effective human-AI collaboration (HAC). HAC entails renewed classroom agencies among teachers, students, and technology (Zhai, 2024b), requiring fair assessment practices, transparency and human oversight of AI uses, and meaningful AI practices that benefit learners (see the REP Framework in Chap. 3).

HAC suggests that, if we view humans and AI as compatible agents, their interactions could be reconceptualized as a form of collaboration. For example, teachers may prepare tests and exams by offering examples of questions and rubrics generated by AI or leveraging AI to provide individualized support, answer questions, and identify resources (UNESCO, 2023). The possibility of collaboration between teachers and AI is underpinned by various theories, such as cultural-historical activity theory, computer-supported collaborative learning theory, and actor-network theory (Lee, Mun, et al., 2025), with varying degrees of explainability. These theories as a whole ensure that both teachers and their students can benefit from AI in the community of learning, and their interactions should be shaped in a manner that allows every participant to contribute. Further research should be done to enhance HAC in science classrooms.

10.3.2 Bias Mitigation Strategies

In addressing the issue of bias in AI scoring models, we begin by recognizing both societal and technical biases and that societal bias is an inherent feature of any society. These inevitabilities are deeply rooted in humankind conventions and traditions, which were shaped by historical and cultural activities. Thus, it is unrealistic to expect an elimination of bias in AI outcomes. However, acknowledging this limitation does not absolve us from responsibilities; instead, it requires us to take meaningful steps toward mitigating AI bias wherever possible. Sociologically, we endeavor to consciously recognize and respond to the existence of bias, not only in how AI models are developed but also in how they are deployed. Technically, strategies and methods (e.g., controlling confounding factors) offer valuable tools for examining and addressing bias, resulting in reduced impact of bias through thoughtful modeling practices. The following strategies can help mitigate AI bias and are aligned with the REP of AI in science assessment.

We may not be able to change how data have been historically accumulated; however, we can change how it will be sampled to build AI scoring models. As technical AI bias could come from the insufficiency and unrepresentativeness of training data, it is important to collect data from diverse demographic groups in terms of gender, race, ethnicity, socioeconomic status, disability, and language. Further, technical bias mitigation strategies, such as adjusting parameter weights, adversarial learning, and setting different threshold values for categories, can be applied (Idowu, 2024). Furthermore, sampling techniques, such as randomization and stratification, can be helpful to minimize bias.

To tackle the issue of training data imbalance, data augmentation techniques may be adopted to generate simulated student responses representing minoritized group categories. These techniques have been adopted in automatic scoring. For example, Fang et al. (2023) found that GPT-4-augmented student responses to chemistry items improved AI scoring model performance. In addition, they reported increased linguistic diversity while retaining scientific semantics in student written responses. Research is needed to study other augmentation methods with regard to both their promises and limitations.

Thoughtful rubric design can help mitigate AI scoring bias (Lee & Zhai, 2025). In science education, researchers have employed both analytical and holistic scoring rubrics to score students' constructed responses. Analytical scoring rubrics break down a student's response into multiple components or criteria (e.g., claim, evidence, reasoning), each of which is evaluated separately to assign scores. In contrast, holistic scoring rubrics assess the response as a whole, assigning a score based on an overall judgement of correctness, plausibility, and coherence. Assigning scores using an analytical scoring rubric is less cognitively demanding for human experts, as it only requires the evaluation of certain aspects of the responses. Therefore, research has reported that analytic scoring shows higher interrater reliability among human experts compared to holistic scoring for science assessment (Klein et al., 1998). Therefore, training data using analytical rubrics tends to be more accurate, which may result in more robust AI scoring performance (Jescovitch et al., 2021), reducing the potential for biased outcomes. However, analytic scoring lacks the process of synthesizing the aspects of the scoring rubric, which may not fit all situations. Researchers and educators need to determine the appropriate type of rubric based on the purpose of the assessment and the type of tasks involved.

Furthermore, recent research reveals that humans and LLMs use significantly different mechanisms in the process of assigning scores (Wu et al., 2025). Although LLMs adapt quickly to scoring tasks, they often rely on shortcuts that bypass the deeper logical reasoning characteristic of human grading. As such, incorporating high-quality rubrics that align with human grading logic can help bridge the gaps and improve LLM scoring accuracy.

When employing humans to label student responses as training data, it is important to ensure that human scorers, including teachers, have sufficient expertise and specific professional learning experiences with respect to the scoring system and assessment tasks. Robust data labeling is essential to exclude new errors from the training data, which could lead to biased scoring outcomes. In other words, we still need teachers to know and be proficient with content and assessment. Capable humans create better AI. In practice, at least two human scorers are required to check the interrater reliability, mitigate subjectivity, and avoid construct-irrelevant variances in student scores.

While there are few reports on how the architecture of machine learning can affect AI scoring bias, it is recommended to try out AI algorithms to optimally reduce potential bias, as is in the typical machine learning model building process. It is also recommended to consciously integrate bias mitigating strategies into AI scoring model building and utilization behaviors. Counteracting the biases in AI scoring models will lead to cutting the reproduction loop of societal biases. Once technical biases are avoided to the extent possible, their systematic influence on the reproduction of societal biases will be similarly decreased (Baldwin et al., 2022). This is also true for scoring machines that use generative AI models that sometimes hallucinate their responses. Once setting hyperparameters (e.g., temperature and top_p in the case of ChatGPT) to get repeatable results, generative models' biases in scoring can be examined and reduced.

10.3.3 Beyond Locality

We mentioned previously that we conceptualize locality as a distinct environment where its components intersect and that the locality and its components can affect how AI is used for science assessment. Addressing locality in AI-based science assessment requires considering its five components, significant but not exhaustive (see Fig. 10.4). Consequently, it can contribute to fair and respectful use of AI in science education (see the REP Framework in Chap. 3).

Standards To address the issue of different standards across educational systems, there is a need to prepare AI models that consider the differences between such standards. This was difficult to realize in traditional natural language processing and machine learning approaches, as it required equipping AI models with standards- (or curriculum-) specific semantics, which may require unfeasible amounts of data from diverse educational contexts. However, recent advances in LLMs resolve this problem. Retrieval Augmented Generation (RAG) supports uploading external document data, which are then embedded in the vector space of LLMs. When a user query is fed to the LLMs, the content of the document is retrieved as appropriate (see Li et al., 2025 for the application of RAG in education). Many recent AI services, such as ChatGPT and Gemini, support this functionality. Thus, RAG can be easily used for AI-based assessment. For curriculum alignment of AI-based assessment, a user can upload or attach standards (or curriculum) documents at various levels (national, regional, state, district, and/or school) to an LLM module that is being used in the assessment process. Then, the automatic scoring machine will refer to the standards (or curriculum) documents as requested by the user in the prompt.

One issue with utilizing the RAG functionality is the confidentiality of curriculum documents. Whether the curriculum documents are available to the public and allowed to be uploaded to the LLM engine should be confirmed. If they are not, a user should turn off the option that grants model training based on their input for proprietary services. If that is the case, the user can consider using open-source LLMs in local environments so that there is no breach of confidentiality.

Language To address the issue of different languages of the learners, it is crucial to ensure that an AI scoring model captures the semantic features of the student responses that are directly relevant to an item's nature, not other linguistic features. If an English-proficient student and a non-English-proficient student respond to an item and their answers imply the same scientific meanings despite their varying English proficiency, their responses should be assessed equivalently (AERA, APA, & NCME, 2014). In this case, a language-correction step in AI assessment can be helpful. For example, Lee et al. (2024) report that ChatGPT sometimes corrects typos in student-written responses during the automatic scoring process, even without explicit order from the researchers. This may help the scoring models process the right word and thus increase accuracy in assessing content knowledge.

Another way to mitigate the matter of language is to train the model with corpora from diverse student groups. Technologically, in principle, LLMs pre-trained with equivalent amounts of corpora for different languages can achieve comparable performances for tasks. For example, ChatGPT's inner engines, such as GPT-4 and the following versions, show better performance in various languages than the initial engine of GPT-3.5. Even in cases where a user cannot pre-train models, they could fine-tune the scoring models for students with linguistic needs.

Culture Cultural backgrounds of respondents should be respected in GenAI-based science assessment. For example, we could consider utilizing GenAI's capabilities in addressing the above-mentioned five tenets of culturally responsive assessment proposed by Nyaaba et al. (2024). They showed that GenAI could be used for generating and revising culturally responsive assessment items with proper prompt engineering methods. Multimodal assessment, in which GenAI can generate and analyze multimodal items and responses, is another way to address the different needs of learners, as it outdoes monomodal assessment that may not embrace learners' diverse needs (Gao et al., 2025; Lee, Shi, et al., 2025).

Besides or even before harnessing GenAI to generate and manage culturally responsive assessment, how cultures impact the perceptions of teachers and learners on AI-based assessment practices needs to be examined. Cross-cultural studies on the ease with which various educational sites from different countries adopt AI for science assessment can provide another avenue for investigating the influence of culture on science assessment (Al-Zahrani & Alasmari, 2024). While there have been policy studies that compare institutional approaches (e.g., for Asia-Pacific countries, see UNESCO, 2024b), empirical studies based on authentic voices of stakeholders, such as in-depth interviews, would be insightful.

Region Although the matter of country-specific driving forces and openness toward AI-based education is beyond the scope of this chapter, we want to say that some regions are more actively seeking AI-driven advances in education compared to other regions. It is important to note that utilizing AI in education, specifically for assessment, is not just for WEIRD countries. Recent responses and recommendations from global organizations, such as UNESCO (2023, 2024a, 2024b), OECD (2023), and the EC (2022), show that AI has the potential to be used to benefit education globally. Also, systematic literature reviews show that non-WEIRD countries, such as China, South Korea, India, Malaysia, and Saudi Arabia, are significantly contributing to AI-integration in science education (Guo, Zheng, et al., 2024; Lee, Yun et al., 2025). As more countries gain an understanding of AI's approachability, more countries could participate and contribute to the innovative science assessment made possible by AI.

Accessibility Addressing the issue of accessibility can be difficult, as it is directly relevant to the distribution of computational resources. For educational initiatives to be effectively implemented, the dissemination of hardware (e.g., PC, mobile devices, smart pads) and related learning materials is critically important for underrepresented groups of learners around the globe (Kim, 2009). However, this is not a new approach specialized for the ever-growing trend of integrating AI for assessment. Previously, computing was typically done by local machines, which meant a user had to have the physical hardware for computing. In contrast, currently, many AI services, such as ChatGPT and Gemini, are run on company servers, not users' hardware. Online computing environments, such as Google Colab, also free users from the constraints of having to possess physical computing environments, significantly enhancing the accessibility of AI for science assessment.

Further, the increasing number of open-source LLMs, starting from Google's BERT family and now Meta's Llama, BigScience's BLOOM, and Mistral AI's Mixtral, also contributes to the accessibility of AI-based science assessment. These open-source LLMs can liberate researchers from paying high fees for AI model subscriptions and API usage and from concerns for student data protection because student data is not shared with the AI model servers. Several studies have already used these models for science assessment research, showcasing the benefits of this advanced approach over typical models (Wu et al., 2025; Wulff et al., 2023).

10.4 Conclusion and Implications

In this chapter, we introduced three potential pitfalls of using AI in science assessment: outsourcing of cognitive tasks in each stage of the assessment process; ignoring societal and technical biases in AI scoring models; and failing to address locality in terms of standards, language, culture, region, and accessibility. To mitigate these potential pitfalls, we presented the following specific strategies: reconceptualizing assessment and the human-AI relationship, avoiding bias in AI scoring models, and pursuing ways to overcome locality.

This chapter offers implications for REP of AI in science assessment. First, AI use in assessment practices needs to benefit students and teachers, and human efforts in the overall assessment process need to be respected. This ties back to responsibly practicing AI-based science assessment. The steps to overcome potential pitfalls of outsourcing, bias, and locality require tremendous efforts, including mobilizing cognitive resources, identifying and suppressing sources of bias, and tackling locality that may limit the effective use of AI in science assessment. While eliminating these potential pitfalls is challenging, it is still important for researchers, teachers, and even learners to take all possible measures to minimize these potential pitfalls and maximize the effectiveness of AI-based science assessment.

Second, the solutions suggested in this chapter highlight the need for more open-source models rather than proprietary models. The availability of more open-source models could awaken teachers and students to think more critically about the reasons behind their choice of a specific model rather than unquestioningly adopting prominent ones and encourage their awareness of AI and HAC competencies. This can help connect responsible practice of using AI to ethical practice. In addition, open-source models tend to be more transparent to users, enabling users to more readily identify potential biases in the models and counteract these biases through retraining and fine-tuning. As previously mentioned, open-source models can also address some components of locality, such as accessibility, as ethical practice. Consequently, the use of open-source models can improve fairness, privacy, and data security in using AI for science assessment (see the REP Framework in Chap. 3).

Finally, frequent oversight of the scorers and assessment process is required (see the REP framework in Chap. 3). Validating and readjusting the assessment process and results compared to the ground-truth score label is the basis for tackling the three potential pitfalls (outsourcing, bias, and locality). One example is highlighted in the article "How ETS Scores the TOEIC® Speaking and Writing Test Responses" (Everson & Hines, 2010). Educational Testing Service (ETS) requires every rater from every region to be certified as a qualified rater (which can tackle locality). Even after raters are certified, they are required to pass daily calibration tests, and if not successful, they are sometimes not allowed to score responses (which can tackle bias). Another example is the tradition of repeated test validation. AERA, APA and NCME (2014) recommend that test validity evidence be evaluated in each specific situation. This principle encourages us to iteratively revisit whether students are cognitively engaged in AI-based science assessment practices as a validity criterion (which can tackle outsourcing).

Declaration of AI Use Parts of this manuscript were prepared with the assistance of generative AI tools. The AI tools were used for language editing or paragraphing of the authors' writing. The authors reviewed, revised, and verified all AI-supported content for accuracy, appropriateness, and originality. No confidential or sensitive data were entered into AI systems. The final responsibility for the content rests solely with the authors.

Acknowledgement This material is based upon work supported by the National Science Foundation under Grant No. 2332964 (PI Zhai). Any opinions, findings, and conclusions or recommendations expressed in this material are those of the author(s) and do not necessarily reflect the views of the National Science Foundation.

References

Al-Zahrani, A. M., & Alasmari, T. M. (2024). Exploring the impact of artificial intelligence on higher education: The dynamics of ethical, social, and educational implications. *Humanities and Social Sciences Communications, 11*(1), 1–12.

American Educational Research Association, American Psychological Association, and National Council on Measurement in Education [AERA, APA & NCME]. (2014). *Standards for educational and psychological testing.*

Ba, S., Zhan, Y., Huang, L., & Lu, G. (2025). Investigating the impact of ChatGPT-assisted feedback on the dynamics and outcomes of online inquiry-based discussion. *British Journal of Educational Technology*. Advance online publication. https://doi.org/10.1111/bjet.13605

Baker, R. S., & Hawn, A. (2022). Algorithmic bias in education. *International Journal of Artificial Intelligence in Education, 32*, 1052–1092.

Baldwin, J. R., Pingault, J. B., Schoeler, T., Sallis, H. M., & Munafò, M. R. (2022). Protecting against researcher bias in secondary data analysis: Challenges and potential solutions. *European Journal of Epidemiology, 37*(1), 1–10.

Beggrow, E. P., Ha, M., Nehm, R. H., Pearl, D., & Boone, W. J. (2014). Assessing scientific practices using machine-learning methods: How closely do they match clinical interview performance? *Journal of Science education and Technology, 23*(1), 160–182.

Bird, K. A., Castleman, B. L., & Song, Y. (2025). Are algorithms biased in education? Exploring racial bias in predicting community college student success. *Journal of Policy Analysis and Management, 44*(2), 379–402.

Chan, K. W., Ali, F., Park, J., Sham, K. S. B., Tan, E. Y. T., Chong, F. W. C., Qian, K., & Sze, G. K. (2025). Automatic item generation in various STEM subjects using large language model prompting. *Computers and Education: Artificial Intelligence, 8*, 100344.

Devlin, J., Chang, M.-W., Lee, K., & Toutanova, K. (2019). BERT: Pre-training of deep bidirectional transformers for language understanding. In J. Burstein, C. Doran, & T. Solorio (Eds.), Proceedings of the 2019 Conference of the North American Chapter of the Association for Computational Linguistics: Human Language Technologies, Volume 1 (Long and Short Papers) (pp. 4171–4186). Association for Computational Linguistics. https://doi.org/10.18653/v1/N19-1423

European Commission [EC]. (2022). *Ethical guidelines on the use of artificial intelligence (AI) and data in teaching and learning for educators*. Publications Office of the European Union. https://education.ec.europa.eu/news/ethical-guidelines-on-the-use-of-artificial-intelligence-and-data-in-teaching-and-learning-for-educators

Everson, P., & Hines, S. (2010). *How ETS scores the TOEIC® speaking and writing test responses*. https://www.ets.org/Media/Research/pdf/TC-10-08.pdf

Fang, L., Lee, G.-G., & Zhai, X. (2023). Using GPT-4 to augment unbalanced data for automatic scoring. *Journal of Educational Measurement*. https://doi.org/10.13140/RG.2.2.20161.74087

Gao, Y., Zhai, X., Li, M., Lee, G., & Liu, X. (2025). A multimodal interactive framework for science assessment in the era of generative artificial intelligence. *Journal of Research in Science Teaching*. Advance online publication. https://doi.org/10.1002/tea.70009

Good, R., Kromhout, R., & Bandler, W. (1986). Human and machine diagnosis of scientific problem-solving abilities. *Journal of Research in Science Teaching, 23*(4), 263–275.

Guo, S., Latif, E., Zhou, Y., Huang, X., & Zhai, X. (2024). Using generative AI and multi-agents to provide automatic feedback. *arXiv Preprint*. https://doi.org/10.48550/arXiv.2411.07407

Guo, S., Wang, Y., Yu, J., Wu, X., Ayik, B., Watts, F. M., Latif, E., Liu, N., Liu, L., & Zhai, X. (2025). Artificial intelligence bias on English language learners in automatic scoring. In *International conference on artificial intelligence in education* (pp. 268–275). Springer Nature.

Guo, S., Zheng, Y., & Zhai, X. (2024). Artificial intelligence in education research during 2013–2023: A review based on bibliometric analysis. *Education and Information Technologies, 29*(13), 16387–16409.

Ha, M., & Nehm, R. H. (2016). The impact of misspelled words on automated computer scoring: A case study of scientific explanations. *Journal of Science Education and Technology, 25*(3), 358–374.

Idowu, J. A. (2024). Debiasing education algorithms. *International Journal of Artificial Intelligence in Education, 34*, 1510–1540.

Jescovitch, L. N., Scott, E. E., Cerchiara, J. A., Merrill, J., Urban-Lurain, M., Doherty, J. H., & Haudek, K. C. (2021). Comparison of machine learning performance using analytic and holistic coding approaches across constructed response assessments aligned to a science learning progression. *Journal of Science Education and Technology, 30*(2), 150–167.

Kim, J. (2024). Leading teachers' perspective on teacher-AI collaboration in education. *Education and Information Technologies, 29*, 8693–8724.

Kim, P. H. (2009). Action research approach on mobile learning design for the underserved. *Educational Technology Research and Development, 57*, 415–435.

Klein, S. P., Stecher, B. M., Shavelson, R. J., McCaffrey, D., Ormseth, T., Bell, R. M., Comfort, K., & Othman, A. R. (1998). Analytic versus holistic scoring of science performance tasks. *Applied Measurement in Education, 11*(2), 121–137.

Krist, C., & Kubsch, M. (2023). Bias, bias everywhere: A response to Li et al. and Zhai and Nehm. *Journal of Research in Science Teaching, 60*, 2395–2399.

Lee, G. G., Ha, H., Hong, H. G., & Kim, H. B. (2018). Exploratory research on automating the analysis of scientific argumentation using machine learning. *Journal of the Korean Association for Science Education, 38*(2), 219–234.

Lee, G. G., Latif, E., Wu, X., Liu, N., & Zhai, X. (2024). Applying large language models and chain-of-thought for automatic scoring. *Computers and Education: Artificial Intelligence, 6*, 100213.

Lee, G.-G., Mun, S., Shin, M.-K., & Zhai, X. (2025, April). Collaborative learning with artificial intelligence speakers: Pre-service elementary teachers' responses to the prototype. *Science & Education, 34*(2), 847–875.

Lee, G. G., Shi, L., Latif, E., Gao, Y., Bewersdorff, A., Nyaaba, M., Guo, S., Liu, Z., Mai, G., Liu, T., & Zhai, X. (2025). Multimodality of AI for education: Towards artificial general intelligence. *IEEE Transactions on Learning Technologies, 18*, 666–683.

Lee, G. G., Yun, M., Zhai, X., & Crippen, K. (2025). Artificial intelligence in science education research: Current states and challenges. *Journal of Science Education and Technology*. Advance online publication. https://doi.org/10.1007/s10956-025-10239-8

Lee, G. G., & Zhai, X. (2025). Nerif: Gpt-4v for automatic scoring of drawn models. *Journal of Science Education and Technology*. https://doi.org/10.1007/s10956-025-10262-9

Lee, U., Lee, J., Bae, J., Jeong, Y., Koh, J., Lee, G., Lee, G., Ahn, T., & Kim, H. (2025). *Pedagogy-R1: Pedagogically-aligned reasoning model with balanced educational benchmark*. https://doi.org/10.48550/arXiv.2505.18467

Li, Z., Wang, Z., Wang, W., Hung, K., Xie, H., & Wang, F. L. (2025). Retrieval-augmented generation for educational application: A systematic survey. *Computers and Education: Artificial Intelligence, 8*, 100417.

National Research Council [NRC]. (2001). *Knowing what students know: The science and design of educational assessment*. National Academies Press.

Nehm, R. H., Ha, M., & Mayfield, E. (2012). Transforming biology assessment with machine learning: automated scoring of written evolutionary explanations. *Journal of Science Education and Technology, 21*(1), 183–196.

Next Generation Science Standards [NGSS] Lead States. (2013). *Next generation science standards: For states, by states*. National Academies Press.

Nguyen, H., & Hayward, J. (2025). Applying generative artificial intelligence to critiquing science assessments. *Journal of Science Education and Technology, 34*(1), 199–214.

Nyaaba, M., Zhai, X., & Faison, M. Z. (2024). Generative AI for culturally responsive science assessment: A conceptual framework. *Education Sciences, 14*(12), 1325.

Organisation for Economic Co-operation and Development [OECD]. (2023). Generative AI in the classroom: From hype to reality? In *Launch meeting of the Schools+ Network, 22–23 May 2023, Boulogne*. https://one.oecd.org/document/EDU/EDPC(2023)11/en/pdf

Pellegrino, J. W., DiBello, L. V., & Goldman, S. R. (2016). A framework for conceptualizing and evaluating the validity of instructionally relevant assessments. *Educational Psychologist, 51*(1), 59–81.

Samuel, Y., Brennan-Tonetta, M., Samuel, J., Kashyap, R., Kumar, V., Krishna Kaashyap, S., Chidipothu, N., Anand, I., & Jain, P. (2023). Cultivation of human centered artificial intelligence: Culturally adaptive thinking in education (CATE) for AI. *Frontiers in Artificial Intelligence, 6*, 1198180.

Sullivan, M., Kelly, A., & McLaughlan, P. (2023). ChatGPT in higher education: Considerations for academic integrity and student learning. *Journal of Applied Learning & Teaching, 6*(1), 1–10.

Tao, W., Zhang, M., & Liu, Y. (2024). Mastering delegation to artificial intelligence creative tools: The concept, dimensions, and development of a scale to measure cognitive outsourcing. *Social Behavior and Personality: An International Journal, 52*(12), 1–15.

United Nations Educational, Scientific and Cultural Organization [UNESCO]. (2023). *Guidance for generative AI in education and research*. https://unesdoc.unesco.org/ark:/48223/pf0000386693_eng

United Nations Educational, Scientific and Cultural Organization [UNESCO]. (2024a). *AI competency framework for teachers*. https://unesdoc.unesco.org/ark:/48223/pf0000391104

United Nations Educational, Scientific and Cultural Organization [UNESCO]. (2024b, February 7). *How generative AI is reshaping education in Asia-Pacific*. https://www.unesco.org/en/articles/how-generative-ai-reshaping-education-asia-pacific

Wu, X., Saraf, P. P., Lee, G., Latif, E., Liu, N., & Zhai, X. (2025). Unveiling scoring processes: Dissecting the differences between LLMs and human graders in automatic scoring. *Technology, Knowledge and Learning*. Advance online publication. https://doi.org/10.1007/s10758-025-09836-8

Wulff, P., Mientus, L., Nowak, A., & Borowski, A. (2023). Utilizing a pretrained language model (BERT) to classify preservice physics teachers' written reflections. *International Journal of Artificial Intelligence in Education, 33*(3), 439–466.

Xu, X., Qiao, L., Cheng, N., Liu, H., & Zhao, W. (2025). Enhancing self-regulated learning and learning experience in generative AI environments: The critical role of metacognitive support. *British Journal of Educational Technology*. Advance online publication. https://doi.org/10.1111/bjet.13599

Yang, J., Latif, E., He, Y., & Zhai, X. (2025). Fine-tuning ChatGPT for automatic scoring of written scientific explanations in Chinese. *Journal of Science Education and Technology, 34*, 719–736. https://doi.org/10.1007/s10956-025-10199-z

Zhai, X. (2024a). AI and machine learning for next generation science assessments. In H. Jiao & R. W. Lissitz (Eds.), *Machine learning, natural language processing and psychometrics* (pp. 201–219). Information Age.

Zhai, X. (2024b). Transforming teachers' roles and agencies in the era of generative AI: Perceptions, acceptance, knowledge, and practices. *Journal of Science Education and Technology*. Advance online publication. https://doi.org/10.1007/s10956-024-10174-0

Zhai, X., Haudek, K. C., Shi, L., Nehm, R. H., & Urban-Lurain, M. (2020). From substitution to redefinition: A framework of machine learning-based science assessment. *Journal of Research in Science Teaching, 57*(9), 1430–1459.

Zhai, X., Haudek, K. C., Stuhlsatz, M. A., & Wilson, C. (2020). Evaluation of construct-irrelevant variance yielded by machine and human scoring of a science teacher PCK constructed response assessment. *Studies in Educational Evaluation, 67*, 100916.

Zhai, X., & Krajcik, J. (2025). Pseudo artificial intelligence bias. In X. Zhai & J. Krajcik (Eds.), *Uses of artificial intelligence in STEM education* (pp. 568–578). Oxford University Press. https://doi.org/10.1093/oso/9780198882077.003.0025

Zhai, X., Krajcik, J., & Pellegrino, J. W. (2021). On the validity of machine learning-based next generation science assessments: A validity inferential network. *Journal of Science Education and Technology, 30*(2), 298–312.

Zhai, X., & Nehm, R. H. (2023). AI and formative assessment: The train has left the station. *Journal of Research in Science Teaching, 60*(6), 1390–1398.

Zhai, X., Nyaaba, M., & Ma, W. (2025). Can generative AI and ChatGPT outperform humans on cognitive-demanding problem-solving tasks in science? *Science & Education, 34*(2), 649–670.

Zhai, X., & Wiebe, E. (2023). Technology-based innovative assessment. In C. J. Harris, E. Wiebe, S. Grover, & J. W. Pellegrino (Eds.), *Classroom-based STEM assessment: Contemporary issues and perspectives* (pp. 83–94). Community for Advancing Discovery Research in Education, Education Development Center.

Zhai, X., Yin, Y., Pellegrino, J. W., Haudek, K. C., & Shi, L. (2020). Applying machine learning in science assessment: A systematic review. *Studies in Science Education, 56*(1), 111–151.

Zhang, B., Xiao, Y., & Luo, J. (2015). Rater reliability and score discrepancy under holistic and analytic scoring of second language writing. *Language Testing in Asia, 5*, 1–9.

Part VI
Educational Outcomes

Chapter 11
Preparing Students for Living in an AI Infused World: Revisiting the Outcomes of STEM Education

Marcus Kubsch, Paul Hur, Knut Neumann, and Juan-Carlos Aguilar

11.1 Introduction

AI is developing fast and projections for how it will develop in the future come with great uncertainties. Some experts envision artificial general intelligence (AGI) which will fundamentally revolutionize and transform our societies around the corner (e.g., AI 2027, 2025; Klein, 2025), while others see AGI as an event in the far future that may well never come (Marcus, 2024; Mitchell, 2024; Stevenson et al., 2025). Given this uncertainty, any discussion of what knowledge and skills students need at the end of K-12 science education to be prepared for their future lives and careers has to remain somewhat speculative.

To account for this uncertainty, we want to make the set of assumptions that have guided our thinking transparent here. First, based on our interactions with leading scientists that use AI in their research, as well as insights gathered from recent Nobel Prizes in Chemistry and Physics for AI-supported work (Nobel Prizes, 2024), and the emerging literature (e.g., Levine & Tu, 2024), we assume that AI will be relevant in all science-related academic fields in the future (See Chap. 2, Ross & Kubsch). Scientists use chatbots like ChatGPT to help them write or translate scientific texts; physicists, for example, have used supervised machine learning to

M. Kubsch (✉) · P. Hur
Department of Physics, Freie Universität Berlin, Berlin, Germany
e-mail: m.kubsch@fu-berlin.de; k.hur@fu-berlin.de

K. Neumann
Leibniz Institute for Science and Mathematics Education (IPN), Kiel, Germany
e-mail: neumann@leibniz-ipn.de

J.-C. Aguilar
Office of Teaching and Learning, Innovative Programs and Research, Atlanta, GA, USA
e-mail: JAguilar@doe.k12.ga.us

X. Zhai, K. J. Crippen (eds.), *Advancing AI in Science Education*, Advances in Technology-Rich Science Education 3,
https://doi.org/10.1007/978-3-032-16871-9_11

separate signal from noise in large experiments (Karagiorgi et al., 2022), and biologists may use unsupervised machine learning to detect patterns in gene expressions (Greener et al., 2022).

Second, based on the increasing ubiquity of AI in the sciences, AI will eventually also be relevant in science-related jobs. For example, healthcare providers are already starting to adopt AI to automate time-consuming, high-volume repetitive tasks (Martin et al., 2024). Based on the AI Index 2025 Annual Report from Stanford University (Masley et al., 2025), we can only expect this trend to increase with steep investments into AI and accelerating usage across industries. Third, AI will be relevant in our everyday lives—outside of work and academia. The *AI Index Report 2025* supports this assumption by documenting how AI is becoming deeply embedded in daily routines: self-driving cars now operate at scale, with services like Waymo providing over 150,000 autonomous rides per week, and AI-enabled medical devices—more than 220 approved by the FDA—are already in use. Millions of people are using AI not just professionally but also for leisure (for example generating songs but also listening to AI generated music (Pelly, 2025), using recommendation algorithms for what movie to watch next, or generating individualized exercise plans), and two-thirds of the global population expect AI-powered products and services to significantly impact their personal lives within the next three to 5 years.

Fourth, perhaps most relevant to the present chapter, we assume that preparing students for the changing educational and vocational landscape requires AI skills and knowledge at the end of K–12. The *AI Index Report 2025* supports this assumption, stating that "expanding AI expertise must begin in K–12 and higher education in order to ensure that students are prepared to be responsible users and developers." These sentiments highlight the projected growth of AI-related roles in areas like data infrastructure, emphasizing the need to expand the computing pipeline starting with early education.

Fifth and lastly, we assume that preparing students for an AI-infused world requires foundational knowledge and skills about AI that K–12 science education can foster. One concrete example is the AI4K12 initiative (AI4K12, 2020), which provides standards around "Five Big Ideas in AI" including ethics, applications, and system design—topics that can be addressed in science classrooms to build critical competencies. As an example, physics knowledge about vectors can help to understand generative AI (Odden & Hansen, 2025).

In this chapter, we treat AI in two distinct but overlapping roles. As a *tool*, AI extends human perception and reasoning, and amplifies inquiry. As an *agent*, however, AI operates with a certain degree of autonomy. Thus, it can help to generate recommendations or explanations that may exhibit hidden biases or, like a human collaborator, make mistakes. Viewing AI through this perspective may help teachers decide when to emphasize the tool use aspect and when to engage in agent oversight.

11.2 The Purpose of Science Education and Educational Outcomes

Beyond assumptions about the influence of AI on science, science education, and society, discussing educational outcomes makes little sense without reflecting on the purposes of science education. Over the past century, the desired outcomes of science education have undergone significant transformations. Initially, science teaching focused on rote memorization of factual knowledge and preparation for higher education or vocational pathways. During the mid-twentieth century, the emphasis shifted toward inquiry-based learning and the development of scientific thinking skills, particularly during the Sputnik-era push for technological advancement. In more recent decades, the focus has expanded to include scientific literacy for all students, aiming to prepare citizens to engage with socio-scientific issues such as climate change, health, and technology (Chu et al., 2021).

Today, science education increasingly incorporates interdisciplinary competencies, critical thinking, and ethical reasoning—particularly in the context of digital and AI-driven transformations. In this context, the OECD's Learning Compass 2030 (OECD, 2019) offers a forward-looking vision by placing student agency at the heart of educational outcomes. Agency, as defined by the OECD, is the capacity to set goals, reflect on one's actions, and shape the world through informed decisions. Science education, guided by this vision, should not only enable students to understand scientific concepts but also empower them to apply these concepts in everyday life situations in both private, professional, and political spheres and navigate complex socio-scientific challenges. Interestingly, the emphasis on the private sphere that, among others, situates science education as a means for personal development and growth reflects a less instrumental perspective on education. This is especially so compared to John Dewey's pragmatic perspective (Riga, 2020), and is one more alike to traditions of Bildung (e.g., Sjöström & Eilks, 2020). Thus, the OECD Learning Compass can be seen as synthesizing different Western traditions regarding the purpose of education.The PISA 2025 Science Framework (OECD, 2023) operationalizes this vision put forward in the OECD Learning Compass through three key competencies that 15-year-old students should be able to demonstrate: (1) explaining phenomena scientifically, (2) constructing and evaluating designs for scientific inquiry and interpreting scientific data and evidence critically, and (3) researching, evaluating, and using scientific information for decision making and action. These competencies support learners in actively engaging with scientific knowledge and in contributing to individual and collective well-being through the anticipation–action–reflection cycle central to the Learning Compass. Thus, science education, guided by this vision, should not only enable students to understand scientific concepts but also empower them to apply these concepts in (everyday life) situations in the private and professional sphere and finally evaluate and reflect science-related information to make decisions in socio-scientific problem spaces. In consequence, when we refer to educational outcomes in this chapter, we mean the knowledge and skills about AI in the context of science that we deem relevant for all students at the end of K-12 science education, irrespective of whether they continue their education in the sciences in higher education, enter vocational training in science related fields, or engage with science as part of their everyday life.

11.3 How AI Is Changing Science, Science-Related Occupations, and Interactions with Science in Our Everyday Lives

AI is increasingly used in science to automate routine tasks such as data collection, processing, and pattern recognition. For example, in climate science, AI models like FireSat enhance wildfire prediction by analyzing satellite data with high temporal resolution, while in molecular biology, AlphaFold 3 has drastically improved the prediction of protein structures via their 3D shapes, earning its creators the Nobel Prize in Chemistry in 2024 (Masley et al., 2025). These examples illustrate that AI does not merely support scientific work—it has begun to become a vital component in producing original scientific insights and accelerating discovery.

Similar developments can be observed in STEM-related industries, where AI is transforming workflows and altering the nature of work itself. For instance, autonomous and electric vehicle specialists, data scientists, and AI engineers are among the fastest-growing occupations worldwide (WEF, 2025). Yet, as argued convincingly in AI Snake Oil by Arvind Narayanan and Sayash Kapoor (2024), while AI is undoubtedly changing the character of jobs—often reshaping how tasks are performed—it is not necessarily eliminating them. Many occupations are being redefined rather than replaced, with AI taking over repetitive or data-intensive subtasks and humans focusing more on judgment, creativity, and interpersonal aspects. In other words, humans may use AI as a tool or supervise AI agents carrying out certain more complex task, making the work done by humans more managerial. This shift is reflected in survey data: by 2030, employers expect a nearly even split between tasks performed by humans, those performed by AI, and those performed collaboratively (WEF, 2025). This underscores the need to prepare students not for a jobless future, but for one in which human-AI collaboration and supervision is the norm.

While these AI-technologies offer significant potential to increase efficiency and uncover new insights, they also introduce new challenges. The application of AI systems is often complex, and their effectiveness depends heavily on the quality of input data, the design of algorithms, and their interpretation. As a result, errors can occur easily. In high-stakes domains like medicine, such errors can have severe consequences. For instance, AI-powered diagnostic tools have been shown to underperform for certain demographic groups due to biased training data, leading to, for example, underdiagnosis of skin conditions in patients with darker skin tones (Masley et al., 2025).

At the same time, AI systems are becoming increasingly capable of generating plausible and often persuasive explanations of scientific phenomena and STEM-related content. Large language models (LLMs) such as GPT-4o and Gemini-1.5 can simulate expert-level discourse, summarize complex studies, or generate educational explanations tailored to different levels of prior knowledge. In medical education, OpenAI's o1 model has reached a new state-of-the-art on the MedQA benchmark, outperforming not only previous AI systems but also human medical professionals in certain diagnostic tasks (Masley et al., 2025). However, these systems are not infallible. They may reproduce inaccuracies, reflect biases in their training data, or fabricate information altogether—a phenomenon often referred to as hallucinations (Alkaissi & McFarlane, 2023).

Moreover, the same technologies that can support learning and communication may also be used to spread misinformation. In particular, bad actors can exploit generative AI systems to deliberately produce and disseminate plausible-sounding but false claims about pressing socioscientific issues such as climate change, vaccine safety, or genetically modified organisms. In 2024 alone, AI-generated misinformation played a documented role in over a dozen election contexts across multiple countries (Masley et al., 2025). The ability of these tools to mimic scientific language and reasoning can make misinformation especially difficult to detect, even for well-informed audiences. New benchmarks such as FACTS and SimpleQA are being developed to test and improve AI systems' factual reliability, but the challenge remains significant (Masley et al., 2025).

Beyond these technical and epistemic concerns, the application of AI can have far-reaching and sometimes disruptive consequences. AI-driven decisions in areas such as healthcare, education, and finance can significantly affect people's lives. These systems may unintentionally reinforce social inequalities or cause harm if not properly designed or regulated. For example, automated loan approval systems or hiring algorithms have been found to perpetuate gender or racial biases (Masley et al., 2025). An often-cited case in healthcare is the underdiagnosis of conditions among marginalized populations due to non-representative training data, which undermines both equity and trust in science-informed technologies.

Furthermore, the environmental footprint of AI development is becoming a growing concern. While AI tools are becoming more efficient and accessible, the energy required to train large models continues to increase. Training GPT-4 in 2023 generated over 5000 tons of CO_2, and Llama 3.1 in 2024 produced nearly 9000 tons—equivalent to the annual carbon footprint of hundreds of individuals (Masley et al., 2025). These environmental costs raise critical questions about the sustainability of AI-driven science and innovation.

Finally, access to AI tools and expertise remains highly unequal. While high-income countries dominate AI development and benefit from early integration into scientific infrastructure, low-income regions often lack the digital infrastructure, technical capacity, and educational programs necessary to participate fully. According to the WEF (2025), while nearly all high-income countries plan to integrate AI literacy into foundational education by 2030, many low-income countries still face barriers such as electricity shortages or limited internet connectivity.

Without targeted investment and international cooperation, this digital divide threatens to exacerbate global disparities in science education and innovation capacity.

11.4 AI Knowledge and Skills for K-12 Science Education

Based on the assessment of how AI is changing science, science related occupations, and interactions with science in our everyday lives in the previous section, it is clear that at the end of K-12 science education, students need to a) understand what AI is, how it works, and how it is used in science, b) be able to use AI tools to explore, analyze, and communicate scientific information, and c) be prepared to critically assess and make judgments about AI's role in science and society. In this section, we will explore what this means in more concrete terms.

11.4.1 Existing Frameworks and Models for AI Knowledge and Skills in K-12

The integration of AI into education has prompted a proliferation of frameworks that seek to define AI literacy and competency. Each of these frameworks offers a unique perspective on what knowledge and skills learners need, and how they can be fostered. Here, we discuss notable frameworks, and by discussing how their perspectives compare, we can highlight their common goals as well as the diverse foci in the area.

First, UNESCO's recent "AI Competency Frameworks for Students and Teachers" (UNESCO, 2025) prioritizes an ethical and human-centered approach and outlines 12 competencies across four key areas: a human-centered mindset, ethics of AI, AI techniques and applications, and AI system design. A parallel framework for teachers details 15 competencies in five domains, and adds a focus on pedagogical adaptation and ongoing professional learning. UNESCO's goals with these frameworks appear to aim as a guidance for nations to develop their own AI education strategies. While the United States does not have a single national policy, with many states issuing their own guidance, the UNESCO frameworks could be a solid start for a comprehensive baseline that complements existing federal initiatives and recommendations from educational organizations.

One of the earlier frameworks, the AI4K12 "Five Big Ideas" developed by Touretzky et al. (2019), is structured around five key conceptual domains: perception, representation and reasoning, learning, natural interaction, and societal impact. These themes provide a foundational structure for understanding the workings and implications of AI systems. However, while the model is conceptually clear and well-aligned with computing education standards, it is relatively abstract in its articulation of learner competencies. It stops short of offering detailed learning progressions or specifying how these ideas translate into classroom tasks or student behaviors.

In contrast, the AI4Future framework by Chiu et al. (2021) emphasizes pedagogical structure and learner development. The model includes layers that address content knowledge, modular and visual scaffolding for teachers, and importantly, roles for learners as users, developers, and ethical designers of AI. This framework moves beyond abstract content to suggest more concrete pedagogical strategies and instructional design principles, such as relevance, co-construction, and contextualization. Still, while the model identifies key instructional goals, it only partially articulates the specific skills students should acquire.

Building upon this work, a subsequent framework by Chiu et al. (2024) introduces a distinction between AI literacy and AI competency. AI literacy is defined as understanding AI systems and their societal impacts, whereas competency includes confidence, the ability to apply knowledge, and reflective judgment. The model structures AI learning around five domains: technology, impact, ethics, collaboration, and self-reflection. This model is more concrete than its predecessors in linking specific dispositions and practices to each domain. It also offers tangible pedagogical pathways, such as project-based learning and community engagement. Nonetheless, while it advances the idea of AI as a subject of inquiry and personal growth, the competencies remain general and would benefit from tighter alignment with subject-specific practices, especially in science education.

A more detailed structure is provided by the framework developed by Chee et al. (2024). This competency model, based on a systematic review, identifies eight main competencies and 18 sub-competencies. These range from data and algorithm literacy to affective dimensions such as confidence and ethical sensitivity. It offers developmental progressions across K-12, higher education, and workforce contexts, and pays special attention to disciplinary differences. Importantly, the model provides a clearer picture of what learners should be able to do, such as recognizing bias in algorithms or collaborating effectively with AI tools. It is arguably the most concrete and comprehensive model reviewed, and aligns well with the needs of a science curriculum that aims to integrate AI both as a topic and as a tool.

The framework developed by Tenório and Romeike (2023) further advances specificity, particularly in post-secondary education for non-technical disciplines. It identifies 52 competencies organized by domain, including machine learning, data handling, and responsible AI use. What makes this framework particularly useful is its grounding in practical, professional contexts. It translates AI literacy into applicable skills for fields like medicine, law, and education. While not developed specifically for K-12, it sets a precedent for articulating competencies that are both context-sensitive and highly concrete.

The most recent addition to this landscape is the German VK:KIWA model by Alles et al. (2025), designed collaboratively by educators for classroom use. This framework introduces four cyclical developmental fields: understanding, applying, reflecting, and shaping. These are described as interdependent phases of a continuous learning process, applicable to both students and teachers. What sets this model apart is its inclusion of three progression levels for each competency field, ranging from foundational awareness to expert practice. This structure allows educators to identify where learners are and what the next step might be. Moreover, the model introduces "AI Leadership" as a central integrative goal, encompassing ethical responsibility, critical use, and participatory design of AI. While the VK:KIWA model does not list detailed competencies in the same way Chee et al. or Tenório and Romeike do, its developmental clarity and adaptability across contexts make it a valuable tool, particularly for curriculum planning and school development.

Another significant contribution comes from Wang and Lester (2023), who offer a component-based definition of AI literacy tailored to K-12 education. Their model emphasizes understanding AI capabilities, applying AI to solve problems, and working with AI in varied sociocultural contexts. Though their work powerfully articulates the rationale for integrating AI literacy into general education and underscores the importance of early, developmentally appropriate instruction, it remains somewhat broad in its educational recommendations. While outlining essential areas of AI knowledge, the framework lacks concrete learning goals or pedagogical strategies aligned with subject-specific standards, such as those in science.

Across these frameworks, several areas of consensus emerge. All models emphasize the importance of understanding how AI works, the ethical and societal implications of AI, and the need for reflective and critical engagement. Most models also acknowledge the value of using AI tools, though the specificity of what that entails varies. There is growing agreement that AI literacy is not just a technical skill but involves social, ethical, and epistemological dimensions.

While all the reviewed frameworks agree on the broad contours of AI literacy—including the importance of understanding how AI works, considering its societal implications, and fostering critical engagement—they diverge significantly in how concretely they define competencies and learning outcomes. The AI4K12 framework remains largely conceptual, offering thematic categories without detailed descriptions of what students should know or be able to do. In contrast, the frameworks by Chee et al. and Tenório and Romeike stand out for their operational clarity, with Chee et al. offering well-defined competencies and Tenório and Romeike providing highly contextualized skill sets for professional domains. The later framework by Chiu et al. and the VK:KIWA model fall in between, aiming to connect conceptual understanding with pedagogical strategies and developmental progression, but still lacking in fine-grained specificity.

Taken together, these models suggest the value of a hybrid approach for science education—one that blends clearly defined conceptual domains with developmental progressions and concrete classroom applications. However, even the most detailed AI literacy frameworks fall short when compared to the level of precision and curricular integration found in the NRC Framework for K-12 Science Education (National Research Council, 2012). That framework distinguishes itself through its articulation of disciplinary core ideas, scientific practices, and crosscutting concepts, all presented with a granularity that enables alignment with instructional planning and assessment. No current AI literacy framework offers an equivalent mapping of competencies to classroom practice.

Furthermore, the reviewed frameworks largely lack domain specificity to science. Although they emphasize the interdisciplinary nature of AI, they offer limited guidance on how AI literacy connects with scientific reasoning, data analysis, or modeling practices fundamental to science education. As a result, while existing models provide valuable guidance, they do not yet equip science educators with the detailed tools needed to integrate AI meaningfully into their instruction. There is a clear need for a domain-specific model for AI in science that combines conceptual coherence with practical utility, modeled after the success of established standards in science education.

11.4.2 Towards a Framework for AI Educational Outcomes in Science Education

In the following, we present a framework of educational outcomes, i.e., skills and knowledge, regarding AI that aims at being concrete and specific to science—addressing what the previous models and frameworks reviewed in the previous section left wanting. We organize these into three categories inspired by the PISA 2025 science framework competencies: knowing about AI in science, applying AI in science, and evaluating and reflecting on AI in science. These three categories reflect a broad consensus across recent AI education frameworks. Moreover, they underscore the need for responsible and ethical AI use, ensuring fairness, transparency, accountability, and the promotion of human dignity in scientific contexts. We emphasize reflection and ethically responsible use of AI which aligns with the Responsible and Ethical Principles (REP) framework described earlier in Chap. 3 of this volume (Crippen, Zhai & Lee). We see REP's responsible principles as guidance for better understanding how AI should be implemented in classrooms, and its ethical principles for why AI is used. We frame our proposed competencies and examples under these distinctions.

Frameworks such as AI4K12 (Touretzky et al., 2019), Chee et al.'s (2024) AI literacy framework, and the VK:KIWA model (Alles et al., 2025) consistently emphasize conceptual understanding, practical engagement, and critical reflection as foundational components. Similarly, Chiu et al. (2024) and Tenório and Romeike (2023) highlight both the importance of knowledge and technical know-how but also include ethical awareness, and self-reflective practice, reinforcing the necessity of prioritizing fairness, transparency, scientific integrity, and human oversight when working with AI. When it comes to applying AI and evaluating and reflecting on AI, we take inspiration from the scientific practices and cross-cutting concepts in the Framework for K-12 Science Education (NRC, 2012).

Our framework (Table 11.1) focuses on educators' need for concrete, science-specific skills by identifying the teachable, granular skills across the three competency groups of "Knowing about AI", "Applying AI", and "Evaluating and Reflecting about AI". We were directly inspired by the PISA 2025 for these groups, but added

Table 11.1 AI knowledge and skills relevant in science for K-12

AI focus	Knowing about AI	Applying AI	Evaluating and reflecting about AI
AI for hypothesis generation, and cues for exploration	– Students know that AI identifies patterns in data and text. They are aware that AI outputs can be used as hypothesis generators and inform scientific explanations, but require validation and testing – Students know that AI insights might lead to generating new questions for scientific inquiry – Students can explain how AI (e.g., LLMs analyzing scientific literature or datasets) can identify patterns, gaps, or other potentially productive areas in scientific knowledge	– Students can use an AI tool (e.g., a research chatbot, automated knowledge mapping tool) to analyze a scientific literature or dataset to generate possible hypotheses, or questions for further investigation – Students can analyze the output of AI data analysis, and propose scientific explanations or mechanisms consistent with the findings and known scientific principles	– Students can evaluate the AI-generated hypotheses/generated questions for relevance, feasibility, and testability to the ongoing science topic – Students are able to weigh their scientific intuition, creativity, critical thinking, and domain knowledge to refine the generations – Students can analyze whether AI-identified patterns provide enough evidence for a scientific explanation, and reflect on the matters of correlation versus causation – Students do not simply accept AI-generated hypotheses as productive, and can think of best ways of rigorously testing them
AI-driven predictive modeling and simulations	– Students know that AI models are approximations – Students can explain that AI (e.g., supervised machine learning models) are trained on data to create predictive simulations (e.g., ecological outlook, chemical reaction yield) that complement or extend existing mathematical or statistical models – Students know that models learn from data, and that data quality and quantity are important for models' predictive accuracy	– Students can interact with AI-driven simulations (e.g., adjusting weights in a virtual physics pulley system, changing inputs like pollution levels for an AI ecosystem) to observe predicted outcomes and explore relationships between variables – Students can compare AI predictions to data from traditional, non-AI models or to real-world observations to the extent possible	– Students can identify core assumptions and potential limitations of specific AI models or simulations (e.g., the scope of training data, algorithmic biases, uninterpretable, black box nature) – Students can evaluate whether the AI models or simulations are more appropriate than a traditional model for the ongoing science topic – Students can reflect on the ethical implications of AI predictive models in sensitive scientific contexts and populations, and consider fairness, accountability, and trust. They can discuss the consequences of model biases

Pattern recognition (unsupervised machine learning)	– Students know that AI uses algorithms for decision-making and optimization based on rules and input data – Students can explain the general nature of how unsupervised learning algorithms (e.g., clustering, dimensionality reduction) identifies latent patterns in data	– Students can use unsupervised machine learning-driven tools (e.g., data visualization software, image sorting) with a dataset to identify groups or anomalies – Students can describe the characteristics of the clusters (e.g., how similar or dissimilar they are to other clusters, what makes each group unique)	– Students can critically evaluate whether the patterns/clusters found by AI are meaningful, or are not valuable for scientific inquiry – Students can explain how the AI's parameters (e.g., number of clusters, distance metrics) can influence the results seen – Students are aware that there are potential data biases for the results, and that these outputs are correlations – Students can discuss the ethical responsibilities of making unsubstantiated assumptions from patterns
Classification and regression (supervised machine learning)	– Students know that AI learns mapping from input features to known labeled target variables – Students can, at a high level, discuss how common supervised learning algorithms (e.g., decision trees, simple neural networks, regression) use input data with labeled targets to train models that can classify new data points or predict outcomes in science	– Students can use pre-trained AI-driven classification or regression tools to classify new scientific examples or make predictions	– Students can interpret outputs of AI-derived predictions, such as understanding classification labels, and performance metrics (e.g., accuracy, confidence values, confusion matrices) – Students can discuss the importance of the interpretability of features (i.e., understand how the model made decisions) when possible – Students can acknowledge how factors like training data bias or data quality can affect the performance observed – Students are strictly vigilant about perpetuating harm or stereotypes based on patterns or predictions observed from AI-derived output

(continued)

Table 11.1 (continued)

AI focus	Knowing about AI	Applying AI	Evaluating and reflecting about AI
AI algorithms and mathematical representation	– Students know that mathematics encodes the underlying information for the algorithms' processes – Students can explain that AI systems rely on a set of instructions (algorithms), driven by mathematical representations of data processed from various methods (e.g., vectorization, feature scaling, normalization) – Students can, at a high level, explain how simple algorithms work in a conceptual sense (e.g., if-then rules, decision trees)	– Students can interact with and create conceptual representations or simpler algorithms, such as basic rule-based systems (e.g., if-then logic), linear regression (e.g., drawing a line of best fit, writing a linear function), and simple sorting tasks	– Students can evaluate the balance between algorithmic complexity and how it may influence performance and interpretability – Students can generally assess how data representation can determine what algorithms can learn – Students can reflect on how AI methods in algorithm design and mathematical data representation can lead to unintentional biases becoming embedded in the models
Informed consideration of AI-derived information as evidence	– Students know that when using AI, they should also evaluate the validity of the source information or data – Students should be aware of uncertainty in the nature of science, and ways it can be influenced by the integrity and validity of data	– Students can incorporate findings from AI tools (e.g., summarized trends from scientific literature, patterns identified in a large quantitative data set, predicted outcomes for an experiment) into an argument – To the extent possible, students are able to state the method and source material of the information and acknowledge potential limitations of AI-derived information	– Students can critique AI-generated data analysis and summaries by questioning (1) the underlying algorithm, (2) the quality, completeness, and biases of training/input data, (3) known limitations of the method or model, and (4) interpretations of the output – Students can assess the trustworthiness of AI-derived evidence and compare it against evidence from direct experimentation or an established scientific theory – Students are aware of their responsibility for rigorous validation when using AI to generate evidence to support scientific claims

AI for information seeking and information literacy	– Students know that AI can improve speed, access, and interpretation of complex scientific concepts, but there is a need for critical evaluation of the AI-generated content – Students are aware how AI tools can be used to synthesize scientific information, but can also generate inaccurate, incomplete, or biased summaries – Students know that AI can perpetuate biases in the training data	– Students can use AI tools (e.g., LLM-powered chatbots) to find and summarize information on a scientific topic from numerous sources or data – Students know how to check AI outputs for validity (e.g., checking original sources, logic and coherence) to verify its facts and claims	– Students are aware of potential unintentional misinformation practices and the importance of transparency in how the information was obtained – Students can critically evaluate the accuracy and completeness of AI-derived information by referencing and weighing the validity of sources – Students can discuss the ethical implications and risks of using AI-generated text, including plagiarism, authorship, and accuracy

the reflection aspect into the third group, as we determined that it was an essential characteristic of AI-related task evaluation, especially when considering the ethical implications, such as the need for transparency, bias mitigation, and reducing potential misinformation propagation. Our framework thus more strongly focuses on the knowledge and related considerations needed for critical AI usage, rather than just the skills needed to interact with the AI applications themselves (i.e., "Applying AI" column is slightly more brief than others for this reason). We do not want to de-emphasize the skills needed to apply AI and AI-powered tools and believe they are important for thriving in the current education landscape. Still, we mapped our framework with what we believe to be a safe assumption that while AI tools and interactions (e.g., prompt engineering, programming/environment-specific syntax) will continue to change, understanding core concepts surrounding AI will remain valuable for maintaining informed perspectives even when faced with future AI tools yet to be created.

We propose what we identified as core AI concepts in the first column ("AI Focus/Concept"), which were partly inspired by the eight essential practices for sciences and engineering identified by the Next Generation Science Standards (NGSS) (Next Generation Science Standards Lead States, 2013). As shown, the final set of seven concepts span across every aspect of science that can be directly mapped to an analogous AI aspect, from ideation ("AI for hypothesis generation, and cues for exploration"), experimentation ("AI-driven predictive modeling and simulations", "Pattern recognition (unsupervised machine learning)", "Classification and regression (supervised machine learning)"), validation ("AI algorithms and mathematical representation"), to interpretation and implications of results ("Informed consideration of AI-derived information as evidence"), as well as including considerations particularly important for AI in "AI for information seeking and information literacy". Throughout these categories, we emphasize the necessity of ethical practices such as respecting privacy, safeguarding autonomy, ensuring the robustness of AI interpretations, and evaluating the trustworthiness of AI-derived evidence in scientific discourse.

We identified concrete, directly related knowledge and skills that mapped across the three competency groups, while keeping in mind suitability for K-12 students. We admit some of these bullet points (e.g., "Students can explain that AI systems rely on a set of instructions/algorithms, driven by mathematical representations of data processed from various methods, such as vectorization, feature scaling, normalization.") may seem quite advanced (and abstract) and perhaps slightly miscalibrated for current classrooms and integration into the broader science curricula. We take a cautiously optimistic stance with our framework, however, and believe that the near future of teacher education and professional development will rapidly integrate AI foundations. We assert that the resulting K-12 graduates with these characteristics, along with a parallel emphasis on ethical and responsible AI engagement to foster scientifically and socially responsible graduates, can be more confident to thrive and be prepared for the uncertainties of the future AI landscape and job market.

When developing our framework, we knew we specifically wanted to emphasize reflection and ethics, as it seemed to be an under-emphasized area across other existing models and frameworks. In relation to REP, we consider reflection as relating responsible (i.e., the how) with ethical (i.e., the why) principles, which can integrate increased transparency and accountability of AI use for the purposes of scientific integrity and human-centric values (Crippen, Zhai & Lee, Chap. 3 of this volume). Thus, we positioned reflection not merely as a final step, but as the central, ongoing practice that makes meaningful ethical evaluation possible. We believe that reflection, as a skill, helps students to go from simply knowing about a potential bias to understanding its implications. Without being mindful of students' reflective use, students may not grow their perspectives beyond surface-level understanding about whether an AI's answer is correct or incorrect, rather than moving toward more consequential inquiries, such as, "What assumptions about the world are embedded in the design of AI tools and systems?", and "How does my interaction with this tool shape my own understanding and potential biases?"

The continuous reflection is directly involved in achieving the ethical dimensions of the framework. For instance, when students use AI for pattern recognition (i.e., third row), a non-reflective student might simply accept the machine-generated clusters as objective groupings. A reflective student, however, would consider an additional ethical implications dimension, and ask questions like: "Who is represented in this dataset, and who is invisibly absent?" or "What are the potential harms of applying these labels to real-world populations". This process, while perhaps effortful for students (and teachers) to engage in, shapes students from passive AI users into more active, critical thinkers who understand that scientific data is not generated without biases. Reflection pushes students to engage with the idea of power and meaning making of AI. Thus, students would be able to ask who built the tool, for what purpose, and whose interests are being served.

Foremost, we hope educators view Table 11.1 as a map for encouraging this reflective capacity, and not necessarily as a checklist. Let us think about a hypothetical lesson workflow. A teacher might begin with an "Applying" task, such as using an AI interactive simulation or game to model population ecosystem changes. Ideally, the teacher would have built in surprising or counterintuitive results from the simulation to create opportunities to trigger reflection for students. Students, with close guidance from the teacher, would practice processes to understand the various "Knowing" points, and to investigate the AI model's underlying algorithms and training data. This, in turn, perhaps leads to a more sophisticated "Evaluating and Reflecting" discussion about the AI's limitations, and the ethical responsibility of using such predictions in real life ecosystems. The pedagogical goal would be to develop intuitive habits to reflect and critically question each application of AI.

The framework was designed to be adaptable across developmental stages. For younger students, reflection might be scaffolded through simple prompts: "Did the tool give you a surprising answer? Why do you think that happened?" For older students, reflection could involve more complex tasks, such as analyzing the source code or logic of a simple algorithm or debating the societal trade-offs of using predictive AI in the real world. The core principle remains the same, however. Students' reflective skills should be given priority in order to navigate the ethical complexities of AI. The outcome, thus, would not be to train every student as an AI developer, but to encourage a generation of scientifically literate citizens who are not only competent users of technology, but are also continuously reflective and ethically responsible.

We believe our three competency groups (knowing, applying, evaluating/reflecting) are not discrete and work synergistically. That is, a student will inevitably develop insights for applying based on knowing the underlying AI concepts, and that evaluating and reflecting about AI will inform future AI knowledge acquisition and usage; simultaneously, longer and active learning about AI, as we outlined in these bullet points, will strengthen students' overall AI competency. As an example, consider the development of competence around "Classification and regression (supervised machine learning)": students will first need to build foundational knowledge (knowing) by understanding that these AI models learn to map input features to known, labeled outputs, as well as be aware of common algorithms like decision trees or simple neural networks, which help to predict or classify objects in science. This foundational knowledge would then enable the student to engage in informed applied practices (applying) such as using a pre-trained classification tool, where they may not necessarily be programming the AI but can still benefit from leveraging known predictive capabilities. We believe, however, that students must also evaluate their AI usage and knowledge (evaluating and reflecting) in order to move beyond getting output and taking them at face value, and toward deeply engaging with the output, questioning classification confidence and accuracy, and thinking about the biases that may be influencing what they see. Importantly, such reflective practices must be rooted in the broader goal of upholding ethical responsibility, promoting fairness, and maintaining scientific integrity when engaging with AI outputs.

Ultimately, what we see as true AI competence and K-12 outcomes regarding AI will likely require a far-seeing and intentional curriculum to meet the competencies in our framework. We believe that it is not only about interacting in an AI-filled world, but producing users who are increasingly continuously reflective about AI's implications for science and society.

11.5 Conclusions

Preparing students for a world deeply infused with artificial intelligence (AI) is one of the biggest challenges and—at the same time—opportunities, facing science education today. As this chapter has outlined, AI is not only transforming scientific research and science-related professions, but is also reshaping how individuals engage with science in everyday life. In this context, defining AI-related educational outcomes for science education becomes critical. However, it must be acknowledged that the significance of knowing about, applying, and reflecting on AI will vary depending on students' post-secondary trajectories—whether they pursue science-related higher education (e.g., studying physics, enter vocational training to become a nurse) or rely on scientific reasoning in everyday contexts (e.g., when making decisions about GMOs or vaccines). For instance, while students aiming for research-intensive careers may require deeper competencies in AI system design and data analytics, those entering vocational pathways might benefit more from practical experience with AI tools in applied settings. Meanwhile, for the general citizenry, the reflective use of AI—especially the ability to critically evaluate AI-generated information—will be essential for informed participation in societal discourse and personal decision-making (Chiu et al., 2024; Chee et al., 2024, see also Crawford, 2021).

To accommodate this diversity, AI educational outcomes should be cultivated progressively throughout K–12 education via thoughtfully constructed learning progressions. Such progressions should ideally integrate into existing frameworks for scientific practices and core disciplinary ideas (National Research Council, 2012). This alignment is crucial not only for curricular coherence but also for ensuring that AI education supports broader goals of science literacy, such as evidence-based reasoning, modeling, and ethical reflection. While existing frameworks like AI4K12 (Touretzky et al., 2019) and Chee et al.'s (2024) competency model provide foundational guidance, further research is needed to map AI-specific learning outcomes onto established science education trajectories. As Wang and Lester (2023) emphasize, this integration must also be age-appropriate and sensitive to developmental stages, especially as AI competencies encompass both technical knowledge and socio-ethical dimensions.

Beyond K–12, the need for aligned pathways into higher education and vocational training becomes equally pressing. The framework by Tenório and Romeike (2023), for example, illustrates how post-secondary students in non-technical disciplines still require robust AI competencies to engage responsibly in their respective fields. These extended pathways should not only build on earlier schooling but also provide opportunities for learners to specialize or deepen their understanding of AI in ways that align with their aspirations and societal roles. The notion of lifelong AI learning, including adult education and professional development, should therefore become a cornerstone of national education policies (Casal-Otero et al., 2023; Chiu et al., 2024).

Importantly, all these educational design efforts must be situated within a landscape of technological uncertainty. While projections about AI's future range from incremental improvements to transformative leaps toward artificial general intelligence (Mitchell, 2024; Marcus, 2024), the reality is that AI's trajectory is neither linear nor fully predictable. Educational frameworks must therefore remain adaptable, fostering not only foundational knowledge but also the metacognitive skills, ethical awareness, and learning dispositions necessary to navigate a rapidly evolving technological landscape. As emphasized by the OECD Learning Compass 2030, cultivating student agency—the ability to set goals, make informed decisions, and reflect on one's actions—will be indispensable in such an environment (OECD, 2019). AI education, accordingly, should prepare learners not just to use or understand current technologies, but to adapt to and shape the AI-infused world they will inhabit.

In sum, the preparation of students for life in an AI-infused world demands a coherent, scaffolded, and flexible approach to AI literacy. It calls for an educational architecture that differentiates between trajectories, aligns across life stages, and remains responsive to ongoing change. Science education, already central to fostering inquiry, reasoning, and engagement with complex systems, is a natural and necessary home for these outcomes. The challenge now is to ensure that educational design, policy, and practice evolve in step with AI—neither outpaced by its advances, nor blind to its implications.

Acknowledgement This material is based upon work supported by the National Science Foundation under Grant No. 2332964 (PI Zhai). Any opinions, findings, and conclusions or recommendations expressed in this material are those of the author(s) and do not necessarily reflect the views of the National Science Foundation.

References

AI 2027. (2025). https://ai-2027.com/.

AI4K12 Initiative. (2020). *AI4K12: Artificial intelligence (AI) for K-12 initiative*. https://ai4k12.org.

Alkaissi, H., & McFarlane, S. I. (2023). Artificial hallucinations in ChatGPT: Implications in scientific writing. *Cureus, 15*(2). https://doi.org/10.7759/cureus.35179

Alles, S., Falck, J., Flick, M., & Schulz, R. (2025). *KI-Kompetenzen für Lehrende und Lernende: Aus der Praxis für die Praxis—eine adaptierbare Basis* [White paper]. Virtuelles Kompetenzzentrum KI und Digitalisierung in der Lehrer:innenbildung (VK:KIWA). https://zenodo.org/records/15001019.

Nobel Prize Outreach. (2024). All nobel prizes 2024. https://www.nobelprize.org/all-nobel-prizes-2024/.

Casal-Otero, L., Catala, A., Fernández-Morante, C., Taboada, M., Cebreiro, B., & Barro, S. (2023). AI literacy in K-12: A systematic literature review. *International Journal of STEM Education, 10*(29). https://doi.org/10.1186/s40594-023-00418-7

Chee, Y. S., Ahn, J., & Lee, Y. (2024). A competency framework for AI literacy: Variations by different learner groups and implications for education. *British Journal of Educational Technology*.

Chiu, T. K., Meng, H., Chai, C. S., King, I., Wong, S., & Yam, Y. (2021). Creation and evaluation of a pretertiary artificial intelligence (AI) curriculum. *IEEE Transactions on Education, 65*(1), 30–39.

Chiu, T. K. F., Ahmad, Z., Ismailov, M., & Sanusi, I. T. (2024). What are artificial intelligence literacy and competency? A comprehensive framework to support them. *Computers and Education Open, 6*, 100171.

Chu, S. K. W., Reynolds, R. B., Tavares, N. J., Notari, M., & Lee, C. W. Y. (2021). *21st century skills development through inquiry-based learning from theory to practice*. Springer.

Crawford, K. (2021). *Atlas of AI: Power, politics, and the planetary costs of artificial intelligence*. Yale University Press.

Greener, J. G., Kandathil, S. M., Moffat, L., & Jones, D. T. (2022). A guide to machine learning for biologists. Nature Reviews Molecular Cell Biology, 23(1), 40–55. https://doi.org/10.1038/s41580-021-00407-0.

Karagiorgi, G., Kasieczka, G., Kravitz, S., Nachman, B., & Shih, D. (2022). Machine learning in the search for new fundamental physics. Nature Reviews Physics, 4(6), 399–412. doi:https://doi.org/10.1038/s42254-022-00455-1.

Klein, E. (2025, March 4). Opinion | the government knows A.G.I. Is coming. *The New York Times*. https://www.nytimes.com/2025/03/04/opinion/ezra-klein-podcast-ben-buchanan.html.

Levine, H., & Tu, Y. (2024). Machine learning meets physics: A two-way street. *Proceedings of the National Academy of Sciences, 121*(27), e2403580121. https://doi.org/10.1073/pnas.2403580121

Marcus, G. (2024, June 5). AGI by 2027? [Substack newsletter]. *Marcus on AI*. https://garymarcus.substack.com/p/agi-by-2027.

Martin, C. P., Lamb, J., Dahab, A., Jones, J., & Bhasker, S. (2024). *Generative AI in healthcare: Current trends and future outlook | McKinsey*. https://www.mckinsey.com/industries/healthcare/our-insights/generative-ai-in-healthcare-current-trends-and-future-outlook#/.

Masley, N., Fattorini, L., Perrault, R., Gil, Y., Parli, V., Kariuki, N., Capstick, E., Reuel, A., Brynjolfsson, E., Etchemendy, J., Ligett, K., Lyons, T., Manyika, J., Niebles, J. C., Shoham, Y., Wald, R., Walsh, T., Hamrah, A., Santarlasci, L., Lotufo, J. B., Rome, A., Shi, A., & Oak, S. (2025). *The AI index 2025 annual report*. AI Index Steering Committee, Institute for Human-Centered AI, Stanford University.

Mitchell, M. (2024). Debates on the nature of artificial general intelligence. Science, 383(6689), eado7069. doi:https://doi.org/10.1126/science.ado7069.

Narayanan, A., & Kapoor, S. (2024). AI snake oil: What artificial intelligence can do, what it can't, and how to tell the difference. In *AI snake oil*. Princeton University Press.

National Research Council (NRC). (2012). *A framework for K-12 science education: Practices, crosscutting concepts, and core ideas*. National Academies Press.

Next Generation Science Standards Lead States. (2013). Appendix F—science and engineering practices in the NGSS. In *In next generation science standards: For states, by states*. The National Academies Press.

Odden, T. O. B., & Hansen, A. S. (2025). Physics intuition about vectors can help us understand generative artificial intelligence. The Physics Teacher, 63(3), 173–176. doi:https://doi.org/10.1119/5.0149555.

OECD. (2019). OECD Future of Education and Skills 2030: Conceptual learning framework. OECD Publishing. https://www.oecd.org/education/2030-project/

Organisation for Economic Co-operation and Development (OECD). (2023). *PISA 2025 science framework (second draft)*. OECD Publishing. https://pisa-framework.oecd.org/science-2025/assets/docs/PISA_2025_Science_Framework.pdf

Pelly, L. (2025). *Mood machine: The rise of Spotify and the costs of the perfect playlist*. Atria/One Signal Publishers.

Riga, F. (2020). Pragmatism—John Dewey. In B. Akpan & T. J. Kennedy (Eds.), *Science education in theory and practice. Springer texts in education*. Springer. https://doi.org/10.1007/978-3-030-43620-9_16

Sjöström, J., & Eilks, I. (2020). The *Bildung* theory—from von Humboldt to Klafki and beyond. In B. Akpan & T. J. Kennedy (Eds.), *Science education in theory and practice. Springer texts in education*. Springer. https://doi.org/10.1007/978-3-030-43620-9_5

Stevenson, C. E., Pafford, A., van der Maas, H. L. J., & Mitchell, M. (2025). *Can large language models generalize analogy solving like people can?* (No. arXiv:2411.02348). arXiv. https://doi.org/10.48550/arXiv.2411.02348.

Tenório, T., & Romeike, R. (2023). AI competencies for non-computer science students: A practice-oriented framework. In Proceedings of the Koli Calling International Conference on Computing Education Research.

Touretzky, D. S., Gardner-McCune, C., Martin, F., & Seehorn, D. (2019). Envisioning AI for K-12: What should every child know about AI? *AI Magazine, 40*(4), 5–14.

UNESCO. (2025). AI competency frameworks for students and teachers.

Wang, N., & Lester, J. (2023). K–12 education in the age of AI: A call to action for K–12 AI literacy. *International Journal of Artificial Intelligence in Education, 33*, 228–232.

World Economic Forum (WEF). (2025, January 7). The future of jobs report 2025. https://www.weforum.org/publications/the-future-of-jobs-report-2025/

Part VII
Conclusion

Chapter 12
Charting the Future of AI-Supported Science Education: A Human-Centered Vision

Xiaoming Zhai and Kent Crippen

12.1 Introduction

We are living in an era of rapid transformation, in which science education must adapt to the evolving demands of a workforce that requires discipline-based competencies in artificial intelligence (AI). Unlike earlier waves of technological change that often built incrementally upon established practices, the present shift is more radical and disruptive, demanding not only new skills but also integrated competence that can thrive with the technology itself. Moreover, the pervasive nature of AI in society implies that students encounter these technologies not only in professional contexts but also in their civic and personal lives, making it imperative for education systems to nurture critical awareness, ethical reasoning, and the ability to question and evaluate AI's role in shaping knowledge and decision-making. To meet this change, our educational system must make substantial adjustment at various levels, ranging from the National AI Strategy to reimagined paradigms of teaching and learning (Shi, 2025). This transition is neither trivial nor paradigmatic, yet it is fundamentally reshaping how we conceive of teaching, learning, and assessment in science education. It challenges educators, researchers, and policymakers to rethink the very purposes and methods of science education in light of these shifts, underscoring the urgency for science education to prepare learners to engage with AI technologies in responsible, ethical, and human-centered ways.

The field of science education is navigating this transformation along three interrelated trajectories, each of which reflects both historical continuities and novel

X. Zhai (✉)
AI4STEM Education Center, University of Georgia, Athens, GA, USA
e-mail: Xiaoming.zhai@uga.edu

K. Crippen
School of Teaching and Learning, University of Florida, Gainesville, FL, USA

X. Zhai, K. J. Crippen (eds.), *Advancing AI in Science Education*, Advances in Technology-Rich Science Education 3,
https://doi.org/10.1007/978-3-032-16871-9_12

possibilities. First, new scholarly communities are emerging to cultivate research and practice at the intersection of AI and science education. A notable example is the Research on AI-involved Science Education (RAISE) special interest group within the NARST organization. Founded in 2022, RAISE has provided a professional home for collaboration and intellectual exchange (Zhai, 2024a, 2024b). Beyond functioning as a network, RAISE has initiated workshops, conference sessions, and mentoring opportunities that have allowed researchers from diverse disciplinary backgrounds to coalesce around shared questions, methods, and values. In this way, the community is actively shaping the identity of a new research subfield rather than merely reacting to technological trends (Lee et al., 2025). Over time, such communities not only produce knowledge but also establish norms, standards, and agendas that shape the trajectory of scholarship and practice for years to come.

Second, a growing body of research is generating foundational knowledge on the potential applications of AI in science education. This scholarly activity extends beyond documenting novel uses of AI technologies and increasingly seeks to theorize how AI reshapes learning processes, instructional design, and teacher professional development. Examples of this activity include the 2021 special issue, titled *Applying Machine Learning in Science Assessment* in the *Journal of Science Education and Technology, which* set forth a new vision for the potential of AI, in particular to advance next-generation science assessment. The volume *Uses of Artificial Intelligence in STEM Education* (Zhai & Krajcik, 2025b), which emerged from the *International Conference on AI-based Assessment in STEM Education* held in Athens, Georgia, in 2022, sets a milestone that exemplifies more than 60 scholar's cutting-edge work in AI-supported science education. These scholarly efforts not only compiled empirical evidence but also articulated visions for integrating AI in sustainable and scalable ways. Together, these contributions mark a transition from early exploratory studies toward more systematic and programmatic lines of inquiry. In addition, newer projects are beginning to develop design principles, empirical frameworks, and methodological innovations that will guide subsequent generations of researchers and practitioners. By doing so, this body of work is gradually moving from curiosity-driven exploration toward the establishment of a coherent, evidence-informed field with its own theoretical foundations.

Next, the community has begun to confront the challenges and ethical implications of AI in science education. While enthusiasm about AI's potential is evident, critical scholarship has emphasized that adoption without deep reflective thought and consideration risks reinforcing inequities, narrowing pedagogical aims, or diminishing the central role of human agency in science learning. Along this line, confusion between AI and the use of AI, misunderstanding of AI and its mechanisms and outcomes have been constantly drawing concerns (Bewersdorff et al., 2023; Zhai & Krajcik, 2025a), not only among practitioners, but also within scholarly work. For example, recent debates published in the *Journal of Research in Science Teaching* (Krist & Kubsch, 2023; Li et al., 2023; Zhai & Nehm, 2023) exemplify several alternative understandings of AI in science education, yet collectively highlight both the promise and the perils of AI. These papers call for greater attention to issues of responsibility, equity, and ethics when using AI in science assessment. These discussions clarify that AI technology is not a neutral tool, but an embedded element in a socio-technical system that includes cultural values, power dynamics, and policy environments. These debates also point to the need for more cross-disciplinary dialogue with ethicists, policymakers, and practitioners, ensuring that the integration of AI into science education is more than driven by technology, but a process that must be informed by broader societal concerns and long-term humanistic goals.

In response to this broadening conversation, the Advancing AI in Science Education (AASE) committee was established under the leadership of the co-- editors, Xiaoming Zhai and Kent Crippen. Funded by National Science Foundation the committee was charged with two primary goals: to identify the transforma (#2332964), tions AI brings to science education and to articulate principles for responsible and ethical practice of AI in science education. These goals were intentionally ambitious, recognizing that addressing them would require not only the synthesis of current research but also the creation of new opportunities for dialogue across disciplinary and institutional boundaries. This work unfolded through multiple virtual meetings, which allowed for the participation of international scholars, and culminated in a two-day in-person conference held in Athens, Georgia, in February 2025. That conference served as both a capstone and a catalyst: it provided an opportunity to refine the committee's shared understandings, generate new ideas, and engage in spirited debate about the implications of AI for the future of science education. The deliberations and collaborations of the committee produced the collective insights that structure this volume, as well as a sense of community that extends beyond the book itself. The book includes 10 chapters authored by the committee members, in addition to introductory and concluding chapters, that together address the committee's charge. In doing so, it aspires not only to document the state of the field but also to serve as a touchstone for future scholarship and practice, a roadmap for teacher educators and policymakers, and an invitation to practitioners who are exploring the promises and pitfalls of AI in their own contexts.

Our approach to this endeavor began with examining the shifts AI has precipitated in contemporary scientific research and considering how these shifts inform science education (see Chap. 2). Building on this foundation, we proposed a framework for responsible and ethical principles (REP) of practicing AI-supported science education (see Chap. 3). This framework composites principles covering eight aspects: *Fairness, Privacy & Data Security, Human Oversight & Autonomy, Beneficial & Meaningful Use, Robustness & Scientific Integrity, Respect for Human Values, Transparency & Explainability,* and *Accountability*. The framework was not conceived in abstraction, but rather, grounded in careful analysis of ongoing case studies and collaborative reflection on best practices. Guided by this framework, the book is organized into five sections, each examining how AI is reshaping core dimensions of science education—*educational goals, pedagogical processes, learning materials, assessment practices,* and *educational outcomes*—and articulate ethical and responsible practice of AI guided by the REP framework. Each section is intended to provide conceptual clarity and practical guidance, demonstrating ways in which AI can be used to expand opportunities for learners while safeguarding human dignity and agency. In what follows, we present the key insights that emerged from the committee's work, charting a vision for the future of AI in science education that is both innovative and human-centered. These insights provide a foundation upon which subsequent scholarship, classroom innovations, and policy initiatives may build and evolve in the years to come.

12.2 Key Insights from the Book's Sections

12.2.1 Educational Goals

The formulation of educational goals must align with the evolving demands of future citizens and the workforce. In light of the transformative influence of AI across all sectors of society, it is evident that the competencies required for individuals to thrive are undergoing profound change. Consequently, educational objectives—particularly those embedded within science education—must be redefined to address these emerging needs. Within this context, two chapters are dedicated to examining the necessary transformations (see Chaps. 4 and 5). The combined insights from these chapters highlight the following:

12.2.1.1 Evolving Purpose of Science Education

Science literacy has expanded beyond content mastery to include applying knowledge in everyday life and exercising agency in addressing societal issues. In the AI era, these goals converge into preparing students as lifelong investigators and responsible citizens.

12.2.1.2 AI as a Transformative Force in Science and Education

Artificial intelligence is reshaping how science is practiced—accelerating inquiry, modeling, experimentation, and data analysis—and requires K–12 education to integrate AI as an authentic component of scientific practice.

12.2.1.3 Science Literacy Plus as a New Educational Goal

A future-ready framework combines science literacy and AI literacy. It emphasizes human-centered mindsets, ethical engagement, technical proficiency, and system design skills, progressing from understanding to applying to creating.

12.2.1.4 Inquiry-Centered, AI-Supported Learning

Science classrooms should move from rote learning toward investigation-centered environments where AI serves as a partner in problem-solving, enabling authentic inquiry cycles that emphasize evidence, iteration, creativity, and reflection.

12.2.1.5 Developing Reflective and Ethical Competencies

Students must cultivate metacognitive, critical, and ethical skills to engage responsibly with AI, addressing risks such as bias, misinformation, inequity, and privacy concerns while promoting transparency, fairness, inclusivity, and environmental responsibility.

12.2.1.6 Personalized and Inclusive Learning with AI

AI tools, such as natural language processing and adaptive guidance, can personalize feedback, make student thinking visible, and build on diverse learner ideas, supporting equitable and meaningful participation in science learning.

12.2.1.7 Collaboration and Partnerships for Sustainable Integration

Equitable and ethical AI integration in education requires partnerships among teachers, researchers, communities, and policymakers to co-develop open-source curricula, resources, and professional learning opportunities.

12.2.1.8 Preparing Autonomous and Resourceful Learners

By embracing AI-supported inquiry and Science Literacy Plus, K–12 education can prepare students to be self-directed, collaborative, and ethically responsible learners capable of addressing urgent scientific and societal challenges.

12.2.2 Educational Procedures

As educational goals evolve, there is an urgent need to reconfigure instructional practices in science classrooms and to prepare teachers with the knowledge and skills necessary to navigate AI-enhanced learning environments (Zhai, 2024a, 2024b). Modern science education is increasingly designed to foster scientific practices and collaborative problem-solving. With the integration of AI, these processes are further amplified through innovative instructional decision-making. AI-supported systems can provide immediate feedback, simulate experimental conditions, and personalize learning pathways, enabling students to generate and test hypotheses, interpret results, and progressively refine their scientific understanding. In this context, teachers' roles are being redefined (Shi & Choi, 2024), and instructional practices are shifting significantly from those of traditional classrooms. Moreover, AI holds the potential to enhance the overall effectiveness of teaching and learning. The two chapters in this section address these transformations, with particular attention to the responsible and ethical use of AI in education.

12.2.2.1 Individualized and Authentic Student Learning

AI enables teachers to tailor instruction through intelligent tutoring systems, exploratory environments, and adaptive curricula, while generative AI can scaffold authentic, self-directed, and agentic learning experiences.

12.2.2.2 Shifting Classroom Dynamics and Roles

The integration of AI requires a move from teacher-centered orchestration to teacher–student–AI co-orchestration, with teachers acting as facilitators, organizers, and critical evaluators of AI contributions.

12.2.2.3 Balancing Autonomy and Guidance

AI can scaffold inquiry and reasoning but also risks undermining independent sense-making if adopted uncritically. Students must develop epistemic autonomy, metacognition, and reflective skills to engage responsibly with AI outputs.

12.2.2.4 Ethical and Equitable Integration

Responsible use of AI requires instructional strategies, collaboration frameworks, and policies that ensure fairness, transparency, inclusivity, privacy protection, and human oversight in both teaching and learning.

12.2.2.5 Building Shared AI Literacy

Both teachers and students must cultivate AI literacy—not only to use AI effectively but also to critically evaluate outputs, recognize limitations, and address ethical challenges in algorithmic mediation.

12.2.2.6 Amplifying Human Aspects of Education

When implemented thoughtfully, AI should enhance—not replace—the human dimensions of science teaching and learning, such as curiosity, inquiry, creativity, and meaningful teacher–student connections.

12.2.2.7 AI as a Tool for Teacher Support and Efficiency

AI can streamline lesson planning, instruction, assessment, and reflection while providing timely insights into student knowledge and struggles, allowing teachers to focus on responsive and targeted instruction.

12.2.2.8 Enhancing Teacher Professional Learning

AI offers practice-based opportunities, personalized feedback, and simulations that strengthen teachers' competencies. Professional development must build AI literacy, model effective use of prompts, and emphasize teacher judgment as central to decision-making.

12.2.3 Learning Materials

If educational goals and procedures are being reshaped by AI, the demand for transformed learning resources in science education becomes evident. This transformation is both profound and multifaceted, emphasizing interactive, adaptive, and multimodal formats. Traditional static textbooks are increasingly being replaced by virtual laboratories, AI-supported simulations, and augmented reality environments that enable learners to engage directly with scientific phenomena. To ensure that science education remains contemporary, content of the learning material must extend beyond conventional methods and reflect the authentic practice employed by scientists. Moreover, AI plays a growing role in supporting educators and researchers in the creation of relevant instructional materials, reducing the demands of time-intensive tasks and allowing curriculum designers to focus more fully on higher-order pedagogical innovation and design. Chapter 8 contributes to the field in several important ways:

12.2.3.1 Embedding Authentic Scientific Practice

AI transforms science learning materials by integrating practices such as data analysis, modeling, simulation, and hypothesis testing, enabling students to engage with science as it is practiced today.

12.2.3.2 Personalization and Adaptivity

AI-supported materials can dynamically respond to learners' needs, prior knowledge, and progress, fostering inclusivity, deeper engagement, and equitable learning opportunities.

12.2.3.3 Interactive and Inquiry-Rich Experiences

AI-supported simulations and tools allow students to become active investigators and even co-creators of models and experiments, providing contextualized, authentic, and inquiry-rich learning experiences.

12.2.3.4 Multimodal and Dynamic Content

Generative AI enables the creation of multimodal resources (text, visuals, audio, video) that make abstract concepts more tangible, but their educational use requires careful validation for scientific accuracy and pedagogical coherence.

12.2.3.5 Accessibility and Cultural Responsiveness

AI enhances accessibility for multilingual, neurodiverse, and differently-abled learners through translation, summarization, and multimodal interaction, while supporting culturally responsive and inclusive learning design.

12.2.3.6 Co-Authoring and Rapid Content Development

Generative AI can serve as a collaborative partner for educators in creating context-sensitive, culturally relevant, and adaptable instructional resources, provided that human oversight ensures quality, integrity, and alignment with learning goals.

12.2.3.7 Ethical and Responsible Integration

The use of AI in learning materials must prioritize transparency, fairness, privacy, inclusivity, and human oversight, ensuring that teacher judgment and student agency remain central.

12.2.3.8 Capacity Building for Educators and Learners

Professional development is essential for teachers to evaluate, adapt, and design AI-supported materials, while students must develop AI literacy to critically interpret, question, and contextualize AI outputs.

12.2.3.9 Reimagining Science Learning Materials

When implemented responsibly, AI can act as a catalyst to make learning materials more authentic, inclusive, and responsive, advancing the broader goals of scientific literacy, critical thinking, and ethical responsibility.

12.2.4 Assessment

AI is also transforming approaches to assessment in science education, shifting emphasis toward more formative, performance-based, and authentic measures of learning. Automated scoring systems can now evaluate students' written explanations, arguments, and drawn models, enhancing efficiency while expanding the scope of constructs that can be assessed. Through AI-supported learning analytics, educators can monitor how students plan investigations, interpret evidence, and construct scientific arguments, providing insight into both the processes and outcomes of science learning. Interactive dashboards supply real-time feedback on student learning progressions of big ideas in science, enabling timely instructional responses. Furthermore, AI-supported longitudinal assessments compile detailed learner profiles, allowing educators to trace the progression of scientific reasoning over extended periods. In this way, assessment becomes a dynamic and integral element of the learning process rather than a concluding event. Two chapters dedicated to this section articulates both the potential and the challenges of AI-supported science assessment.

12.2.4.1 Expanding Assessment Constructs for the AI Era

Science assessment must evolve to capture competencies relevant to AI-supported learning, including collaboration with AI, critical evaluation of AI outputs, ethical reasoning, and higher-order problem-solving.

12.2.4.2 Designing AI-Resilient and Authentic Assessments

Traditional recall-based tasks are increasingly vulnerable to AI-supported shortcuts. Assessments should emphasize performance-based, inquiry-driven, and higher-order thinking tasks that preserve authentic student engagement and align with standards.

12.2.4.3 Balancing Efficiency with Integrity

While AI can streamline scoring and feedback, overreliance by students (outsourcing responses) or teachers (outsourcing item design and scoring) risks undermining meaningful learning, assessment literacy, and instructional judgment.

12.2.4.4 Ensuring Validity, Fairness, and Oversight

AI-supported scoring requires transparent rubrics, continuous validation, and human-in-the-loop oversight to safeguard interpretability and fairness across diverse populations. Frequent recalibration is essential to maintain educational value.

12.2.4.5 Mitigating Bias through Technical and Societal Strategies

Both societal and technical biases must be addressed. This includes auditing systems for equity, using balanced and diverse training datasets, designing analytic rubrics, and adopting open-source or transparent AI tools to improve accountability.

12.2.4.6 Accounting for Locality and Context

Assessment practices must consider standards, language, culture, region, and accessibility. Strategies such as retrieval-augmented generation for curriculum alignment, multilingual model training, and culturally responsive design are necessary for inclusivity.

12.2.4.7 Transforming Classroom Assessment into Formative Practice

AI can enable ongoing, formative assessment by providing real-time feedback, multimodal data analysis, and learner-driven engagement, fostering more personalized and responsive instruction.

12.2.4.8 Fostering AI Literacy and Ethical Engagement

Both teachers and students need AI literacy to understand its affordances, limitations, and ethical implications. Human judgment, reflection, and responsibility must remain central to all AI-supported assessment practices.

12.2.4.9 Positioning AI as a Catalyst, Not a Substitute

AI should be leveraged to enrich assessment by deepening inquiry, supporting fairness, and amplifying human expertise—rather than as a shortcut for efficiency or replacement of authentic teacher and student roles.

12.2.5 *Educational Outcomes*

In the end, shifts in goals, procedures, learning resources, and assessment practices reshape what counts as achievement in science education. Success is no longer measured solely by memorizing facts or completing lab tasks; instead, it highlights originality in designing experiments, teamwork across disciplines, persistence in tackling challenges, and thoughtful consideration of the social and ethical implications of science and technology, including AI. Graduates of AI-supported science education are imagined as flexible, lifelong learners, prepared to confront both scientific and societal issues responsibly—for example, by advancing renewable energy or addressing climate change. This reimagining ensures that science education goes beyond delivering subject knowledge, fostering the broader capabilities needed to drive innovation in an AI-infused world.

12.2.5.1 Preparing Students for an AI-Infused World

As AI transforms science, work, and daily life, K–12 science education must ensure students develop the knowledge, skills, and dispositions needed to navigate and thrive in AI-supported environments.

12.2.5.2 Core Competencies for AI in Science Education

Educational outcomes should emphasize three interconnected competencies: (a) Knowing about AI: understanding its concepts, functions, and role in science and society, (b) Applying AI: using AI tools for inquiry, modeling, analysis, and problem-solving, and (c) Evaluating and Reflecting on AI: critically assessing AI's outputs, biases, and ethical implications.

12.2.5.3 Reflection and Ethical Responsibility

Students must be equipped to move beyond surface-level tool use, cultivating habits of questioning assumptions, recognizing limitations, and addressing societal and ethical impacts embedded in AI systems.

12.2.5.4 Learning Progressions and Pathways

Coherent, developmentally appropriate progressions should integrate AI outcomes into existing science frameworks (e.g., NRC Framework, NGSS), while accounting for multiple trajectories—deep technical competencies for research, applied skills for vocational fields, and reflective capacities for all citizens.

12.2.5.5 Scientific Literacy as the Ultimate Goal

The aim is not to train all students as AI developers but to cultivate scientifically literate citizens who can thoughtfully and ethically engage with AI to contribute to science, work, civic life, and global challenges.

12.3 Recommendations for Future Research

Building on the insights articulated in this volume, future research must take deliberate steps to advance a human-centered vision of AI in science education. As AI continues to reshape scientific practice, pedagogy, and learning environments, research must both leverage these transformations and remain grounded in the *REP* framework introduced in this book. The following directions highlight how future work can respond to the changes AI brings while safeguarding fairness, transparency, human oversight, and respect for human values. To deepen the agenda, each recommendation is elaborated to show both the opportunities created by AI and the ethical guardrails required to ensure responsible adoption.

12.3.1 Develop Coherent Conceptual and Theoretical Frameworks for AI-Supported Science Education

Future research should refine constructs such as *Science Literacy Plus* and *AI literacy* to reflect the ways AI redefines scientific inquiry, knowledge production, and classroom learning. Theoretical work must not only account for AI's potential to accelerate data analysis, modeling, and collaboration but also integrate REP principles such as fairness, autonomy, and meaningful use. By articulating conceptual frameworks that balance innovation with responsibility, scholars can provide educators and policymakers with clear guidance for ethically grounded practice. This work should also consider how AI disrupts traditional notions of scientific authority and democratizes knowledge, ensuring that new frameworks explicitly safeguard transparency, integrity, and the preservation of human values. Expanding this line of research could involve mapping how AI transforms epistemic practices across different scientific domains, investigating how students' conceptions of science evolve when AI becomes a partner in inquiry, and theorizing how REP values interact with curricular standards and policy mandates. Such scholarship should also link conceptual development to design-based research that tests how theoretical models play out in classrooms, allowing theories to remain dynamic and responsive to emerging challenges and opportunities.

12.3.2 Advance Methodological Innovation and Build Robust Evidence in AI Contexts

AI enables the collection of rich, multimodal data and the use of advanced analytics to study learning in unprecedented detail. Future research should harness these capabilities to design studies that explain *why* and *under what conditions* AI enhances equity, creativity, and scientific reasoning. At the same time, methodologies must embody REP values by ensuring privacy protections, transparency in analytics, and accountability in interpretation. Research that blends human-centered inquiry with AI-supported tools will be essential to building trustworthy, evidence-based insights. Methodological expansion should also include longitudinal designs that track learners across educational stages, participatory approaches that include teacher and student voices, and cross-context comparisons to reveal how AI can responsibly adapt to varied learning environments. Further growth in this area might involve the creation of cross-institutional research consortia that share anonymized datasets while adhering strictly to REP safeguards, as well as the development of open-source methodological toolkits that empower educators and researchers worldwide to study AI-supported learning responsibly. By embedding ethical protocols into every stage of methodological design, scholars can ensure that evidence-building advances both scientific understanding and human dignity.

12.3.3 Prioritize Ethical, Equitable, and Human-Centered Research Agendas

As AI tools shape assessment, personalization, and instructional decision-making, it is imperative to investigate how they influence equity and agency. Future studies should focus on reducing algorithmic bias, protecting student data, and amplifying diverse voices in classrooms. Guided by REP principles of fairness, transparency, and respect for human values, research must explore not only the risks of inequity but also strategies for designing AI systems that actively promote inclusivity. Investigating how ethical commitments are enacted in real-world settings will ensure AI adoption remains human-centered. Additional attention should be devoted to contexts outside formal schooling, such as community learning hubs and informal science spaces, where AI tools may reach vulnerable populations. By doing so, researchers can better illuminate how ethical practices scale across diverse ecosystems. Expanding this agenda requires comparative work that documents how ethics is interpreted across cultural contexts, the development of rubrics for assessing ethical dimensions of AI integration, and collaborations with ethicists and community stakeholders who can help translate REP commitments into culturally responsive practices. Such expanded efforts will position ethics not as an afterthought but as a central driver of innovation.

12.3.4 Strengthen Teacher Roles and Professional Learning for Ethical AI Integration

AI is altering classroom dynamics, shifting teachers from sole orchestrators to co-facilitators alongside students and AI systems. Research should explore how educators can critically evaluate AI tools, balance guidance with student autonomy, and retain authority over ethical decision-making. Professional learning must cultivate both technical fluency and moral judgment, aligning with REP commitments to human oversight and accountability. Studies should also examine how teachers' professional identities and agency evolve in AI-supported environments, ensuring that educators remain empowered shapers of ethical practice. Future work should additionally investigate how professional development can model ethical AI use, provide teachers with case-based simulations, and create international communities of practice that exchange lessons on balancing AI innovations with human-centered pedagogy. Expanding this research could include studying how pre-service teacher preparation programs embed AI literacy, how cross-generational mentorship models support ethical practice, and how professional learning networks can act as watchdogs ensuring adherence to REP. Such expanded work will clarify how teachers can remain both pedagogical leaders and ethical stewards in AI-supported classrooms.

12.3.5 Innovate Learning Materials and Assessment Practices with Responsibility

AI has enabled adaptive, generative, and multimodal resources that can transform science learning materials and assessments. Future research should investigate how such resources can embody authenticity, accessibility, and cultural responsiveness while aligning with REP principles of transparency, robustness, and scientific integrity. Assessments must be designed to capture higher-order competencies—such as critical evaluation of AI outputs and ethical reasoning—while resisting overreliance on automation. By prioritizing validity, fairness, and human oversight, research can ensure that AI-supported assessments foster authentic learning and reflective engagement. In addition, future agendas should examine how generative AI reshapes textbook creation, how adaptive assessments can be audited for equity, and how student reflection on AI use can itself become a learning outcome, reinforcing REP commitments through practice. Expanding this focus might involve the co-design of culturally sustaining digital resources with community partners, empirical studies testing the reliability of AI-supported grading across diverse populations, and design experiments that integrate student self-assessment of AI outputs into formative feedback cycles. Through such expansion, the field can ensure that learning resources and assessments both leverage AI's strengths and safeguard against its risks.

12.3.6 Foster Interdisciplinary and Cross-Sector Collaboration Guided by Responsible and Ethical Principles

Because AI integration in education is embedded in larger socio-technical systems, future research should encourage partnerships across disciplines, sectors, and communities. Collaborative projects must not only advance technological innovation but also ensure alignment with REP commitments. For example, open educational resources and shared data infrastructures should be designed with transparency, accountability, and inclusivity in mind. Interdisciplinary teams—including educators, technologists, ethicists, and policymakers—can co-create solutions that are both innovative and human-centered. Research should also examine governance models for such collaborations, exploring how collective responsibility can be distributed fairly, how policies can enforce REP, and how voices from underrepresented communities can have meaningful influence in shaping the direction of AI in education. Further expansion could include studies of international governance structures for educational AI, analyses of how policy frameworks operationalize REP commitments, and experimental collaborations that test new forms of participatory design with teachers and students. By scaling up collaborative research while embedding REP safeguards, the field can chart models of AI governance that are democratic, inclusive, and future-ready.

12.3.7 Embrace Global and Comparative Research Agendas to Ensure Inclusive AI Futures

AI adoption is unfolding differently across nations and cultures, with varied implications for equity, pedagogy, and policy. Future research should compare these diverse pathways to identify practices that balance innovation with REP commitments to fairness, cultural responsiveness, and respect for human dignity. Global collaborations can generate repositories of open-source AI tools, ethical guidelines, and multilingual resources that reflect both universal principles and local adaptations. Special attention should be given to voices from the global South and marginalized communities, ensuring that the human-centered vision for AI in science education is inclusive and just. Further expansion in this area should involve joint cross-country projects, UNESCO-level policy dialogues, and knowledge exchanges that enable equitable participation of countries with fewer technological resources, thereby preventing the reproduction of global inequities. To extend this agenda, scholars might explore how indigenous knowledge systems intersect with AI tools, and how resource-constrained contexts innovate responsible low-tech AI solutions.

12.4 Conclusion

The integration of AI into science education represents not simply a technological innovation but a fundamental transformation of educational purposes, practices, and outcomes. This volume highlights how AI can enrich inquiry, personalize learning, and expand opportunities, yet its adoption must remain firmly anchored in Responsible and Ethical Principles. Guided by fairness, transparency, privacy, and respect for human values, AI should serve as a catalyst to amplify—rather than replace—human curiosity, creativity, and collaboration. We thus call for a human-centered vision of AI in science education serving learners who are scientifically literate, ethically grounded, and capable of critical engagement with AI. Teachers, as ethical stewards of practice, and policymakers, as architects of equitable systems, play essential roles in ensuring that AI enhances rather than diminishes human agency. By aligning innovation with responsibility and ethics, science education can transform the disruptive potential of AI into a force for human flourishing and civic responsibility.

Declaration of AI Use Parts of this manuscript were prepared with the assistance of generative AI tools. The AI tools were used for language editing or paragraphing of the authors' writing. The authors reviewed, revised, and verified all AI-supported content for accuracy, appropriateness, and originality. No confidential or sensitive data were entered into AI systems. The final responsibility for the content rests solely with the authors.

Acknowledgement This material is based upon work supported by the National Science Foundation under Grant No. 2332964 (PI Zhai). Any opinions, findings, and conclusions or recommendations expressed in this material are those of the author(s) and do not necessarily reflect the views of the National Science Foundation.

References

Bewersdorff, A., Zhai, X., Roberts, J., & Nerdel, C. (2023). Myths, mis- and preconceptions of artificial intelligence: A review of the literature. *Computers and Education: Artificial Intelligence, 4*(100143), 1–11. https://doi.org/10.1016/j.caeai.2023.100143

Krist, C., & Kubsch, M. (2023). Bias, bias everywhere: A response to Li et al. and Zhai and Nehm. *Journal of Research in Science Teaching, 60*, 2395–2399.

Lee, G., Yun, M., Zhai, X., & Crippen, K. (2025). Artificial intelligence in science education research: Current states and challenges. *Journal of Science Education and Technology*, 1–18.

Li, T., Reigh, E., He, P., & Adah Miller, E. (2023). Can we and should we use artificial intelligence for formative assessment in science? *Journal of Research in Science Teaching*. https://doi.org/10.1002/tea.21867

Shi, L. (2025). Global perspectives on AI competence development: Analyzing national AI strategies in education and workforce policies. *Human Resource Development Review*, 15344843251332360.

Shi, L., & Choi, I. (2024). A systematic review on artificial intelligence in supporting teaching practice: Application types, pedagogical roles, and technological characteristics. In X. Zhai & J. Krajcik (Eds.), *Uses of artificial intelligence in STEM education* (pp. 321–347). Oxford University Press.

Zhai, X. (2024a). *Conclusions and foresight on AI-based STEM education: A new paradigm*. Oxford University Press. https://doi.org/10.1093/oso/9780198882077.003.0026

Zhai, X. (2024b). Transforming teachers' roles and agencies in the era of generative AI: Perceptions, acceptance, knowledge, and practices. *Journal of Science Education and Technology*. https://doi.org/10.1007/s10956-024-10174-0

Zhai, X., & Krajcik, J. (2025a). Pseudo artificial intelligence bias. In *Uses of artificial intelligence in STEM education* (pp. 568–578). Oxford University Press. https://doi.org/10.1093/oso/9780198882077.003.0025

Zhai, X., & Krajcik, J. (2025b). *Uses of artificial intelligence in STEM education*. Oxford University Press.

Zhai, X., & Nehm, R. (2023). AI and formative assessment: The train has left the station. *Journal of Research in Science Teaching, 60*(6), 1390–1398. https://doi.org/10.1002/tea.21885

The manufacturer's authorised representative in the EU is Springer Nature Customer Service Centre GmbH, Europaplatz 3, 69115 Heidelberg, Germany. If you have any concerns regarding our products, please contact ProductSafety@springernature.com

Printed and bound by CPI Group (UK) Ltd, Croydon, CR0 4YY
21/09/2026
02219988-0007